Enzymes Biotechnology

Enzymes Biotechnology

Ms Kusum Chopra
B.Tech (Biotechnology) (MIET) U.P. Tech.

Ms Hiru Ranabhatt
M.Sc. Plant Biotechnology from TERI University, New Delhi

WOODHEAD PUBLISHING INDIA PVT LTD

New Delhi • Cambridge • Oxford • Philadelphia

CRC Press
Taylor & Francis Group
Boca Raton London New York

CRC Press is an imprint of the
Taylor & Francis Group, an **informa** business

First published 2026
by CRC Press
4 Park Square, Milton Park, Abingdon, Oxon OX14 4RN

and by CRC Press
2385 NW Executive Center Drive, Suite 320, Boca Raton FL 33431

CRC Press is an imprint of Informa UK Limited

British Library Cataloguing-in-Publication Data
A catalogue record for this book is available from the British Library

Print edition not for sale in South Asia (India, Sri Lanka, Nepal, Bangladesh, Pakistan or Bhutan)

ISBN: 9781041217503 (hbk)
ISBN: 9781041231141 (pbk)
ISBN: 9781003735403 (ebk)

DOI: 10.1201/9781003735403

Typeset in Times New Roman
by Y Square Multilingual Services, Chennai

WPI

Contents

Preface

Natural proteins produced in tiny quantities by all living organisms (bacteria, plants and animals) and functioning as highly selective biochemical catalysts in converting one molecule into another. Enzymes are essential to life because they speed up metabolic reactions to a very great extent, but do not undergo any change in themselves. In the industry they are used for degrading oil spills and wastes into harmless compounds, in cleaning fat (food) stains and in fermentation processes to make alcoholic beverages. Enzymes are very sensitive to environmental conditions and function best only within a narrow range of temperature and acidity (pH) levels.

This book on Enzymes Biotechnology is divided into six sections and contains 24 chapters.

Section I discusses general considerations & engineering aspects. Chapter 1 is devoted to basic concepts of enzymes. Chapter 2 deals with engineering enzymes for biocatalysts. Since enzymes are proteins, enzyme engineering is a part of the larger activity of protein engineering. Enzyme engineering utilises r-DNA technology to introduce the desired changes in amino acid sequences of enzymes.

Section II discusses classification, nomenclature and production of enzymes. Chapter 3 focuses on enzymes nomenclature and classification. Chapter 4 deals with immobilised. Immobilisation of enzymes is a useful tool to meet cost targets and to achieve technological advantages. Immobilisation enables repetitive use of enzymes and hence significant cost savings. From the technological point of view, immobilised enzymes can easily be separated from the reaction liquid and make labourious separation steps unnecessary. Chapter 5 concentrates on carbohydrase. Carbohydrase is an enzyme that catalyses the break down of carbohydrates into simple sugars. Carbohydrases are produced in the pancreas but act in the stomach breaking down carbohydrates. Chapter 6 is devoted to protease which are enzymes that digest proteins. Various types of proteases such as plant, animal, microbial, serine, etc., are discussed in detail. Chapter 7 deals with lipase. Lipase is a physiologically necessary enzyme. It occurs in many plants and animal organisms, as well as in micro-organisms. However, its richest source are bacteria, fungi and yeast. Chapter 8 concentrates on oxidoreductase which are enzyme catalysing a reaction in which two molecules of a compound interact so that one molecule is oxidised and the other reduced with a molecule of water entering the reaction. Various types of oxidoreductases along with their general characteristics are discussed. Chapter 9 deals with production of industrial enzymes. Fermentation is a method of generating enzymcs for industrial purposes. Fermentation involves the use of micro-organisms, like bacteria and yeast to produce the enzymes.

Section III discusses marine, microbial, fungal and soil enzymes. Chapter 10 deal with marine microbial enzymes. Marine micro-organisms take active part in the mineralisation of complex organic matter through degradative pathways of their metabolism in the marine environments and contribute to the secondary production in sea. Chapter 11 is devoted to microbial enzymes in bioremediation of pollution. Biological methods or bioremediation promises to provide an ecofriendly solution to this menacing problem. Biodegradation refers to the breakdown of substances or molecules by biological system into smaller units.

Chapter 12 concentrates on fungal laccase enzyme for biotechnological application. Laccase belongs to the small group of enzymes called the blue multi copper oxidases. Chapter 13 deals with soil enzymes. Soil enzyme activities are of paramount importance to evaluate functions of micro-organisms, cycling of nutrients and carbon-sources decomposition. Chapter 14 is devoted to enzyme catalysis for polymer function and degradation. Enzyme catalysed reactions are green alternative approaches to functionalise polymers compared to conventional methods. This technique is especially advantageous due to the high selectivity, high efficiency, milder reaction conditions and recyclability of enzymes.

Section IV discusses industrial applications of enzymes. Chapter 15 focuses on applications of oxidoreductase enzymes which are widely distributed among microbial, plant and animal organisms. They catalyse the exchange of electrons or redox equivalents between donor and acceptor molecules, in reactions involving electron transfer, proton abstraction, hydrogen extraction, hydride transfer, oxygen insertion, or other key steps. Chapter 16 deals with enzymes in dairy, meat, fish and wine industry. Dairy enzymes are enzymes used for the production of cheese and yoghurt as well as other milk products. The function of these enzymes varies widely from coagulants, which are used to make cheese, to bioprotective enzymes to enhance the shelf life and safety of dairy products. For decades, enzymes such as malt and fungal alpha-amylases have been used in bread-making. Feeding enzymes to poultry is one of the major nutritional advances in the last sixty years. Some of the enzymes used in poultry are – β-glucanases, xylanases, β-galactosidases, phytases, proteases, lipases, amylases. Some of the enzymes used in meat industry are – papain, bromelain, ficin, *bacillus* protease, aspartic protease. Some of the enzymes used in fish processing are – protease, glucose oxidase, catalase. Some of the enzymes used in wine industry are – polygalacturonase, pectin esterase, pectin lyase, arabinanase, galactanase, rhamnogalacturonase. Chapter 17 concentrates on enzymes in textile industry. The use of enzymes in the textile industry is an example of white/industrial biotechnology, which allows the development of environmentally friendly technologies in fibre processing and strategies to improve the final product quality. Some of the enzymes used in textile industry are – amylases, pectinases, cellulases, serine proteases, cysteine proteases, etc. Chapter 18 focuses on enzymes in detergent industry. Lipases are added to detergents such as household and industrial laundry and in household dishwashers, where their function is in the removal of fatty residues and cleaning clogged drains. Some of the enzymes used in detergent industry are – proteases, amylases, lipases, cellulases.

Chapter 19 deal with enzymes in green chemistry and organic synthesis. Enzymes are important tools in green chemistry as they act as catalysts which accelerate the chemical reactions, works without using harsh conditions and chemicals and are reusable. Chapter 20 is devoted to enzymes in clinical applications. It is shown that many diseases associate with many components of the enzyme metabolism systems are now widely applied in clinical examinations as special markers for diseases.

Section V discusses enzymes of nanoparticles. Chapter 21 concentrates on enzymes in biosynthesis of nanoparticles. While a large number of microbial species are capable of producing metal nanoparticles (NPs), mechanism of nanoparticle biosynthesis is very important. Chapter 22 deals with nanoparticles in enzyme immobilisation. Chitosan nanoparticles due to their highest specific surface area are much proper for immobilisation of higher amount of enzymes.

Section VI discusses risk assessment and safety regulation of enzymes. Chapter 23 is devoted to risk assessment guidance for enzyme containing products. Risk assessment is important for enzyme-containing consumer products since it helps ensure the continued safety of currently marketed products. Chapter 24 concentrates on safety regulation of food enzymes. Safety concerns for consumers largely focus on toxic by-products that might be inadvertently present in the enzyme preparation and might thereby be transferred to foodstuff.

Proofreading and Editing done by S. C. Bhatia. We are thankful to the editorial team of Woodhead Publishing India Pvt. for their wholehearted cooperation in bringing out the book in time.

It may not be wrong to hold that the present reference textbook of Enzymes Biotechnology is a complete treatise on this subject. It is essential reading for B. Tech. (Biotechnology/Microbiology/Food Technology/Environmental) and students pursuing B.Sc/M.Sc course in Biotechnology and Microbiology. Besides students, this book will prove useful to industrialists and consultants in the respective fields.

It has been prepared with meticulous care, aiming at making the book error-free. Constructive suggestions are always welcome from users of this book.

Ms Kusum Chopra

Ms Hiru Ranabhatt

Section I

General considerations and engineering aspects

1

Basic concepts of enzymes

1.1 Introduction

Enzymes are nature's sustainable catalysts. They are biocompatible, biodegradable and are derived from renewable resources. Enzymes constitute a large biological globular protein molecule responsible for thousands of metabolic processes that sustain life and function as catalysts to facilitate specific chemical reactions within the cell. These reactions are essential for the life of the organism. The living cell is the site of tremendous biochemical activity called metabolism. This is the process of chemical and physical changes which go on continually in the living organism; enzyme facilitate life processes in essentially all life-forms from viruses to man. Enzymes have been naturally tailored to perform under different physiological conditions. Build-up of new tissues, replacement of old tissues, conversion of food into energy, disposal of toxic materials, reproduction almost all the activities that can be characterised as 'life.' Enzymes act as life catalysts, substances that accelerate the rate of a chemical reaction. By reducing the activation energy (E_a) necessary to initiate the reaction, thus dramatically increasing the rate of reaction. Enzymes do not initiate reactions that would not naturally occur but they accelerate any reaction that is already underway. Enzymes enable the reaction to take place more rapidly at a safer, relatively low temperature that is consistent with living systems. During an enzyme-mediated reaction, the substrate physically attaches to the enzyme at its active site, allowing the substrate(s) to be converted to new product molecule(s). Most enzyme reaction rates are millions of times faster than those of comparable uncatalysed reactions. Enzymes are neither consumed by the reactions they catalyse, nor do they alter the equilibrium of these reactions. Enzymes are known to catalyse about 4000 biochemical reactions.

1.2 Proteinaceous nature of enzymes

Enzymes are proteins and are nature's own biocatalyst and their function is determined by their complex structure. With the exception of a small group of catalytic RNA molecules, all enzymes are proteins which are made up of amino acids linked together by peptide bonds. By the early 1800s, the proteinaceous nature of enzymes had been recognised. Knowledge of the chemistry of proteins drew heavily on improving techniques and concepts of organic chemistry in the second half of the 1800s, it culminated in the peptide theory of protein structure, usually credited to authors Fischer and Hofmeister. However, methods that had permitted the separation and synthesis of small peptides were unequal to the task of purifying enzymes. Indeed, there was no consensus that enzymes were proteins. After isolation of a series of crystalline proteolytic enzymes beginning with pepsin by Northrop and others in 1930, the proteinaceous nature of enzymes was established. They are high molecular weight compounds made up principally of chains of amino acids linked together by peptide bonds. Their catalytic activity depends on the integrity of their native protein conformation. If an enzyme is denatured or dissociated into its subunits, catalytic activity is usually lost. Thus the primary, secondary, tertiary and quaternary structures of protein enzymes are essential to their catalytic activity.

1.2.1 Structure of enzymes

Enzymes are proteins and are agreeable to structural analysis by the methods of protein chemistry, molecular biology and molecular biophysics. Like all proteins, enzymes are composed mainly of the 20 naturally occurring amino acids. The structures of enzymes can be elucidating by the physical methods such as spectroscopic methods, X-ray crystallography and more recently, multidimensional NMR methods. On the basis of arrangement of amino acids enzyme structure can be classified into following types.

Primary structure

The structure and reactivity of a protein are defined by the identity of the amino acids that make up its polypeptide chain, this amino acid sequence of the peptide chains is the primary structure of the enzyme.

Secondary structure

Secondary structure is due to the interaction of amino acids with each other in the same chain of protein. As a result the protein chain can fold up on itself in two ways, namely α-helix or β-sheet resulting secondary structures.

Tertiary structure

The arrangement of secondary structure elements and amino acid side chain interactions that define the three-dimensional structure of the folded protein, so that specific contacts are made between amino acid side chains and between backbone groups. The resulting folded structure of the protein is referred to as its tertiary structure.

Quaternary structure and domains

Many enzymes consist of more than one polypeptide chain (or subunit) that aggregate to confer catalytic activity. In some enzymes the subunits are identical, in others they differ in sequence and structure. This description of subunit arrangement in such enzymes is called the quaternary structure. A typical enzyme is not an entity completely folded as a whole, but may consist of apparently autonomous or semiautonomous folding units called domains. The 3D structure of enzymes is given in Fig. 1.1.

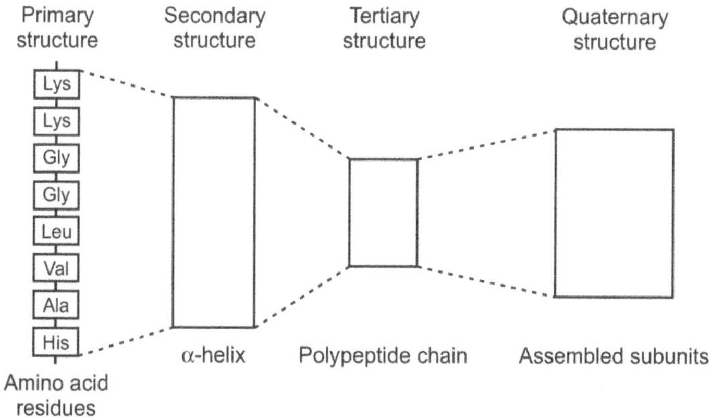

Figure 1.1: 3D structure of enzymes.

1.2.2 Chemical nature of enzymes

Enzymes are proteins that are produced by all living organisms. They are high molecular weight compounds made up principally of chains of amino acids linked together by peptide bonds. They have molecular weights ranging from 10000 to 2000000.

The important terminologies related to enzymes are given below:

Co-factor: A non-protein chemical component required for proteins biological activity are called co-factor.

Apoenzyme: The protein part of an active enzyme is called apoenzyme.

Holoenzyme: The active enzyme is composed of Apoenzyme and a co-factor is termed as holoenzyme.

Co-enzyme: Co-enzyme is a non-protein compound or substance that is necessary for an enzyme to initiate the function of the enzyme.

Prosthetic group: A co-enzyme or metal ion that is very tightly or even covalently bound to the protein component of the enzyme is called a prosthetic group.

1.2.3 Specificity of enzymes

Enzymes are the most remarkable and highly specialised proteins, they have a high degree of specificity for their substrates and they accelerate chemical reactions tremendously.

In general, four types of behaviour can be described:

1. Absolute specificity: Catalyse only one reaction.
2. Group specificity: Catalyses a particular type of functional group, which can occur in a variety of substrate.
3. Linkage specificity: Catalyses a particular type of chemical bond regardless of the rest of the molecular structure.
4. Stereochemical specificity: The enzyme will act on a particular steric or optical isomer.

1.3 Sources of enzymes

Enzymes occur in all living organisms and catalyse biochemical reactions necessary to support life. A wide array of enzymes are extracted from plant sources, they have many advantages including cost of production and stability of products. An ample range of sources are used for commercial enzyme production from a broad spectrum of plant species. Non-microbial sources provide a larger proportion of these, at the present time. Microbes are preferred to plants and animals as sources of enzymes because:

1. They are generally cheaper to produce.
2. Their enzyme contents are more predictable and controllable.
3. Regular supply due to absence of seasonal fluctuations and rapid growth of micro-organisms on inexpensive media.
4. Plant and animal tissues contain more potentially harmful materials than microbes, including phenolic compounds (from plants), endogenous enzyme inhibitors and proteases.
5. Microbial enzymes are also more stable than their corresponding plant and animal enzymes and their production is more convenient and safer.

About fifty years ago, enzymes were being extracted strictly from animals like pig and cow from their pancreases. Animal enzymes were multifold, they were not very stable at the low pH environment so that the enzyme product was destroyed before doing the job. To overcome this problem plant enzymes were discovered, most important one is extraction of peroxidase from horseradish roots occurs on a relatively large scale because of the commercial uses of the enzyme. Peroxidase can also be extracted from soyabean, it is also having the common features with horseradish peroxidase. Some plants like *Cruciferous* vegetables, including *broccoli*, cabbage, kale and collard and turnip greens and papaya are rich in catalase.

Wheat sprouts contain high levels of catalase and vegetarian sources of catalase include apricots, avocados, carrots.

Catalase is also present in some microbes and bacteria, *Aspergillus niger* culture also produces catalase enzyme.

1.4 Mechanism of enzyme action

Enzymes are macro-molecules that help to accelerate (catalyse) chemical reactions in biological systems. Some biological reactions in the absence of enzymes may be as much as a million times slower. Any chemical reaction converts one or more molecules, called the substrate, into different molecule(s), called the product. Most of the reactions in biochemical processes require chemical events that are unfavourable or unlikely in the cellular environment, such as the transient formation of unstable charged intermediates or the collision of two or more molecules in the precise orientation required for reaction. In some of the reactions like, digestion of food, send nerve signals, or contract a muscle simply do not occur at a useful rate without catalysis.

Enzyme overcomes these problems by providing a specific environment within which a given reaction can occur more rapidly. Enzymes are usually proteins – each has a very specific shape or conformation. Within this large molecule is a region called an active site, which has properties allowing it to bind tightly to the substrate molecule(s).

As proposed by Charles-Adolphe Wurtz, an active site is a three dimensional cleft or crevice formed by groups that come from different parts of the amino acid sequence - residues far apart in the amino acid sequence may interact more strongly than adjacent residues in the sequence. The active site encloses a substrate and catalyses its chemical transformation. The enzyme substrate complex was first discovered in 1880, is central to the action of enzymes. The enzyme–substrate (ES) interactions can be explained by the following theories.

1.4.1 Lock-and-key theory

The lock-and-key theory explains the high specificity of enzyme activity. Enzyme surfaces accommodate substrates having specific shapes and sizes, so only specific substances 'fit' in an active site to form an ES complex (Fig. 1.2). A limitation of this theory is that it requires enzymes conformations to be rigid. Research suggests that instead enzymes are at least somewhat flexible.

1.4.2 Induced-fit theory

1. A modification of the lock-and-key theory called the induced-fit theory proposes that enzymes have flexible conformations that may adapt to incoming substrates (Fig. 1.3).
2. The active site adopts a shape that is complementary to the substrate only after the substrate is bound.

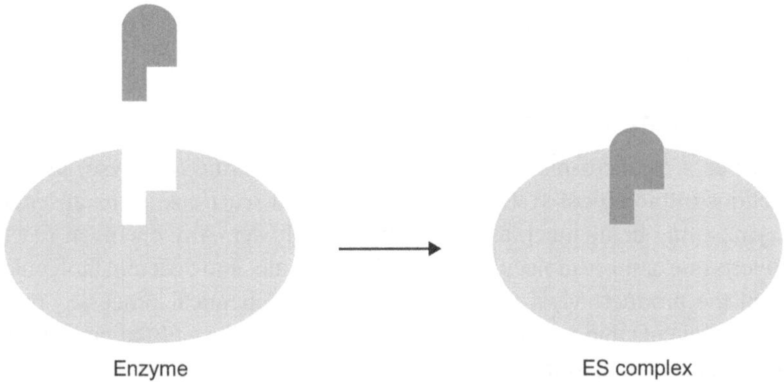

Figure 1.2: Enzyme surfaces accommodate substrates.

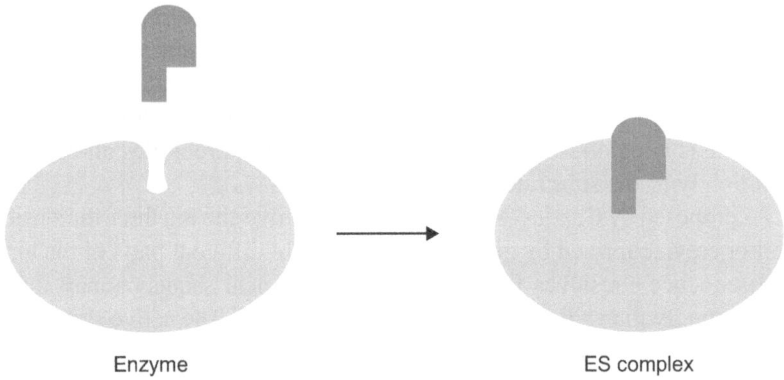

Figure 1.3: Enzymes showing flexible conformations.

1.5 Enzyme assay

In recent years enzyme assays have greatly advanced in their scope and in the diversity of detection principles employed. Enzyme assays are laboratory methods to visualise enzyme activities. An enzyme assay consists of mixing the enzyme with a substrate in a solution of controlled pH with any additional substance whose effect is to be tested, incubating the reaction mixture at an appropriate temperature for the required time, stopping the reaction precisely and then measuring the amount of reaction that has occurred. The amount of reaction that has taken place may be quantified by two ways, in terms of the disappearance of substrate or the appearance of product. In recent years a large variety of enzyme assays have been developed to assist the discovery and optimisation of enzymes, in particular for 'white biotechnology' where selective enzymes are used with great success for economically viable, mild and environmentally benign production processes.

1.5.1 Types of enzyme assay

All enzyme assays measure either the consumption of substrate or production of product over time. The Fig. 1.4 shows enzyme assay and different analytical techniques.

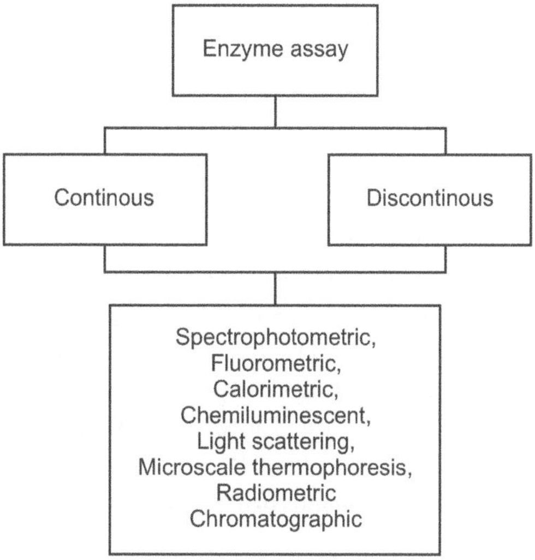

Figure 1.4: Flow diagram of enzyme assay classification.

A large number of different methods of measuring the concentrations of substrates and products exist and many enzymes can be assayed in several different ways. Biochemists usually study enzyme-catalysed reactions using four types of experiments:

1. Initial rate experiments: When an enzyme is mixed with a large excess of the substrate, the enzyme-substrate intermediate builds up in a fast initial transient. Then the reaction achieves a steady-state kinetics in which enzyme substrate intermediates remains approximately constant over time and the reaction rate changes relatively slowly. Rates are measured for a short period after the attainment of the quasi-steady state, typically by monitoring the accumulation of product with time. Because the measurements are carried out for a very short period and because of the large excess of substrate, the approximation free substrate is approximately equal to the initial substrate can be made. The initial rate experiment is the simplest to perform and analyse, being relatively free from complications such as back-reaction and enzyme degradation. It is therefore by far the most commonly used type of experiment in enzyme kinetics.

2. Progress curve experiments: In these experiments, the kinetic parameters are determined from expressions for the species concentrations as a function of time. The concentration of the substrate or product is recorded in time after the initial fast transient and for a sufficiently long period to allow the reaction to approach equilibrium. We note in passing that, while they are less common now, progress curve experiments were widely used in the early period of enzyme kinetics.
3. Transient kinetics experiments: In these experiments, reaction behaviour is tracked during the initial fast transient as the intermediate reaches the steady-state kinetics period. These experiments are more difficult to perform than either of the above two classes because they require rapid mixing and observation techniques.
4. Relaxation experiments: In these experiments, an equilibrium mixture of enzyme, substrate and product is perturbed, for instance by a temperature, pressure or pH jump and the return to equilibrium is monitored. The analysis of these experiments requires consideration of the fully reversible reaction. Moreover, relaxation experiments are relatively insensitive to mechanistic details and are thus not typically used for mechanism identification, although they can be under appropriate conditions.

Enzyme assays can be split into two groups according to their sampling method: continuous assays, where the assay gives a continuous reading of activity and discontinuous assays, where samples are taken, the reaction stopped and then the concentration of substrates/products determined.

Continuous assays

Continuous assays are most convenient, with one assay giving the rate of reaction with no further work necessary. There are many different types of continuous assays.

Spectrophotometric: In spectrophotometric assays, you follow the course of the reaction by measuring a change in how much light the assay solution absorbs. If this light is in the visible region you can actually see a change in the colour of the assay, these are called colorimetric assays. The MTT assay, a redox assay using a tetrazolium dye as substrate is an example of a colorimetric assay.

UV light is often used, since the common co-enzymes NADH and NADPH absorb UV light in their reduced forms, but do not in their oxidised forms. An oxidoreductase using NADH as a substrate could therefore be assayed by following the decrease in UV absorbance at a wavelength of 340 nm as it consumes the co-enzyme.

Direct versus coupled assays: Even when the enzyme reaction does not result in a change in the absorbance of light, it can still be possible to use a spectrophotometric assay for the enzyme by using a coupled assay. Here, the product of one reaction is used as the substrate of another, easily-detectable reaction.

Fluorometric: Fluorescence is when a molecule emits light of one wavelength after absorbing light of a different wavelength. Fluorometric assays use a difference in the fluorescence of substrate from product to measure the enzyme reaction. These assays are in general much more sensitive than spectrophotometric assays, but can suffer from interference caused by impurities and the instability of many fluorescent compounds when exposed to light.

An example of these assays is again the use of the nucleotide co-enzymes NADH and NADPH. Here, the reduced forms are fluorescent and the oxidised forms non-fluorescent. Oxidation reactions can therefore be followed by a decrease in fluorescence and reduction reactions by an increase. Synthetic substrates that release a fluorescent dye in an enzyme-catalysed reaction are also available, such as 4-methylumbelliferyl-β-D-galactoside for assaying β-galactosidase.

Calorimetric: Calorimetry is the measurement of the heat released or absorbed by chemical reactions. These assays are very general, since many reactions involve some change in heat and with use of a microcalorimeter, not much enzyme or substrate is required. These assays can be used to measure reactions that are impossible to assay in any other way.

Chemiluminescent: Chemiluminescence is the emission of light by a chemical reaction. Some enzyme reactions produce light and this can be measured to detect product formation. These types of assay can be extremely sensitive, since the light produced can be captured by photographic film over days or weeks, but can be hard to quantify, because not all the light released by a reaction will be detected.

The detection of horseradish peroxidase by enzymatic chemiluminescence (ECL) is a common method of detecting antibodies in western blotting. Another example is the enzyme luciferase, this is found in fireflies and naturally produces light from its substrate luciferin.

Light scattering: Static light scattering measures the product of weight-averaged molar mass and concentration of macro-molecules in solution. Given a fixed total concentration of one or more species over the measurement time, the scattering signal is a direct measure of the weight-averaged molar mass of the solution, which will vary as complexes form or dissociate. Hence the measurement quantifies the stoichiometry of the complexes as well as kinetics.

Light scattering assays of protein kinetics is a very general technique that does not require an enzyme.

Discontinuous assays

Discontinuous assays are when samples are taken from an enzyme reaction at intervals and the amount of product production or substrate consumption is measured in these samples.

Radiometric: Radiometric assays measure the incorporation of radioactivity into substrates or its release from substrates. The radioactive isotopes most frequently used in these assays are ^{14}C, ^{32}P, ^{35}S and ^{125}I. Since radioactive isotopes can allow the specific labelling of a single atom of a substrate, these assays are both extremely sensitive and specific. They are frequently used in biochemistry and are often the only way of measuring a specific reaction in crude extracts (the complex mixtures of enzymes produced when you lyse cells). Radioactivity is usually measured in these procedures using a scintillation counter.

Chromatographic: Chromatographic assays measure product formation by separating the reaction mixture into its components by chromatography. This is usually done by high-performance liquid chromatography (HPLC), but can also use the simpler technique of thin layer chromatography. Although this approach can need a lot of material, its sensitivity can be increased by labelling the substrates/products with a radioactive or fluorescent tag. Assay sensitivity has also been increased by switching protocols to improved chromatographic instruments (e.g., ultra-high pressure liquid chromatography) that operate at pump pressure a few-fold higher than HPLC instruments.

1.5.2 Nature and origin of diversity in enzyme assay

The major consideration in the design of an enzyme assay method is that the amount of enzyme present should be the only variable which affects the reaction rate. This condition is more likely to be met if the assay procedure ensures optimal substrate concentration, pH, temperature, etc., and thus many of the modifications that have from time to time been made to assay methods have been motivated by the desire to achieve optimal conditions. Unfortunately, it has not always been appreciated that such modifications almost inevitably result in an alteration in the enzyme activity as measured by the method.

1.5.3 Measures of enzyme activity

In order to express the activity of an enzyme in absolute terms it is necessary to ensure that the assay procedure used is measuring the true initial velocity and that this is proportional to the enzyme concentration. Under these conditions the ratio will be a constant that can be used to express the activity of an enzyme quantitatively. This can be valuable for comparing data obtained with the enzyme from different laboratories, assessing the effects of physiological and pharmaceutical challenges on cell or tissues, monitoring the extent of purification of enzymes and comparing the activities of different enzymes, or of the same enzyme from different sources or with different substrates.

1.5.4 Units of enzyme and specific activity

The activity of an enzyme may be in any convenient units, such as absorbance change per unit time per mg of enzyme protein, but it is preferable to have a more standardised unit in order to facilitate comparisons. The most commonly used quantity is the units, sometime referred to as the international unit or Enzyme unit. One unit of an enzyme activity is defined as that catalysing the conversion of 1 μM substrate in 1 min. The specific activity of an enzyme preparation is the number of units per mg of protein.

If the relative molecular mass of an enzyme is known it is possible to express the activity as the molecular activity, which is defined as the number of units per μmol of enzyme, in other words the number of mol of product formed, substrate used per mol of enzyme per minute. This may not correspond to the number of mol substrate converted per enzyme active site per minute since an enzyme molecule may contain more than one active site. If the number of active site per mol is known the activity may be expressed as the catalytic centre activity, which corresponds to mol substrate used, or product formed per minute per catalytic center. The nomenclature commission of the International Union of Biochemistry recommended *katal* as the enzyme unit which is abbreviated as kat, in this the unit of time is expressed in terms of second rather than minute with international system of unit (SI units).

One katal corresponds to the conversion of 1 mol of substrate per second. It is larger quantity than the enzyme unit.

$$1 \text{ kat} = 60 \text{ mol min}^{-1} = 6 \times 10^7 \text{ Units}$$

$$1 \text{ Unit} = 1 \text{ μmol min}^{-1} = 16.67 \text{ nkat}$$

Specific activity

Specific activity is a term used in measuring enzyme kinetics. It is defined as the amount of substrate the enzyme converts (reactions catalysed), per mg protein in the enzyme preparation, per unit of time or in other words the enzymatic activity per unit mass of enzyme. Specific activity is a measure of enzyme purity. The value becomes larger as an enzyme preparation becomes more pure, since the amount of protein (mg) is typically less, but the rate of reaction stays the same.

1.5.5 Michaelis–Menten kinetics

As enzyme-catalysed reactions are saturable, their rate of catalysis does not show a linear response to increasing substrate. If the initial rate of the reaction is measured over a range of substrate concentrations (denoted as $[S]$), the reaction rate (v) increases as $[S]$ increases, as shown in Fig. 1.5. However, as $[S]$ gets higher, the enzyme becomes saturated with substrate and the rate reaches V_{max}, the enzyme's maximum rate.

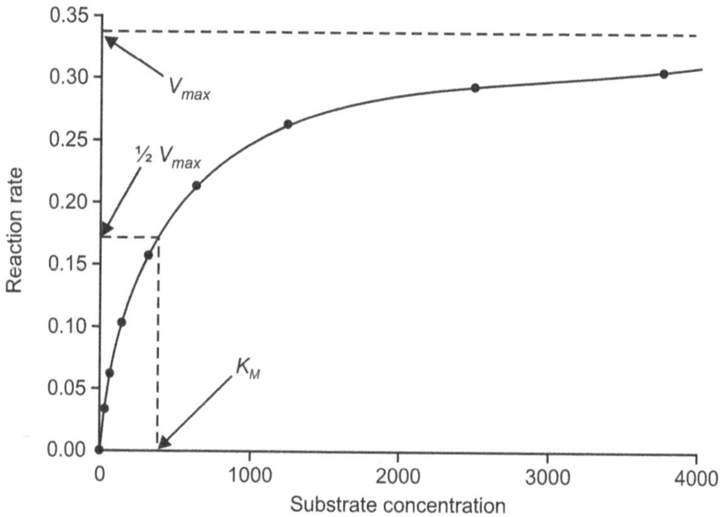

Figure 1.5: Saturation curve for an enzyme showing the relation between the concentration of substrate and rate.

The Michaelis–Menten kinetic model of a single-substrate reaction is shown in Eq. 1.1.

$$\underbrace{E + S \underset{k_{-1}}{\overset{k_1}{\rightleftharpoons}}}_{\text{Substrate binding}} ES \underbrace{\xrightarrow{k_2}}_{\text{Catalytic step}} E + P \qquad \ldots(1.1)$$

There is an initial bimolecular reaction between the enzyme E and substrate S to form the enzyme–substrate complex ES. Although the enzymatic mechanism for the unimolecular reaction $ES \xrightarrow{k_{cat}} E + P$ can be quite complex, there is typically one rate-determining enzymatic step that allows this reaction to be modelled as a single catalytic step with an apparent unimolecular rate constant k_{cat}. If the reaction path proceeds over one or several intermediates, k_{cat} will be a function of several elementary rate constants, whereas in the simplest case of a single elementary reaction (e.g., no inter-mediates) it will be identical to the elementary unimolecular rate constant k_2. The apparent unimolecular rate constant k_{cat} is also called turnover number and denotes the maximum number of enzymatic reactions catalysed per second.

The Michaelis–Menten equation describes how the (initial) reaction rate v_0 depends on the position of the substrate-binding equilibrium and the rate constant k_2.

$$v_0 = \frac{V_{\max}[S]}{K_m + [S]}$$

with the constants

$$K_m = \frac{k_2 + k_{-1}}{k_1} \approx K_D$$

$$V_{\max} = k_{cat}[E]_{tot}$$

This Michaelis–Menten equation is the basis for most single-substrate enzyme kinetics. Two crucial assumptions underlie this equation (apart from the general assumption about the mechanism only involving no intermediate or product inhibition and there is no allostericity or co-operativity). The first assumption is the so-called quasi-steady-state assumption (or pseudo-steady-state hypothesis), namely that the concentration of the substrate-bound enzyme (and hence also the unbound enzyme) changes much more slowly than those of the product and substrate and thus the change over time of the complex can be set to zero $d[ES]/dt = 0$.

The second assumption is that the total enzyme concentration does not change over time, thus $[E]_{tot} = [E] + [ES] = \text{constant}$. A complete derivation is discussed below.

The Michaelis constant K_m is experimentally defined as the concentration at which the rate of the enzyme reaction is half $V_{\max}$, which can be verified by substituting $[S] = K_m$ into the Michaelis–Menten equation and can also be seen graphically. If the rate-determining enzymatic step is slow compared to substrate dissociation ($k_2 << k_{-1}$), the Michaelis constant K_m is roughly the dissociation constant K_D of the ES complex. If $[S]$ is small compared to K_m then the term $[S]/(K_m + [S]) \approx [S]/K_m$ and also very little ES complex is formed, thus $[E]_0 \approx [E]$. Therefore, the rate of product formation is:

$$v_0 \approx \frac{k_{cat}}{K_m}[E][S] \qquad \text{if}\,[S] \ll K_m$$

Thus the product formation rate depends on the enzyme concentration as well as on the substrate concentration, the equation resembles a bimolecular reaction with a corresponding pseudo-second order rate constant k_2/K_m. This constant is a measure of catalytic efficiency. The most efficient enzymes reach a k_2/K_m in the range of 10^8–10^{10} M^{-1} s^{-1}. These enzymes are so efficient they effectively catalyse a reaction each time they encounter a substrate molecule and have thus reached an upper theoretical limit for efficiency (diffusion limit), these enzymes have often been termed perfect enzymes.

1.6 Factors affecting enzyme activity

Velocity or rate of enzymatic reaction is assessed by the rate of change in concentration of substrate or product at a given time duration. Various factors which affect the activity of enzymes include:

1. Substrate concentration.
2. Enzyme concentration.
3. Effect of temperature.
4. Effect of pH.
5. Effect of time.
6. Concentration of co-enzymes.
7. Concentration of metal ion activators.
8. Presence of inhibitors.
9. Product concentration.

1.6.1 Effect of substrate concentration

Reaction velocity of an enzymatic process increases with constant enzyme concentration and increase in substrate concentration. The velocity *(V)* is expressed in micromoles of substrate converted per minute. As the concentration of substrate increases, the velocity of the reaction increases. Continued increase in substrate concentration may lead to a reduction in rate of the reaction and leads to flattened curve.

The maximum velocity obtained from a enzymatic reaction is called as V_{max}. V_{max} represents the maximum reaction rate possible in the presence of excess substrate (Fig. 1.6).

1.6.2 Effect of enzyme concentration

As there is optimal substrate concentration, rate of an enzymatic reaction or velocity *(V)* is directly proportional to the enzyme concentration (Fig. 1.7). Presence of excess substrate and an increase in the enzyme concentration may result in some limitations. It is worthy of note that the enzymes are rarely saturated with substrates under physiological conditions.

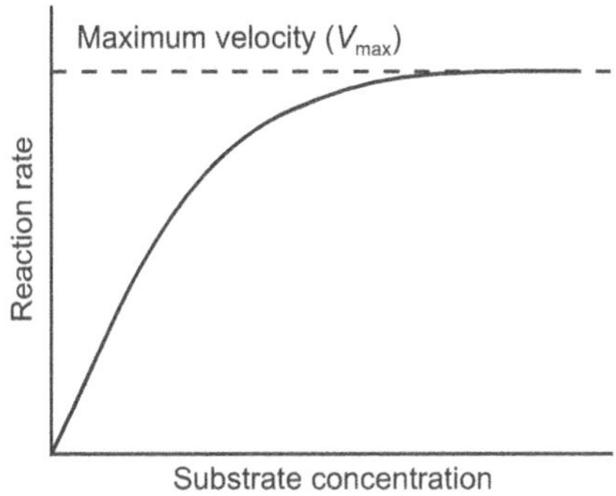

Figure 1.6: Effect of substrate concentration.

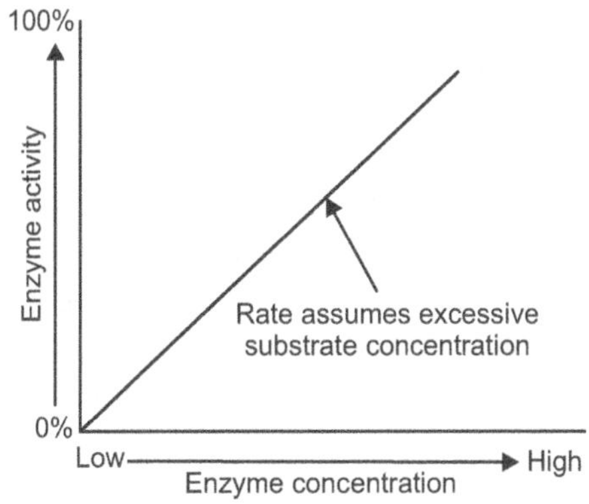

Figure 1.7: Effect of enzyme concentration

1.6.3 Effect of temperature

At very low temperature, enzymes are inactive. Enzyme activity increases gradually with the rise in temperature until a temperature at which the enzyme attains its maximal activity, this temperature is called optimum temperature (Fig. 1.8) which lies between 37–40°C in humans.

Optimum temperature is the temperature at which the enzyme attains its maximal activity. The rise in temperature from low temperature to optimum temperature causes an increase in the rate of reaction due to:

1. The rise in temperature increases the initial energy of substrate leading to a decrease in the activation energy and lower the energy barrier of the reaction.
2. Also, the rise in temperature increases collision of the molecules, i.e., more molecules become in the bond forming or bond breaking distance.

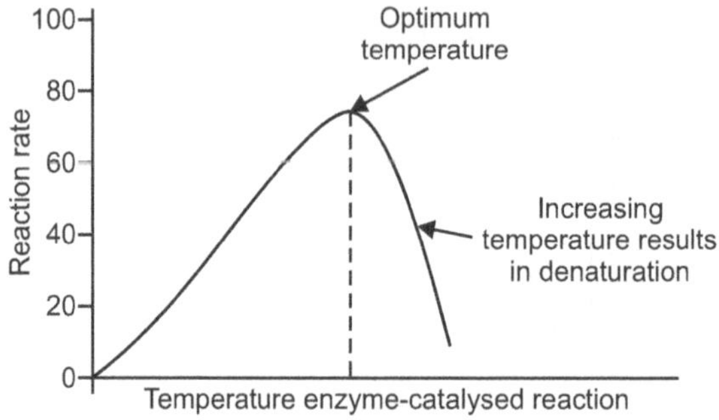

Figure 1.8: Enzyme catalysed reaction and optimum temperature.

The rise in temperature above the optimum temperature leads to a decrease in the rate of enzyme activity. At higher temperature (60–65°C in humans) irreversible loss of enzyme activity occurs due to denaturation of the enzymes, which are protein in nature.

1.6.4 Effect of pH

Each enzyme has an optimum pH at which it attains its maximal activity (Fig. 1.9). Any change of pH below or above the optimum pH decreases the rate of enzyme action due to:

1. Changes in pH lead to changes in the ionisation state of the substrate or the enzyme or both.
2. Also, extreme changes in pH lead to denaturation of the enzyme that is protein in nature.

Some exceptions such as pepsin (with optimum pH 1–2), alkaline phosphatase (with optimum pH 9–10) and acid phosphatase (with optimum pH 4–5) are even noticed.

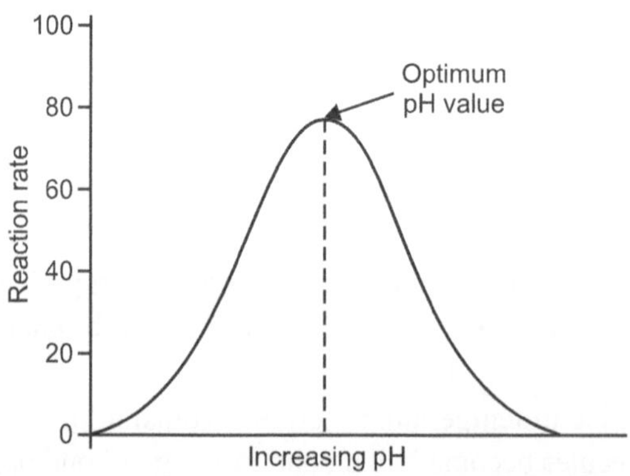

Figure 1.9: Optimum pH value and rate of enzyme.

Effect of time

At the beginning, the rate of reaction increases but by time the rate of reaction decreases due to: (i) depletion of substrate, (ii) accumulation of end products and (iii) change in pH of the reaction, which becomes different from the optimum pH of the enzyme.

1.6.5 Concentration of co-enzymes

In the conjugated protein enzymes that need co-enzymes, the increase in the co-enzyme concentration causes an increase in the rate of enzyme action.

1.6.6 Concentration of metal ion activators

The increase in metal ion activators increases the rate of enzyme action. Many enzymes are activated by metal ions, e.g., (i) chloride ions activate salivary amylase and (ii) calcium ions activate thrombokinase enzyme.

1.6.7 Presence of inhibitors

Inhibitors decrease or even abolish enzyme activity. Enzyme inhibitors may be: (i) competitive inhibitors and (ii) non-competitive inhibitors.

1.6.8 Effect of product concentration

In case of a reversible reaction catalysed by a enzyme, as per the law of mass action the rate of reaction is slowed down with equilibrium. So, rate of reaction is slowed, stopped or even reversed with increase in product concentration.

This phenomena can be better explained by the equation:

$$A \xrightarrow{E1} B \xrightarrow{E2} C \xrightarrow{E3} \| \!\rightarrow D$$

In the above equation, in case of absence of the enzyme E3, the product C will accumulate. Enzymatic activity of E2 will be inhibited with accumulation of the product C. In such inborn error of one enzyme will block the whole pathway.

2

Engineering enzymes for biocatalysts

2.1 Introduction

Improvement in the activity and usefulness of an existing enzyme or creation of a new enzyme activity by making suitable changes in its amino acid sequence is called enzyme engineering. When this approach is used to modify the properties of any protein, whether enzyme or nonenzyme, it is termed as protein engineering. Since enzymes are proteins, enzyme engineering is a part of the larger activity of protein engineering. Enzyme engineering utilises r-DNA technology to introduce the desired changes in amino acid sequences of enzymes. In addition, the level of production of an enzyme may be increased by introducing more copies of the gene into the concerned organism.

The chief objective of enzyme engineering is to produce an enzyme that is more useful for industrial and other applications.

The various properties of an enzyme that may be modified to achieve this objective are as follows:

1. Improved kinetic properties.
2. Elimination of allosteric regulation.
3. Enhanced substrate and reaction specificity.
4. Increased thermostability.
5. Alteration in optimal pH.
6. Suitability for use in organic solvents.
7. Increased/decreased optimal temperature.

2.2 Principles of enzyme engineering

The structure and function of an enzyme molecule, for that matter of any protein molecule, are chiefly determined by its amino acid sequence, i.e., its primary structure. Therefore, any change in the properties of an enzyme is always reflected in its primary structure. Conversely, a change in the amino acid sequence should alter the properties of the enzyme. But, this is not always the case because the enzymatic properties, etc., are changed only when amino acid changes are introduced in certain critical regions of the protein. Therefore, it is of great importance to know the critical regions for the various functions of an enzyme and to be able to predict the effect of specific amino acid changes in these areas on the various functions. However, the present knowledge of the relationships between amino acid sequences, 3D structure of protein and properties of enzymes, obtained from a large database is only partially operative. It allows an explanation of the changes in structure and function on the basis of the changes in amino acid sequence, but it does not allow a dependable prediction of the influences of specific amino acid changes on the structure and function of the enzymes (Fig. 2.1).

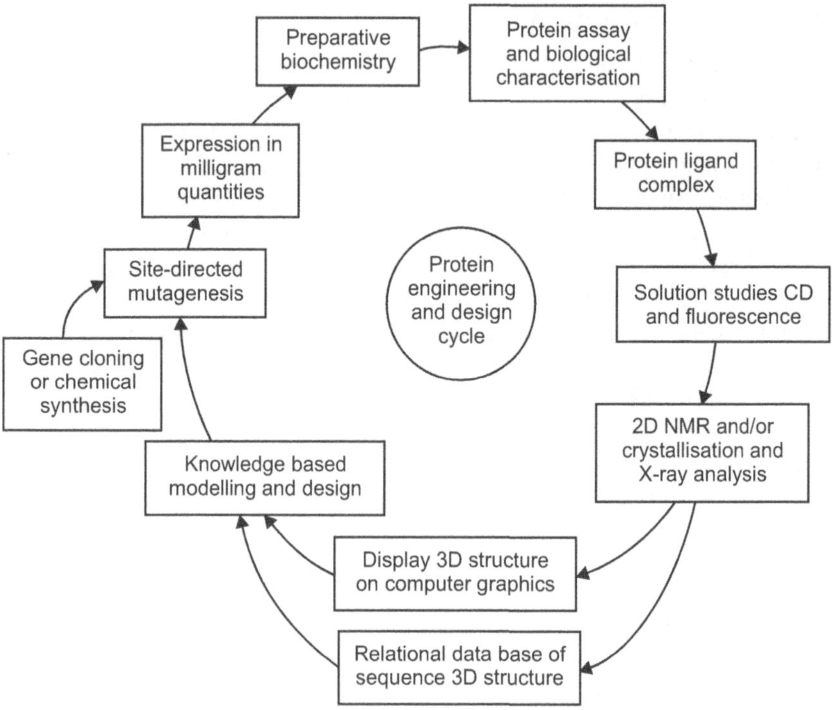

Figure 2.1: Protein engineering and design cycle.

It may, however, be reasonable to anticipate that as more elaborate databases and improved softwares become available, it should become possible to predict with a far greater confidence the structural and functional changes in enzymes produced by the specified changes in their amino acid sequences. The effectiveness of enzyme engineering will be greatly enhanced then and this activity may have a tremendous influence on enzyme technology.

2.3 Steps in enzyme engineering

The strategies for enzyme engineering and their theoretical considerations are quite involved. The steps involved in enzyme engineering (Fig. 2.2) are briefly described in as under.

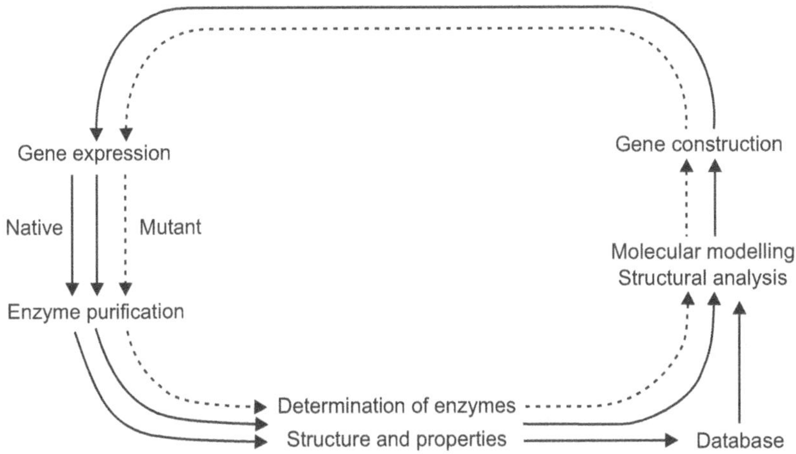

Figure 2.2: Steps in enzyme engineering.

1. The first step consists of isolation of the concerned enzyme and determination of its structure and properties. Both amino acid sequence and the 3D structure are usually obtained from X-ray diffraction, nuclear magnetic resonance (NMR), etc.
2. The data so obtained are analysed together with the database of known and putative structural effects of amino acid substitutions on enzyme structure and function. Molecular modelling is performed to determine a possible change in amino acid sequence for the desired improvement in the structure/function of an enzyme.
3. The next step consists of constructing a gene that will encode the amino acid sequence specified at the end of step 2. This is best achieved by isolation and cloning of the endogenous gene encoding the concerned enzyme and using this gene for site directed mutagenesis.
4. Once the appropriated gene is constructed, it is introduced and expressed in a suitable host, e.g., *E. coli*.
5. The recombinant or mutant enzyme so produced is isolated, purified and used for determination of its structure and properties. The information so obtained is added to the database. If the enzyme structure and function are not altered as desired, the next cycle of experimentation (step 2–5) is undertaken.

2.4 Engineering enzymes for biocatalysis

Natural enzymes are capable of catalysing reactions with up to 10^{17}-fold rate accelerations and with exquisite control of regio- and stereochemistry. Such control makes the use of enzymes very attractive as an alternative to traditional catalysts in the synthesis of complex and highvalue molecules, especially where chemical routes are difficult to implement. Enzymes have evolved over billions of years to operate most effectively under physiological conditions, on a narrow range of natural substrates and usually at concentrations in the low mM range. By contrast, an efficient industrial synthetic process may require a biocatalytic enzyme to operate on non-natural substrates and under conditions in which the enzyme becomes unstable or inactive, such as extremes of temperature, pH and pressure, after repeated or prolonged use, or in the presence of organic solvents which facilitate substrate solubility or product extraction. As a result of these requirements, by far the majority of naturally occurring enzymes are not suitable for industrial-scale biocatalytic processes without further modification of the enzyme itself. Many examples of the engineering of enzymes for their improved use as biocatalysts have been published including: increased activity on novel substrates, altered enantioselectivity, enhanced stability at extreme temperatures, pH or in organic solvents and the preference for alternative enzyme cofactors.

Two broad approaches have been taken to engineer the amino-acid sequences of enzymes, namely rational design and directed evolution. Rational design applies the growing knowledge of enzyme structure and function to make one or more amino-acid changes that are predicted to elicit the desired improvements to enzyme function. Such knowledge is often based upon the bioinformatical analysis of protein sequences or amino-acid propensities, generalised rules derived by characterising the effect of mutations upon enzyme properties and by implementation of molecular potential functions that enable the effect of mutations upon structure to be predicted.

On the other hand, directed evolution techniques mimic natural evolution processes such as random mutagenesis and sexual recombination. Just as nature has produced its enzymes by the process of evolution without using any 'knowledge' of enzyme structure and function, directed evolution similarly permits us to engineer enzymes without understanding them in great detail. Computational methods also provide a valuable tool for the evolutionary design of proteins. Such tools can potentially be used to obtain useful information from protein sequences and structures that can guide the random mutagenesis of proteins making directed evolution more efficient.

Alternatively, the use of molecular potential functions can be used to predict the effects of mutations on protein structure and stability for libraries of enzyme variants generated *in silico*. The latest developments and applications of these tools will also be discussed.

Recent advances in both the rational and evolutionary design of enzymes will be outlined below. However, these two general approaches are not mutually exclusive and techniques for improving directed evolution strategies may include the knowledge-based design of enzyme variant libraries, as will also be discussed. Overall, this section discusses the protein engineering of enzymes for biocatalysis and associated new patents. While many of the methods are of general applicability to protein engineering, only those of interest to modifying enzymes for biocatalysis are included. The final section brings together examples of the applications of protein engineering to various biocatalytic enzymes, particularly those appearing as published patents.

2.4.1 Rational design strategies

The degree of success of rational design studies depends largely on the enzyme property to be altered. For example, the expression level of a protein is readily improved by ensuring the use of the DNA codons for which the host cell has the greatest preference. Many successful studies have applied rational design to the improved conformational stability of enzymes, through the introduction of disulphide bonds, proline residues or the mutation of protein sequences towards the consensus for a given enzyme family. An alternative approach that uses combinatorial mutation towards the consensus sequence is discussed later as a directed evolution strategy. Since these successful applications, progressively more challenging goals have been achieved using rational design.

An ability to improve the aqueous solubility of membrane-bound enzymes is of potential use in the application of these enzymes as biocatalysts and also for determining their three-dimensional structures which would critically enable further enzyme design work. An automated computational design strategy has recently been demonstrated to effectively solubilise two membrane-associated proteins, phospholamban and KcsA.

In the example of phospholamban (PLB), the researchers first predicted the sites for mutagenesis, in the absence of a PLB structure. They used the experimentally observed effects of mutations upon pentameric protein structure formation, to define a perturbation index as a sinusoidal function of residue position in helices. While one residue, F32, was mutated to a more chromogenic tyrosine residue to assist experimental characterisation, the remaining ten were subjected to an energy minimisation designed to optimise the net free energy, α-helical propensity and water solubility of the protein. Sequence diversity was introduced using a sequence entropy term in the potential function. In the PLB example, the new water-soluble variant was found to retain the pentameric form and degree of helical structure and also to be stabilised upon phosphorylation, in the same manner as the wild-type PLB. More recently, the formation of this water-soluble variant of PLB has subsequently enabled the crystal structure of its tetrameric form to be obtained, providing information about the PLB structure that was previously difficult to obtain. While the techniques described are limited to the solubilisation of helical membrane-associated proteins, there is a clear potential for the application of this approach to the solubilisation of membrane associated enzymes with highly α-helical content.

Some striking examples of the *de novo* design of protein structures have been achieved based on existing knowledge and using computational design algorithms. Computational design has been successfully used to introduce binding functionality into natural protein structures, such as the design of TNT, L-lactate, D-lactate and serotonin binding-sites into *Escherichia coli* periplasmic binding proteins (PBP) and also to redesign the DNA-sequence specificity of an endonuclease.

Furthermore, computational design has been used to add binding functionality, such as calcineurin-binding, to protein scaffolds that were themselves obtained previously by rational design.

Enzyme functionality has also been introduced into protein scaffolds using computational design methods, demonstrating the potential application of such methods to biocatalysis. For example, the ability to hydrolyse paranitrophenyl acetate ester has been engineered into the catalytically inert thioredoxin protein scaffold. In a spectacular example, Hellinga and coworkers introduced triose phosphate isomerase (TIM) activity into the bacterial ribose-binding protein (RBP), a periplasmic receptor that has no known catalytic activity, using the computational design algorithm described above. The nascent enzyme functionality was then further improved by experimental directed evolution to achieve a rate enhancement of 3.4×10^5 over the uncatalysed reaction.

The *de novo* design of a protein structure that also has partial enzyme functionality has been demonstrated for a four-helix bundle with stoichiometric ferridoxase activity, obtained by introducing a diiron-binding site. However, the creation of an enzyme functionality that has an efficiency comparable with natural enzymes, remains elusive using rational design techniques alone, most likely due to the greater complexity of the problem and the reliance on static protein structure models and transition-state theory for predicting enzyme structure-function relationships. These models omit the potentially significant contributions of enzyme dynamics and quantum tunnelling towards catalysis. Recent efforts to study the function of enzymes using molecular dynamics simulations and experimental probes of structural dynamics are expected to lead to improved rational design methods for enzyme functionality. While the rational design of a highly efficient catalytic activity into an inert protein scaffold remains challenging, it is worth noting that the improvement of an existing enzyme activity using computational design methods has been recently demonstrated. In this example, the activity of chorismate mutase from *E. coli* was improved by 50% despite modelling an *ab initio* calculated transition-state structure bound into only a static active-site configuration of the enzyme.

2.4.2 Directed evolution strategies

While rational design methods seek to design beneficial mutations or protein sequences by applying empirically derived rules or theoretical models, directed evolution uses a combinatorial approach to create libraries of enzymes from which enhanced variants can be identified experimentally. The two most important aspects to consider when devising directed evolution programmes are: (i) the availability of suitable screening or selection-based methods for identifying 'hits' with improvements of the desired enzyme property and (ii) an appropriate mutagenesis strategy given the extent of change required in the target property and the existing availability of information relating to the target enzyme structure and function. These two aspects are discussed below along with recent related innovations that impact on the application of directed evolution to enzyme biocatalysis.

2.4.3 Screening and selection methods

For directed evolution to be successful and efficient, it is essential that good screening or selection tools are employed for accurately identifying enzymes with improvements in the desired properties. Furthermore, the screening or selection based method available for the target activity or enzyme property critically dictates the enzyme variant library size that can be practically tested and therefore, the most appropriate strategy for library design. In general, the ability to isolate variants from larger libraries will improve the chances of success in finding an enzyme with the desired enhanced properties. Screening methods analyse individual colonies cultured on agar plates, in microtitre plates, or in microfluidic chips and are typically used to screen in the range of 10^3–10^6 individual enzyme variants.

However, selection-based methods can use phage display, cell surface display, ribosomal display, or plasmid display, to isolate high-performing enzymes from libraries ranging up to 10^{14} variants. While selection-based methods enable much greater sequence diversity to be explored, they rely upon indirect selection for catalytic activity such as binding to transition-state analogues, whereas screening methods bring the advantage of direct evaluation of catalytic proficiency. Screening methods also permit enzyme variants to be assessed in accurately controlled and non-physiological environments which may otherwise prohibit the use of selection-based techniques.

The gap between library sizes that can be practicably analysed by screening and by selection-based techniques is gradually being diminished, for example by using fluorescence-activated cell-sorting (FACS) of cell surface displayed enzymes, capable of sorting up to 10^7 enzyme variants per hour. A recent high-throughput screening technique has been developed in which water-in-oil-in-water emulsion droplets are sorted by FACS as outlined in Fig. 2.3, to

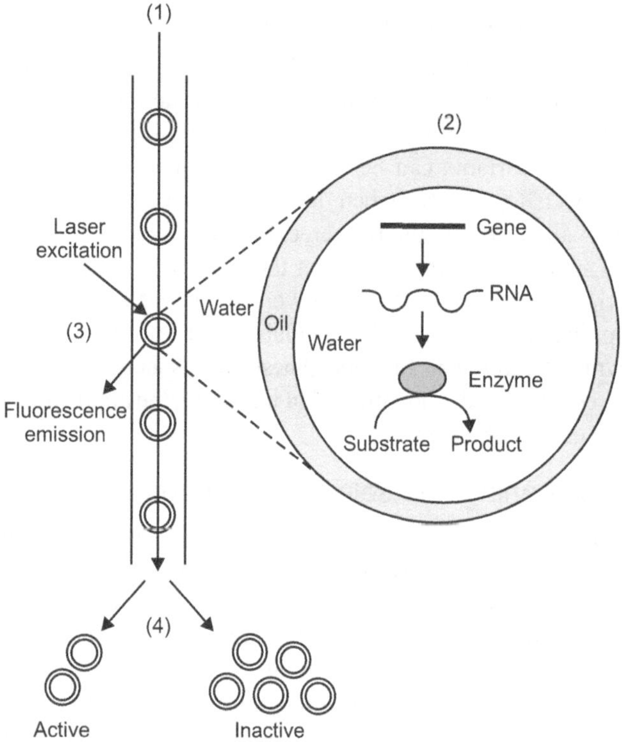

Figure 2.3: Selection of active enzymes in double emulsion microdroplets by fluorescence-activated cell sorting (FACS): (1) water-in-oil-in-water emulsion droplets are passed through a fluorescence-activated cell sorter, (2) each of these double emulsion droplets contains a variant gene which is transcribed and translated to a mutant enzyme. Active enzymes convert fluorogenic substrate into fluorescent product. (3) laser excitation produces fluorescence for double emulsion droplets containing active enzyme and (4) droplets are sorted by FACS into those containing active and inactive enzyme variants.

identify the best enzyme variants. This technique takes advantage of *in vitro* compartmentalisation, in which library encoding DNA and an *in vitro* transcription/translation reaction mixture create enzyme variants within an emulsion droplet. The droplets also contain a fluorogenic enzyme substrate, such that enzyme variants can then be assessed and sorted by FACS. The DNA that encodes the highest performing variants can then also be isolated from the same droplet and used for sequencing. In an alternative version of this technique, single bacterial cells expressing enzyme variants cytoplasmically or displayed on the cell surface, can be encapsulated in water-in-oil-in-water emulsions and screened by FACS.

2.4.4 Library construction

The first methods of directed evolution employed mutagenesis that was targeted across entire genes by error-prone PCR or recombination of homologous genes by DNA shuffling. While these techniques provided the ground breaking benefits of being able to enhance the properties of enzymes without the need for any knowledge of the enzyme structure or its relationship to enzyme function, they still suffer from a number of deficiencies that present a barrier to their wider application. For example, the gains in enzyme activity or other desired properties obtained by these techniques, may not always sufficiently improve an industrial biocatalytic process such that an investment in them is seen as worthwhile. On the other hand, performing directed evolution experiments require a degree of experience sufficient to overcome the difficulties in creating large libraries with techniques that depends on inefficient DNA ligation reactions.

Since the early directed evolution experiments a number of advances have been made that help to overcome these issues, in particular, the development of mutagenesis techniques that avoid the need for DNA ligations. One example was the introduction of mutator strains which enable mutations to accumulate in plasmids harboured within a bacterial strain deficient in DNA repair mechanisms. However, such strains were of limited use as the chromosomal DNA of the host strain was also susceptible to mutagenesis, resulting in long-term instability in cell viability. A recent technique significantly reduces this problem by using a highly error-prone DNA polymerase that preferentially mutates ColE1 plasmids. The errorprone DNA polymerase was created by introducing three point mutations based upon homology with Taq polymerase I. It has been shown to introduce an 80000-fold greater level of mutagenesis on average, than wild-type DNA polymerase I and targets plasmid genes with a 400-fold preference over chromosomal gene mutagenesis.

Many developments have also been made that improve the potential gains obtained by directed evolution strategies. While the ability to isolate variants from larger libraries will lead to a greater chance of success in finding an enzyme with the desired enhanced properties, there may be limitations imposed by the library size that can be practicably assessed by the available screening or selection-based methods. It is, therefore, recognised that an ability to improve the quality of enzyme libraries would be beneficial, for example, by promotion of the diversity or uniqueness of clones, or by reducing the number of enzyme residues subjected to random mutagenesis, without incurring any loss in evolutionary potential. Most DNA shuffling methods utilise the homology between parent sequences to reassemble fragments that have been generated using either restriction-enzyme digests, or short bursts of PCR.

Consequently it is difficult to recombine sequences with less than approximately 70% DNA sequence similarity, thus potentially limiting the range of genetic diversity and therefore, enzyme functionality that can be explored. Recent methods have enabled the non-homologous random recombination (NRR) of DNA sequences. While most NRR methods typically achieve only a single cross over, a few enable multiple cross overs to occur, including a technique in which DNA hairpins are added to a mixture of DNA fragments before reassembly by ligation with T4 DNA ligase. The use of T4 DNA ligase enables the recombination of DNA fragments independently of their homology, whereas the addition of the DNA hairpins, which can only ligate at one end, permits the range of DNA sequence lengths obtained to be carefully controlled.

In an alternative approach to NRR, a method that mimics exon shuffling has been described in an example which recombined two distantly related polymerase genes, to identify novel DNA polymerases. The approach, called structure-based combinatorial protein engineering (SCOPE), requires structural knowledge or sequence analysis of the parent enzymes to design PCR primers that enable recombination to occur at defined structural boundaries.

Amino-acid mutations introduced by traditional error-prone PCR tend to be biased towards those that can be obtained by single-base substitutions, typically accessing only 5.7 alternative residues on average. Greater sequence diversity can be obtained by applying saturation mutagenesis to one or more specific sites using degenerate oligonucleotides that encode for potentially all twenty amino-acids. In one approach, every single site in a protein can be systematically mutagenised independently to identify the optimal residues at each position. The best mutations can then be recombined with the aim of combining the enhancements endowed by each individual mutation.

Another recent method enables sequence saturation mutagenesis (SeSaM) to occur with less bias than error-prone PCR, at random positions within the targeted sequence, yet with only a single 1-3bp mutation appearing in each variant. The method creates random-length DNA gene fragments by PCR doped with α-phosphothioate nucleotides (dNTPaS) and selective cleavage of the sites of dNTPaS incorporation using iodine. The 3′ ends of each DNA fragment are then 'tailed' with typically 1–3 units of the universal base deoxyinosine (dITP), before elongation to the full length genes by PCR with a single-stranded template. A subsequent PCR to amplify the product allows promiscuous incorporation of standard bases at only the sites of dITP in the template DNA.

The incorporation of random saturation mutations at specific multiple sites provides the potential of identifying pairs of mutations that result in synergistic effects on enzyme properties. This type of library can be achieved using oligonucleotide PCR primers that encode for potentially all twenty amino-acids at each position, in a modification of the original Quikchange® protocol as supplied by Stratagene. The original protocol employs Pfu DNA polymerase to extend mutagenic primers annealed to a template plasmid. This then creates a non-methylated nicked plasmid progeny containing mutations defined by the oligonucleotide primers. Subsequent digestion with the methylated-DNA specific DpnI restriction enzyme then removes the methylated parental plasmid DNA that is obtained from cell lysates, leaving only the mutated plasmid DNA progeny. The improved method significantly enhances the ability to introduce mutations at multiple sites by including a blend of Pfu DNA polymerase with Taq polymerase, Pfu flap endonuclease (PfuFEN-1) and Taq DNA ligase in the Quikchange® PCR reaction. Adaptation of the Quikchange methods are also available for performing site-directed mutagenesis of genes in plasmids, with improved transformation directly into *Bacillus* strains and without the need for an intermediate transformation of *E. coli.*

Mutagenesis at specific sites of a plasmid DNA generally makes use of PCR for amplification of mutagenised DNA, as discussed in the examples above. However, the use of mismatch endonucleases, mixed with a proof-reading polymerase, dNTPs and a ligase, potentially obviates the PCR step by simply excising and repairing mismatched DNA in the heteroduplex formed between parental template plasmid and a mutagenic oligonucleotide. This new method presents an efficient alternative for site-directed mutagenesis and also for the recombination of point mutations from several gene variants.

The functionality represented by the standard twenty amino-acids may be limiting in terms of engineering novel enzyme functions. For example, the incorporation of nonnatural amino-acids into proteins may potentially permit enzyme chemistries that are not observed in nature. It is now possible, through ground breaking work, to incorporate nonnatural amino-acids into proteins at specific sites during their expression from bacterial cells. This was achieved by the directed evolution of a tyrosyl t-RNA synthetase to enable it to charge the t-RNA which is complementary to the little used amber codon TAG, with a non-natural amino-acid.

2.4.5 Identifying target sites for mutagenesis

Alongside the development of techniques for creating different types of library as discussed above, the methods available for identifying key residues within an enzyme to target with mutagenesis have also advanced considerably. Decreasing the redundancy of enzyme libraries by targeting mutagenesis has the potential to allow more useful sequence space to be searched and makes more efficient use of the practical limits often imposed by the throughput of analytical tools. Two general approaches are taken. In the first, aligned sequence information from homologous enzymes can be used to identify residues that vary in nature, which may be linked to various enzyme properties. It has recently been observed that the consensus protein sequence for a particular enzyme has greater stability in many cases than the wild-type enzymes. A recent method for improving enzyme stability takes advantage of this observation by creating libraries of enzymes that contain variations only at sites that deviate from the consensus sequence. Interestingly, a similar method has also recently been used to invert the cofactor specificity of a lactate dehydrogenase enzyme from NAD^+ to $NADP^+$.

The second general approach for identifying key residues uses computational techniques, often though not necessarily, the same as those applied to rational design. In these examples, the conformational stabilities of enzyme variants generated *in silico*, are estimated. The degree of loss in conformational stability estimated by such a method has been shown to correlate well with mutagenic hot-spots that were determined experimentally and could therefore potentially be used for identifying amino acids in a protein that are tolerant to mutagenesis.

A similar concept has been used for pre-screening libraries *in silico* in a technique dubbed protein design automation (PDA), which eliminates sequences that are incompatible with the protein fold of the target enzyme. In this example, the 7×10^{23} potential variants created by *in silico* mutagenesis of nineteen active-site residues of β-lactamase, were reduced down to just 17,2800. The remaining variants were designed and screened experimentally in a single round of directed evolution to produce entirely novel enzyme variants with a 1280-fold increase in cefotaxime resistance.

2.4.6 Examples of enzyme engineering for biocatalysis

Many enzymes have been modified by protein engineering methods and some recently patented examples are summarised below.

The targeted random mutagenesis of seven amino-acid residues of human butyrylcholinesterase expressed in mammalian cells, were recently used to evolve an increased activity towards cocaine hydrolysis. Such an enzyme has therapeutic potential for treating severe cases of cocaine toxicity. Seven residues that line the active-site gorge when aligned to the available structure of the homologous acetylcholinesterase, were chosen for mutagenesis. Previous biochemical data that identified sites important for cocaine hydrolysis were also used to guide the choice of the seven target sites. Several enzyme variants were found that had up to 100-fold greater activity towards cocaine hydrolysis than the wild-type butyrylcholinesterase.

Using a range of protein engineering techniques, including error-prone PCR, saturation mutagenesis and DNA shuffling, mutants of toluene 4-mono-oxygenase, toluene-oxylene mono-oxygenase and toluene-4-mono-oxygenase from various *Pseudomonads* have been obtained with improved activities towards benzene, toluene, nitrobenzene, nitrophenols, catechols, *o*-methoxyphenol and *o*-cresol. Such biocatalytic reactions can be used in the synthesis of phenol, catechol, nitrophenols, nitrocatechols, 3-methoxy-catechol, 1,2,3-trihydroxybenzene, methoxyhydroquinone, methylhydroquinone, 4-methylresorcinol, methylhydroquinone and pyrogallol.

B12-dependent dehydratases with improved reaction kinetics have been obtained using error-prone PCR and oligonucleotide-directed mutagenesis to target the DhaB1 gene, encoding the α-subunit of glycerol dehydratase. Such enzymes are useful in the production of glycerol and 1,3-propanediol.

Nitrilases can be used in biocatalytic processes to convert nitriles to carboxylic acids. Recently, error-prone PCR has been used to generate mutants of an *Acidovorax facilis* nitrilase having improved activity towards 3-hydroxynitriles. Isolated enzyme variants were capable of up to 1.9-fold improved activities when used to convert 3-hydroxynitrile to 3-hydroxycarboxylic acid, 3-hydroxyvaleronitrile to 3-hydroxyvaleric acid, or 3-hydroxybutyronitrile to 3-hydroxybutyric acid.

In another example highlighting the general applicability of error-prone PCR, two carotenoid biosynthesis genes were simultaneously mutated randomly to improve the production of astaxanthin from beta-carotene. Targeting of the two enzymes, CrtO ketolase and CrtZ hydroxylase, simultaneously resulted in mutations within both genes and a net increase in the astaxanthin biosynthesis yield of up to 20% above that obtained with the wild-type enzyme pathway.

Error-prone PCR has also been used to improve the activity of 5′-xanthylic acid (XMP) aminase, which is potentially useful for the biocatalytic synthesis of 5′-guanylic acid (GMP) from XMP in microbial fermentations. Six improved mutants were found, with between two and six accumulated mutations and the best mutant having 3.8-fold greater activity resulting from three mutations. Xylanases are useful enzymes added to animal feed to aid the digestion of feed grains containing xylan. The growing need to sterilise the feed grain requires a xylanase enzyme that can retain activity after heat treatment, as well as being active at physiological digestive conditions (pH 3.5–5.0, 40°C). A recent patent has combined an engineered disulphide bond, previously introduced into a homologous xylanase enzyme and the Q162H mutation previously reported for xylanases.

Previously, the disulphide bond was shown to only marginally increase the xylanase thermostability, whereas the Q162H mutation had no effect on thermostability. However, the two mutations combined together result in a xylanase enzyme with a thermal denaturation midpoint increase of 14°C and which also retains 40% of the original activity after a 30 min treatment at 70°C and at least 30% activity at pH 3.5–6 and 40–60°C. A similar product, 5′-inosinic acid (IMP), is synthesised in an industrial process that uses mutant acid phosphatases from *E. blattae*, for which the phosphomonoesterase activity has been reduced to less than 40% of the wild-type enzyme due to the mutations G74D or I153T. Decreased esterase activity allows the mutant enzymes to synthesise IMP from inosine.

Another enzyme that can be used as a supplement to animal feeds is phytase. This enzyme is useful during digestion for releasing inorganic phosphate from phytate, which contains over 70% of the phosphate in plant material. A variant of the *E. coli* pH 2.5 acid phosphatase has been shown, rather unexpectedly, to exhibit increased phytase activity upon digestion of the enzyme with a mixture of trypsin and pepsin.

Alpha-amylase is an enzyme that is commonly used for the degradation of starch during textile and paper desizing and also as a component of household detergents. Several examples of engineered alpha-amylases have been recently patented, including variants with improved activity and/or thermostability, solvent stability and resistance to multimerisation.

Polysaccharide lyases have potential uses in polysaccharide sequencing and degradation. A recent patent describes rationally designed variants of the polysaccharide lyase, chondroitinase B and also potential uses in the inhibition of anticoagulation, angiogenesis and maternal malarial infection. Active-site residues involved in catalysis and substrate binding were targeted by site-directed mutagenesis. While mutagenesis of the catalytic residues were damaging to enzyme activity, the mutation of a substrate binding arginine (R364) to alanine resulted in an altered product profile to that of wild-type enzyme after the digestion of dermatan sulphate.

2.5 Current and future developments

In recent years, significant advances have been made in the understanding of protein structure and function that have also enabled rational design techniques to become more successful and further reaching than previously. In particular, the use of computational design algorithms with advanced features has made it possible to introduce nascent enzyme functionality into proteins and in at least one example, to improve the efficiency of an existing enzyme activity. Advances in rapid screening techniques, new methods for creating genetic diversity and computational tools for predicting hot-spots in protein sequences, have also pushed forward the boundaries that can be reached using directed evolution. The challenge now resides in refining computational methods for more accurate prediction of the effect of mutations on enzyme activity, potentially including the effects of protein dynamics. Meanwhile, improved understanding of natural enzyme evolution and better prediction of important residues within enzymes will enable directed evolution techniques to become much more powerful.

Section II

Classification, nomenclature and production of enzymes

3

Enzymes nomenclature and classification

3.1 Introduction

Enzymes are protein catalysts that increase the velocity of a chemical reaction and are not consumed during the reaction they catalyse. All enzymes contain a protein backbone. In some enzymes this is the only component in the structure. However there are additional non-protein moieties usually present which may or may not participate in the catalytic activity of the enzyme. Covalently attached carbohydrate groups are commonly encountered structural features which often have no direct bearing on the catalytic activity, although they may well effect an enzyme's stability and solubility. Other factors often found are metal ions (cofactors) and low molecular weight organic molecules (coenzymes). These may be loosely or tightly bound by noncovalent or covalent forces. They are often important constituents contributing to both the activity and stability of the enzymes. This requirement for cofactors and coenzymes must be recognised if the enzymes are to be used efficiently and is particularly relevant in continuous processes where there may be a tendency for them to become separated from an enzyme's protein moiety.

[Note: Some types of RNA can act like enzymes, usually catalysing the cleavage and synthesis of phosphodiester bonds, RNAs with catalytic activity are called ribozymes and are much less commonly encountered than protein catalysts.]

Enzyme composition: Enzymes can have molecular weights ranging from about 10000 to over 1 million. A small number of enzymes are not proteins, but consist of small catalytic RNA molecules. Often, enzymes are multiprotein complexes made up of a number of individual protein subunits. Many enzymes catalyse reactions without help, but some require an additional non-protein component called a cofactor. Cofactors may be inorganic ions such as Fe^{2+}, Mg^{2+}, Mn^{2+}, or Zn^{2+}, or consist of organic or metallo-organic molecules known as co-enzymes.

Enzymes are stable and work in mild conditions: Enzymes work at low temperature and moderate pH and are far more stable catalysts than other chemicals or biological molecules. For this reason enzymes are the most environmentally-friendly solution for industrial manufacturing. Enzymes are not living organisms, they are simply biological molecules. They do not therefore live or die. They just keep on working until they are dissolved, usually by other enzymes. That enzymes are catalysts also means that they do not become part of the final product of the biochemical reaction which they are catalysing. When the biochemical reaction is over, the product of the reaction leaves the enzyme. The enzyme is then ready to effect the same reaction on another molecule again and again. Given the right conditions, the enzyme can go on and on for as long as needed. In some production processes, this lowers costs.

Enzymes are biodegradable and environmentally friendly: Most manufacturing produces industrial waste that can present a threat to nature if it includes chemicals. Enzymes can usually do the same job cheaper and do not threaten the environment. Enzymes are fully biodegradable. When industrial enzymes leave a production plant with the wastewater, the retired enzymes do not last long in the surrounding environment. Nature has many micro-organisms that easily break down enzymes into single amino acids, which are used to build up life around us.

Enzymes consist of long strings of amino acids: Like all other proteins, enzymes are made of amino acids. Each enzyme is made of between a hundred and up to a million amino acids placed like pearls on a string. Each amino acid is bonded to the next by chemical bonds. Each enzyme has its own unique sequence of amino acids, which is determined by the genes in the cells.

The vast majority of enzymes are made of only 20 different kinds of amino acid. The structure and function of the enzyme is determined by the order of the amino acids.

Enzymes have a three-dimensional structure: Enzymes consist of millions of amino acids placed one after the other like pearls on a long string, but most enzymes do not look like a long string. In most enzymes, the string is coiled and folded thousands of times to form a highly complex three-dimensional structure, which is unique to each enzyme. It is the chemical interactions between the amino acids that force the enzymes into their three-dimensional structure, which is held together by the many different links between the different amino acids.

Arrangement of amino acids determines the enzymes function: It is the unique three-dimensional structure of each enzyme that determines the function of the enzyme. Even slight changes in the sequence of the amino acids on the string have a huge impact on the structure and function of the protein. With just one or perhaps a few amino acids replaced or switched, an enzyme may not only look different, but also act differently and convert to working on other biological molecules or treating them differently.

Enzymes have active sites that make them highly specific: Enzymes are large molecules with hundreds of amino acids, but only a small part of the enzyme participates in the catalysis of biochemical reactions. This is called the active site. The three-dimensional structure of the enzyme determines the appearance of the active site. The active site precisely accommodates the shape of the biological substrate. The enzyme and substrate fit together like a key in a lock and only substrates with the right shape will be transformed by the enzyme. This is what makes enzymes specific in their action.

3.2 Biological catalysts

This section discusses shared properties with chemical catalysts and differences between enzymes and chemical catalysts.

1. Shared properties with chemical catalysts:
   (a) Enzymes are neither consumed nor produced during the course of a reaction.
   (b) Enzymes do not cause reactions to take place, but they greatly enhance the rate of reactions that would proceed much slower in their absence. They alter the rate but not the equilibrium constants of reactions that they catalyse.
2. Differences between enzymes and chemical catalysts:
   (a) Enzymes are proteins.
   (b) Enzymes are highly specific and produce only the expected products from the given reactants, or substrates (i.e., there are no side reactions).
   (c) Enzymes may show a high specificity toward one substrate or exhibit a broad specificity, using more than one substrate.
   (d) Enzymes usually function within a moderate pH and temperature range.

3.2.1 Functions of enzymes

Enzymes allow many chemical reactions to occur within the homeostasis constraints of a living system. Enzymes function as organic catalysts. A catalyst is a chemical involved in, but not changed by, a chemical reaction. Many enzymes function by lowering the activation energy of reactions. By bringing the reactants closer together, chemical bonds may be weakened and reactions will proceed faster than without the catalyst. The use of enzymes can lower the activation energy of a reaction. Enzymes can act rapidly, as in the case of carbonic anhydrase (enzymes typically end in the -ase suffix), which causes the chemicals to react 10^7 times faster than without the enzyme present. Carbonic anhydrase speeds up the transfer of carbon dioxide from cells to the blood. There are over 2000 known enzymes, each of which is involved with one specific chemical reaction. Enzymes are substrate specific. The enzyme peptidase (which breaks peptide bonds in proteins) will not work on starch (which is broken down by human-produced amylase in the mouth).

Enzymes are proteins. The functioning of the enzyme is determined by the shape of the protein. The arrangement of molecules on the enzyme produces an area known as the active site within which the specific substrate(s) will 'fit'. It recognises, confines and orients thc substrate in a particular direction.

3.2.2 Functions of enzymes in cells

Enzymes are proteins that do the everyday work within a cell. Their basic function is to speed up the process and efficiency of a reaction without themselves being consumed in the process. Enzymes are responsible for moving large parts of a cell's internal structure, such as pulling chromosomes apart when a cell divides. Enzymes make the energy molecules that are constantly needed for the cell to survive. And they break down molecules, recycle the old parts and make new molecules that allow the cell to grow.

Catalysts for change: Enzymes are catalysts, meaning they speed up the rate at which reactants interact to form products in a chemical reaction, while not being consumcd in the reaction. They physically combine chemical reactants in a way that lowers the energy required for bonds to break and new bonds to form, making the formation of a product much faster. They lower what is called the activation energy of the reaction, or the amount of energy required for a hybrid of the reactants and products to form. The hybrid then becomes the product. Without enzymes, these chemical reactions would proceed at a rate that is hundreds to thousands of times slower.

Making energy: Living organisms store the energy required for daily life in the form of chemical energy. Adenosine triphosphate, or ATP, is the main form of chemical energy. ATP is a charged battery that can be discharged to release energy that powers the movement of enzymes. Enzymes are also required to make ATP, however. The main enzyme that produces ATP is called ATP Synthase, which is part of the electron transport chain in the mitochondria of cells. For every molecule of glucose that is broken down for energy, ATP Synthase makes about 32 to 34 ATP molecules.

Molecular motors: Enzymes are the protein machines that perform the day-to-day functions within cells. They deliver packages from one part of the cell to another. They pull chromosomes apart when the cell undergoes mitosis. They pull cilia, which are like the oars of a cell, to help cells move or to help cells move mucus up our airway into our throat. Common motor proteins are myosins, kinesins and dyneins. These families of motor proteins catalyse the breakage of ATP into ADP (adenosine diphoshphate) to get the energy they need to do their grunt work.

Breaking and building: The cells that comprise organisms obtain energy by breaking down organic carbon compounds such as sugar, protein and fat.

Breaking these molecules down into smaller parts is called catabolism, while building new molecules from these recycled smaller parts is called anabolism. Enzymes perform these functions at every step of the way. Energy sources such as glucose, a simple sugar, store a lot of energy. But the cell cannot access that energy to make ATP unless it is able to break the bonds within the glucose molecule.

3.3 Classification of enzymes

Enzymes are divided into six major classes with several subclasses (Table 3.1 and Fig. 3.1).

Table 3.1: Six major classes of enzymes and examples of their subclasses.

Classification	*Distinguishing feature*
1. Oxidoreductases	$A_{red} + B_{ox} \rightarrow A_{ox} + B_{red}$
Oxidases	Use oxygen as an electron acceptor but do not incorporate it into the substrate
Dehydrogenases	Use molecules other than oxygen (e.g., NAD+) as an electron acceptor
Oxygenases	Directly incorporate oxygen into the substrate
Peroxidases	Use H_2O_2 as an electron acceptor
2. Transferases	A–B + C → A+B– C
Methyltransferases	Transfer one-carbon units between substrates
Aminotransferases	Transfer NH_2 from amino acids to keto acids
Kinases	Transfer PO_3~ from ATP to a substrate
Phosphorylases	Transfer $PO_{3\sim}$ from inorganic phosphate (P) to a substrate
3. Hydrolases	A–B + H_2O → A–H + B–OH
Phosphatases	Remove PO_3~ from a substrate
Phosphodiesterases	Cleave phosphodiester bonds such as those in nucleic acids
Proteases	Cleave amide bonds such as those in proteins
4. Lyases	A(XH)–B → A–X+B–H
Decarboxylases	Produce CO_2 via elimination reactions
Aldolases	Produce aldehydes via elimination reactions
Synthases	Link two molecules without involvement of ATP
5. Isomerases	Isomers have the same molecular formula but differ in their structural formula
Racemases	Interconvert L and D stereoisomers
Mutases	Transfer groups between atoms within a molecule
6. Ligases	A+B+ATP → A–B+ADP+Pi
Carboxylases	Use CO_2 as a substrate
Synthetases	Link two molecules via an ATP-dependent reaction

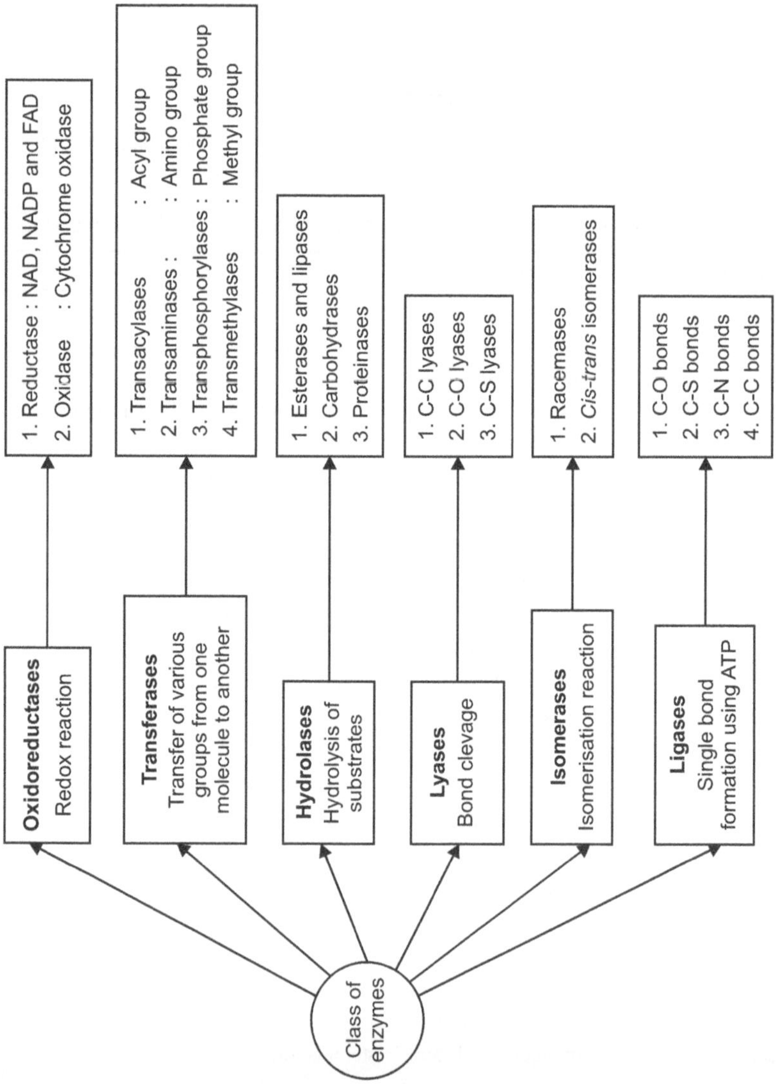

Figure 3.1: Classification of enzyme and examples of related enzyme classes.

EC 1 oxidoreductases: Catalyse the transfer of hydrogen or oxygen atoms or electrons from one substrate to another, also called oxidases, dehydrogenases, or reductases. Note that since these are 'redox' reactions, an electron donor/acceptor is also required to complete the reaction.

EC 2 transferases: Catalyse group transfer reactions, excluding oxido-reductases (which transfer hydrogen or oxygen and are EC 1). These are of the general form:

$$A\text{–}X + B \leftrightarrow BX + A$$

EC 3 hydrolases: Catalyse hydrolytic reactions. Includes lipases, esterases, nitrilases, peptidases/proteases. These are of the general form:

$$A\text{–}X + H_2O \leftrightarrow X\text{–}OH + HA$$

EC 4 lyases: Catalyse non-hydrolytic (covered in EC 3) removal of functional groups from substrates, often creating a double bond in the product, or the reverse reaction, i.e., addition of function groups across a double bond.

EC 5 isomerases: Catalyses isomerisation reactions, including racemisations and *cis-tran* isomerisations.

EC 6 ligases: Catalyses the synthesis of various (mostly C-X) bonds, coupled with the breakdown of energy-containing substrates, usually ATP.

3.4 Nomenclature of enzymes

Each enzyme is assigned two names. The first is its short, recommended name, convenient for everyday use. The second is the more complete systematic name, which is used when the enzyme must be identified without ambiguity.

3.4.1 Recommended name

Most commonly used enzyme names have the suffix '*-ase*' attached to the substrate of the reaction, for example, glucosidase, urease, sucrase, or to a description of the action performed, for example, lactate dehydrogenase and adenylate cyclase.

[Note: Some enzymes retain their original trivial names, which give no hint of the associated enzymic reaction, for example, trypsin and pepsin.]

3.4.2 Systematic name

The International Union of Biochemistry and Molecular Biology (IUBMB) developed a system of nomenclature in which enzymes are divided into six major classes, each with numerous subgroups. The suffix '*-ase*' is attached to a fairly complete description of the chemical reaction catalysed, for example D-*glyceraldehyde 3-phosphate: NAD oxidoreductase.* The IUBMB names are unambiguous and informative, but are sometimes too cumbersome to be of general use.

A skeletal outline of the IUB system is presented below:

1. Reactions and the enzymes that catalyse them form six classes, each having 4–13 subclasses.
2. Thc enzyme name has two parts. The first names the substrate or substrates. The second, ending in *-ase,* indicates the type of reaction catalysed.
3. Additional information, if needed to clarify the reaction, may follow in parentheses, e.g., the enzyme catalysing

$$\text{L-malate} + NAD^+ = \text{Pyruvate} + CO_2 + NADH + H^+$$

   is designated 1.1.1.37 L-malate: NAD^+ oxidoreductase (decarboxylating).
4. Each enzyme has a code number (EC) that characterises the reaction type as to class (first digit), subclass (second digit) and subsubclass (third digit). The fourth digit is for the specific enzyme.

Thus, EC 2.7.1.1 denotes class 2 (a transferase), subclass 7 (transfer of phosphate), subsubclass 1 (an alcohol is the phosphate acceptor).

The final digit denotes hexokinase, or ATP:D-hexose 6-phosphotransferase, an enzyme catalysing phosphate transfer from ATP to the hydroxyl group on carbon 6 of glucose.

3.4.3 Co-enzymes

Enzymes may be simple proteins, or complex enzymes. A complex enzyme contains a non-protein part, called as prosthetic group (co-enzymes). Co-enzymes are heat stable low molecular weight organic compound. The combined form of protein and the co-enzyme are called as holo-enzyme. The heat labile or unstable part of the holo-enzyme is called as apo-enzyme. The apo-enzyme gives necessary three dimensional structures required for the enzymatic chemical reaction. Co-enzymes are very essential for the biological activities of the enzyme. Co-enzymes combine loosely with apo-enzyme and are released easily by dialysis. Most of the co-enzymes are derivatives of vitamin B complex group of substance. One molecule of the co-enzyme with its enzyme is sufficient to convert a large group of substrate.

Co-enzymes are further divided into two groups. The first groups of co-enzymes are a part of reaction catalysed by oxidoreductase by donating or accepting hydrogen atoms or electrons.

The first group of co-enzymes are also called as co-substrates or secondary substrates. Because they are involved in counter-balance in change occurring in the substrate.

The second group of co-enzymes involves in reactions transferring groups other than hydrogen.

3.4.4 Difference between catalysts and enzymes

When compared to man made catalysts, enzymes have several unique properties.

1. They are extremely specific. Enzymes will catalyse reactions involving only one substrate or a very small group of structurally related substrates. Enzymes show:
   (a) Absolute specificity - working upon only a single substrate.
   (b) Group specificity - working upon a related group of molecules containing a specific functional group.
   (c) Linkage specificity - working on molecules that contain a specific type of chemical bond.
2. Enzymes are stereospecific. If a molecule exists as a pair of enantiomers, the enzyme will use only one of the pair as substrate and produce only one of the pair as the product. For example, the enzymes that are involved in amino acid metabolism and/or protein synthesis will only utilise the L-amino acids as substrates.
3. Reactions catalysed by enzymes produces only one product. Wasteful side reactions do not occur during enzyme catalysed reactions.
4. Enzymes are very much faster than man made catalysts. The best man made catalysts increase reaction rates about 10^7 fold, on average man made catalysts increase reaction rates 10^2 to 10^4 fold. Enzymes can enhance the rate from 10^{17} to 10^{20} fold when compared to the uncatalysed reaction.

4

Immobilised

4.1 Introduction

Immobilised enzymes are widely used for variety of applications. Based on the type of application, the method of immobilisation and support material can be selected. The immobilised enzymes can be separated from the reaction mixture and reused and also immobilised in order to prevent the enzyme from being exposed to harsh conditions, high temperature, surfactants and oxidising agents, etc. The immobilised enzymes are also widely used in food industry, pharmaceutical industry, bioremediation, detergent industry, textile industry, etc. Enzyme immobilisation improves the operational stability and is also due to the increased enzyme loading which causes the controlled diffusion. Several hundreds of enzymes are immobilised and used for various large scale industries. Immobilisation technique reduces the effluent treatment costs.

An enzyme derived from an organism or cell culture that catalyses metabolic reaction in living organisms and/or substrate conversions in various chemical reactions. The enzymes are the potential catalyst which works in mild temperature, pressure, pH, substrate specificity under suitable reaction conditions and for the production of desired products without any intermediate products as contaminations for these advantages enzyme are used in variety of application such as cosmetics, paper industry, textile industry, food industry, pharmaceutical industry, laundry and in detergents, etc. The biotechnological method of producing enzyme is expensive, hence new methods have been implemented to reduce the cost.

The enzymes have various other limitations such as low stability, highly sensitive to the process conditions and these problems can be overcome by the immobilisation techniques. Immobilised enzymes are being used since 1916, when Nelson and Griffin discovered that invertase when absorbed to charcoal has the ability to hydrolyse the sucrose. The possibility of immobilised enzyme for its reuse and stability was identified by Grubhofer and Schelth, who reported the covalent immobilisation of several enzymes. The repeated assay can be done with the immobilised enzyme which reduces the cost of assay and the reuse of enzyme process is also very simple and it can be attained through ultrafiltration technique. Irreversible enzyme immobilisation includes covalent binding and entrapment. Reversible enzyme immobilisation includes adsorption, ionic binding, affinity binding and metal binding.

Immobilisation of enzyme by enzymatic process is recently identified by researchers in order to avoid harsh immobilisation.

4.2 Enzyme immobilisation methods

4.2.1 Covalent binding

Covalent binding is a conventional method for immobilisation, it can be achieved by direct attachment with the enzyme and the material through the covalent linkage. The covalent linkage is strong and stable and the support material of enzymes includes polyacrylamide, porous glass, agarose and porous silica. Covalent method of immobilisation is mainly used when a reaction process does not require enzyme in the product, this is the criteria to choose covalent immobilisation method. This covalent binding of the enzyme with the support material involves two main steps such as, the activation of the support material by the addition of the reactive compound and the second one is the modification of the polymer backbone to activate the matrix (Fig. 4.1).

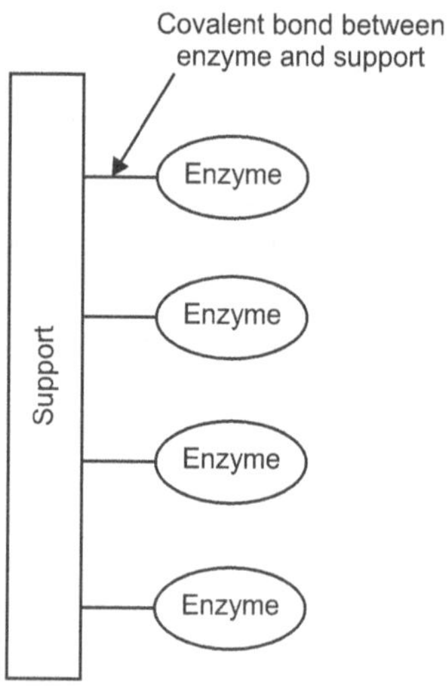

Figure 4.1: Covalent binding.

The activation step produces the electrophilic group on the support material, so that the support material couples/reacts with the strong nucleophiles on the proteins. For example glutaraldehyde is the activation method, in this reaction the amine group reacts with the activated matrix. The covalent binding is normally formed between the functional group in the support matrix and the enzyme surface that contains the amino acid residues. The amino acid residues involved in the covalent binding are the sulphydryl group of cysteine, hydroxyl group of serine and threonine. The attachment between the enzyme and the support material can be achieved either through direct linkage or through the spacer arm. The potentiality of using the spacer arm is that it provides the greater degree of the mobility to the enzymes hence the enzymes show the higher activity when compared to the direct attachment.

Advantages of covalent coupling

1. The strength of binding is very strong, so, leakage of enzyme from the support is absent or very little.
2. This is a simple, mild and often successful method of wide applicability

Disadvantages of covalent coupling

1. Enzymes are chemically modified and so many are denatured during immobilisation.
2. Only small amounts of enzymes may be immobilised (about 0.02 grams per gram of matrix).

4.2.2 Entrapment

Enzymes are occluded in the synthetic or natural polymeric networks, it is a permeable membrane which allows the substrates and the products to pass, but it retains the enzyme inside the network, the entrapment can be achieved by the gel, fibre entrapping and microencapsulation. The advantage of entrapment of enzyme immobilisation is fast, cheap and mild conditions required for reaction process. The disadvantage is that limitation in mass transfer. The support matrix protects the enzymes from microbial contamination, proteins and enzymes in the micro environment.

Microencapsulation method is that the enzyme molecules are capsulated within spherical semipermeable membranes with a selective controlled permeability. This method provides the large surface area between polymeric material and the enzyme. The drawback of this method is inactivation of enzyme during encapsulation.

Advantages and disadvantages of entrapment

Advantages of entrapment: Loss of enzyme activity upon immobilisation is minimised.

Disadvantages of entrapment: (i) The enzyme can leak into the surrounding medium, (ii) another problem is the mass transfer resistance to substrates and products and (iii) substrate cannot diffuse deep into the gel matrix.

4.2.3 Adsorption

This is a simple method of preparing an immobilised enzymes and the materials used for adsorption are activated charcoal, alumina, ion exchange resins, this method is cheap and easy for use and the disadvantage is a weak binding force between the carrier and the enzyme. This method comes under carrier bound immobilisation and the process of immobilisation is reversible. Adsorption (Fig. 4.2) is the easiest and oldest immobilisation techniques. The interaction between the enzyme and the surface of the matrix through weak forces by salt linkage, hydrogen bonds, hydrophobic bonds, ionic bonds and van der waals forces. Based on the charges of the matrix and the protein arrangements the strongly bound, but not distorted enzyme will be formed. The advantage of enzyme adsorption is minimum activation step and as a result of minimum activation, no reagents required. It is cheap and easy way of immobilisation.

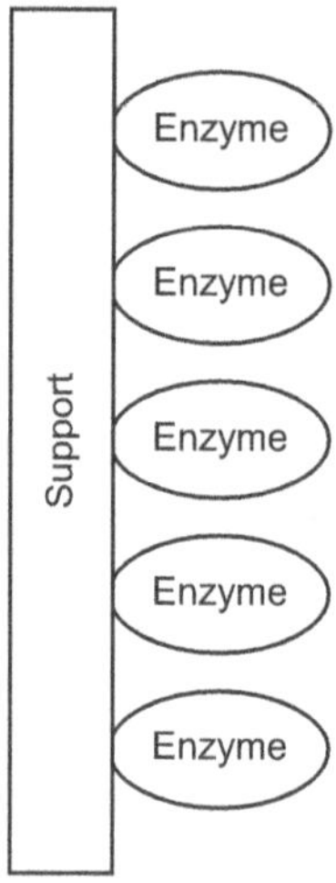

Figure 4.2: Adsorption.

Advantages of adsorption

1. Little or no confirmation change of the enzyme.
2. Simple and cheap.
3. No reagents are required.
4. Wide applicability and capable of high enzyme loading.

Disadvantages of adsorption

1. Desorption of the enzyme protein resulting from changes in temperature, pH and ionic strength.
2. Slow method.

4.2.4 Ionic binding

The bonding involved between the enzyme and the support material is salt linkages. In non-covalent immobilisation the process will be reversed by changing the temperature polarity and ionic strength conditions. This principle is similar to protein-ligand interactions principles used in chromatography.

4.2.5 Affinity binding

The immobilisation of enzyme linked to the matrix through the specific interactions. The two methods are being followed in affinity immobilisation. The first method is the activation of the support material which contains the coupled affinity ligand, so that the enzyme will be added. The advantage of this method is the enzyme is not exposed to any harsh chemicals conditions. The second method, the enzyme modified to another molecule which has the ability to bind towards a matrix.

4.2.6 Metal linked immobilisation

In metal linked enzyme immobilisation, the metal salts are precipitated over the surface of the carriers and it has the potential to bind with the nucleophilic groups on the matrix. The precipitation of the metal ion on the carrier can be achieved by heating. This method is simple and the activity of the immobilised enzymes is relatively high (30–80%). The carrier and the enzyme can be separated by decreasing the pH, hence it is a reversible process. The matrix and the enzyme can be regenerated, by the process.

4.3 Properties of immobilised enzymes

The stability of the immobilised enzymes is based on the temperature and time. The activity of the enzyme is retained throughout series of cycles. Due to immobilisation, the properties of enzymes will be altered such as catalytic activity with respect to the support matrix. The change in the enzyme properties in the immobilised enzyme is due to the enzyme and the substrate reacts in the micro-environment which is different from the enzyme substrate reaction in the bulk solution environment. The change is also due to the change in the three dimensional conformation of the protein when linked with the support matrix. These conformational changes are to a lesser extent and these changes occur in the limited enzyme systems. Enzyme immobilisation improves the operational stability and is also due to the increased enzyme loading which causes the controlled diffusion. The stabilisation as a result of, number of bonds formed between the enzyme and the support matrix. When the immobilised enzyme acts on the macro-molecular substrates, active site of the enzymes does not able to access with the substrates, hence the enzyme loses its activity.

4.4 Application of the immobilised enzymes

4.4.1 Food industry

In food industry, the purified enzymes are used but during the purification the enzymes will denature. Hence the immobilisation technique makes the enzymes stable. The immobilised enzymes are used for the production of syrups. Immobilised beta-galactosidase used for lactose hydrolysis in Whey for the production of bakers yeast. The enzyme is linked to porous silica matrix through covalent linkage. This method is not preferably used due to its cost and the other technique developed by Valio in 1980, the enzyme galactosidase was linked to resin (food grade) through cross-linking. This method was used for the various purposes such as confectionaries and icecreams.

4.4.2 Biodiesel production

Biodiesel is monoalkyl esters of long chain fatty acids. Biodiesel is produced through triglycerides (vegetable oil, animal fat) with esterification of alcohol (methanol, ethanol) in the presence of the catalyst. The production of catalyst is a drawback of high energy requirements, recovery of glycerol and side reaction which may affect the pollution. Hence the biological production of liquid fuel with lipases now-a-days has a great consideration with a rapid improvement. Lipase catalyses the reaction with less energy requirements and mild conditions required. But the production of lipase is of high cost, hence the immobilisation of lipase which results in repeated use and stability. The immobilisation of lipase includes several methods entrapment, encapsulation, cross-linking, adsorption and covalent bonding. Adsortion method of immobilisation is widely used in recent years when compared to covalent bond, entrapment and cross-linking. In the biological production of biodiesel the methanol inactivates the lipase, hence the immobilisation method is an advantage for the biodiesel production. The low cost of lipase, candida sp., as origin is of more industrial use. The nanostructured carriers are with high porosity, natural material activated carbon, celite, zeolite. The carriers for lipase immobilisation by covalent attachment of olive pomace, resins, polyurethane foam, chitosan, silica and magnetic nanostructures, When compared to the natural support material chitosan is used for enzyme binding, the immobilised lipase retains its stability for 10 cycles of pomace oil, esterification, while maintaining 80% residual activity.

4.4.3 Wastewater treatment

The increasing consumption of fresh water and water bodies are mixed up with polluted industrial wastewater and the wastewater treatments are necessary at present. The sources of dye effluents are textile industry, paper industry, leather industry and the effluents are rich in dye colourants. These effluents are threat to the environment and even in low concentration it is carcinogenic. Now-a-days enzymes are used to degrade the dye stuffs. The enzymes used in the wastewater treatments are preoxidases, laccase, azo reductases. These enzymes due to harsh conditions like extreme temperature, low or high pH and high ionic strength may lose its activity, to overcome this problem immobilised enzymes are used. The Horse radish peroxidases are entrapped in calcium alginate beads.

The immobilised laccase enzyme has the ability to degrade dyes anthracinoid dye, Lancet blue and Ponceau red. Adsorption method is widely used because of its easy regeneration. During the covalent method of immobilisation the conformational change in the enzyme occurs which will affect the activity of the enzyme. In single enzyme nanoparticle (SEN), the enzyme is protected by a nanometer thick substance as it provides the large surface area. SEN has the ability to retain its activity during the extreme conditions. SEN is also used in the removal of heavy metals from the wastewater.

Lipase has the ability to hydrolyse oil and fats to long chain fatty acid and glycerol. The immobilised lipase is of high interest for the hydrolysis of oils and fats for treating the wastewater from the food industry. The drawback of the conventional treatment methods is slow biodegradability, oil and fats are absorbed on the surface of sludge. Researchers are now focusing on the treatment with immobilised lipase. Lipase immobilised on the sol gel/calcium alginate with the size of 82 μm, immobilised lipase. Immobilised lipase operated for 100 days in continuous sludge without any problem, does not produce foam in the reactor.

4.4.4 Textile industry

The enzymes derived from microbial origin are of great interest in textile industry. The enzymes such as cellulase, amylase, liccase, pectinase, cutinase, etc., are used for various textile applications such as scouring, biopolishing, desizing, denim finishing, treating wools, etc. Among these enzymes cellulase has been widely used from the older period to till now. The textile industries now turned to enzyme process instead of using harsh chemical which affects the pollution and cause damage to the fabrics. The processing of fabrics with enzymes requires high temperatures. Hence, enzyme immobilisation for this process is able to withstand at extreme and able to maintains its activity for more than 5–6 cycles. Endoglucanase is a component of cellulase enzyme, endoglucanase is microencapsulated with Arabic gum is a natural polymer with the biodegradable property is used as a matrix for encapsulation of endoglucanase. Encapsulation of endoglucanase prevented it to retain its activity in the presence of detergents.

4.4.5 Detergent industry

The detergent industry also employs enzymes for removal of stains. The enzymes used in detergent industry are protease which is used to remove the stains of blood, egg, grass and human sweat. Amylase used to remove the starch based stains like potatoes, gravies, chocolate. Lipase used to remove the stains of oil and fats and also used to remove the stains in cuffs and collars. Cellulase is used for cotton based fabrics in order to improve softening, colour brightening and to remove soil stains. Now-a-days Biotech cleaning agents are widely used in the detergent industries. When compared to synthetic detergents the biobased detergents have good cleaning property.

The enzymes based detergents can be used in low quantity when compared to the synthetic detergent, it has increased biodegradability, does not affect the environment works well in low temperature and these are the advantages of enzymes in detergent industry. Proteases hydrolyse the proteins and protease cannot be used for keratin based fabric wool and silk which cause adverse damage to the garment. So protease directly cannot be used for wool and silk garments, protease loses its stability in the presence of surfactants and oxidising agents, hence protease is immobilised by covalently linking with Eudragit S-100 using carbodiimide coupling. The immobilised protease treated with wool for 72 hr with 100U at 40°C the free enzyme was degraded the wool but the immobilised enzyme retained 76% of the tensile strength of wool. Protease also immobilised by entrapment method with polyacryamide gel, the enzyme retained its activity for about 6 cycles with incubation time of 20 min at 55°C for each cycles, by this immobilisation method, protease activity is retained for about 83% of the initial activity after six cycles. The lipase immobilisation is carried in order to prevent lipase from protease action and surfactant inhibition lipase is immobilised on acrylamine glass beads coated with zirconia with the size of 55 nm. The Maximum activity was found for immobilised lipase at the pH of 6.5 and for free enzyme it is 7.5.

4.4.6 Biomedical industry

Immobilised enzymes are used for diagnosis and treatment of diseases in the medical field. The inborn metabolic deficiency can be overcome by replacing the encapsulated enzymes (i.e., enzymes encapsulated by erythrocytes) instead of waste metabolites which act as a carrier for the exogenous enzyme drugs and the enzymes are biocompatable in nature, hence there is no immune response. The enzyme encapsulation through the electroporation is a easiest way of immobilisation in the biomedical field and it is a reversible process for which enzyme can be regenerated. The enzymes when combined with the biomaterials provides biological and functional systems. The biomaterials are used in tissue engineering application for repair of the defect. The advantage of the enzyme immobilisation in biomedical is that the free enzymes are consumed by the cells and not active for prolonged use, hence the immobilised enzymes remains stable, to stimulate the growth and to repair the defect. The cancer therapy is delivery of enzymes to the oncogenic sites have been improved with new methods. The nanoparticles and nanospheres are often used as enzyme carriers for the delivery of therapeutic agents.

4.5 Nanoparticles as immobilisation matrix

Nanoparticles act as very efficient support materials for enzyme immobilisation, because of their ideal characteristics for balancing the key factors that determine biocatalysts efficiency, including specific surface area, mass transfer resistance and effective enzyme loading. Diffusion problem is more relevant when we are dealing with the macro-molecular substrates, for such systems the nano-particles are the ideal candidates. Moreover, the enzyme bound nanoparticles show Brownian movement, when dispersed in aqueous solutions showing that the enzymatic activities are comparatively better than that of the unbound enzyme. In addition, magnetic nanoparticles possess additional advantage, can be separated easily using an external magnetic field. Studies have shown that immobilisation of enzymes to the nanoparticles can reduce protein unfolding and can improve stability and performance.

Enzymatic immobilisation on Au and Ag nanoparticles have been studied using either as whole cells or isolated enzymes, which include lysozyme, glucose oxidase, aminopeptidase, as well as alcohol dehydrogenase. Cruz and others reported the Immobilisation of enzymes *S. Carlsberg* and *Candida antarctica* lipase B (CALB) on fumed silica nanoparticles for applications in nonaqueous media and they observed catalytic activities were remarkably high. Won and others immobilised acetylcholinesterase onto magnetic glasses based on iron oxide/silica, for paraoxon sensing. Ganesana and others performed the immobilisation of acetylcholinesterase on nickel nanoparticles and obtained a highly sensitive detection method for organophosphate pesticides.

Uygun and others employed magnetic poly (2-hydroxyethyl methacrylate-*N*-methacryloyl-(l)-phenylalanine) to immobilise α-amylase. They reported a substrate affinity increases upon enzyme immobilisation and showed that a specific activity of 85% was maintained after 10 reuses. Khoshnevisan and others immobilised cellulase on magnetic nanoparticles obtaining a smaller activity than for the free enzyme, but at 80°C the immobilised enzyme showed slightly greater activity. Lee and others used amino-functionalised silica-coated magnetic nanoparticles to immobilise trypsin. They applied this system to a pressure-assisted digestion for proteome analysis. It was observed for each of the experiments in which the magnetic nanoparticles were employed an increased number of protein identification in comparison with the experiment with free trypsin. Recently Smith and others have reported the immobilisation of enzymes (Peroxidase, cellulase, trypsin and alpha amylase) on TiO_2 nanoparticles. The immobilised enzymes show higher activity than free enzymes. It also showed enhanced thermal stability compared to its soluble counterpart at higher temperature.

All the advantages of immobilised enzymes on micron-sized particles are inherited when nanomaterials are used as solid supports. Broadly there are four main approaches to link a protein or enzyme to the nanoparticles.

Electrostatic adsorption: The most widely used linkage approach consists of electrostatic adsorption. This is the simplest approach and is already used routinely as an electron dense marker in histology. The interaction between the nanoparticle and protein may be modulated by the pH or charge screening by controlling the ionic strength of the medium.

Covalent attachment to the surface modified nanoparticles: Another general method for nanoparticle–protein conjugation is covalently linking a protein to the nanoparticle ligand. This approach has been greatly advanced by extreme control over the surface chemistry of the nanoparticles. For example, a variety of organic functional groups can be introduced to the surface using mild conditions. The popular labelling chemistry utilises the covalent binding of primary amines with sulpho-NHS esters or R-COOH groups via reaction with EDC. Nanoparticles labelled with NHS esters can react to form covalent bonds with the primary amine of lysine on a protein. In addition, nanoparticles coated with maleimide groups can react with the thiol of cysteine on a protein. Oxide nanoparticles (TiO iron oxide, Copper oxide, silver and gold oxide) can be easily modified by Silanisation yielding a modified surface exhibiting amino groups, which can be used as adsorbent or as coupling sites for linking various proteins.

Conjugation using specific affinity of protein: Nanoparticle–protein conjugation can also be achieved by using specific labelling strategies.

Direct conjugation to the nanoparticles surface: A direct reaction of a chemical group on the protein without the use of a linker is usually desired if the particle is used as a biosensor electron transfer is used. For Au and Ag nanoparticles, this can be achieved by the Au-thiol or Ag- thiol chemistry where a protein with a cysteine covalently bonds to an Au or Ag nanoparticle. The conjugation requires incubation of the protein and the nanoparticle together as the Au–S or Ag-S bond is strongly favoured. Similarly, for sulphur containing nanoparticles such as ZnS/CdSe, cysteine can directly form a disulphide bridge with the surface S atom. Some important new consequences arise when the size of the carrier approaches nanodimensions. Mostly, these all work out in the favour of using nanosized materials.

Present studies shows that different types of nanomaterial are available for enzyme immobilisation for examples carbon nanotubes (CNTs), nanoparticles, magnetic nanoparticles, mesoporous media, nanofibres, nanocomposites, nanorods and sol–gel materials containing nanometer-size particles and single-enzyme nanoparticles. However, the major problem for their application is their high cost and complex supports preparation. Strategies or protocols for synthesising the nanoparticles should be developed which are low cost, ecofriendly and can be used for large scale synthesis.

Thus, immobilisation process has been used for enhancing enzyme activity and stability in aqueous and non-aqueous media. Selecting and designing the support matrix are important in enzyme immobilisation. Recently, the use of nanoparticles has emerged as a versatile tool for generating excellent supports for enzyme stabilisation due to their small size and large surface area. It has been observed that the stability and activity of enzymes increases when immobilised on such materials. The nanomaterials are key components in the future market of high technology. Nanoparticles strongly influence the mechanical properties of the material like stiffness and elasticity and provide biocompatible environments for enzyme immobilisation.

4.6 Summary of advantages and disadvantages of enzyme immobilisation

4.6.1 Advantages of enzyme immobilisation

1. Multiple or repetitive use of a single batch of enzymes.
2. Immobilised enzymes are usually more stable.
3. Ability to stop the reaction rapidly by removing the enzyme from the reaction solution.
4. Product is not contaminated with the enzyme.
5. Easy separation of the enzyme from the product.
6. Allows development of a multienzyme reaction system.
7. Reduces effluent disposal problems.

4.6.2 Disadvantages of enzyme immobilisation

1. It gives rise to an additional bearing on cost.
2. It invariably affects the stability and activity of enzymes.
3. The technique may not prove to be of any advantage when one of the substrate is found to be insoluble.
4. Certain immobilisation protocols offer serious problems with respect to the diffusion of the substrate to have an access to the enzyme.

5

Carbohydrase

5.1 Introduction

Certain enzymes, found in many products of biological origin, catalyse the hydrolysis and synthesis of glycosides, oligosaccharides and polysaccharides. As a group, they are usually termed 'carbohydrases' or 'glycosidases.'

Carbohydrase is an enzyme that catalyses the breakdown of carbohydrates into simple sugars. Carbohydrases are produced in the pancreas but act in the stomach breaking down carbohydrates. Carbohydrate is usually a compound consisting of carbon, hydrogen and oxygen. Carbohydrase is a protein/enzyme that catalyses carbohydrates to break down the compound carbohydrate so the products can be simple sugar which is a monosaccharide.

Example:

Maltase reduces maltose into glucose: $\underset{\text{Maltose}}{C_{12}H_{22}O_{11}} + \underset{\text{Water}}{H_2O} \rightarrow \underset{\alpha\text{-glucose}}{2C_6H_{12}O_6}$

5.2 Amylases

Amylases are starch degrading enzymes. They are widely distributed in microbial, plant and animal kingdoms. They degrade starch and related polymers to yield products characteristic of individual amylolytic enzymes. Initially the term amylase was used originally to designate enzymes capable of hydrolysing α-1,4-glucosidic bonds of amylose, amylopectin, glycogen and their degradation products. They act by hydrolysing bonds between adjacent glucose units, yielding products characteristic of the particular enzyme involved.

In recent years a number of new enzymes associated with degradation of starch and related polysaccharides structures have been detected and studied. The enzymes of actual or potential commercial importance of microbial origin that split α-1,4 and α-1,6 bonds in these structures, may be divided in the following six classes.

1. Enzymes that hydrolyse α-1,4 bonds and by-pass α-1,6 linkages, e.g., α-amylase (endoacting amylases).
2. Enzymes that hydrolyse α-1,4 and cannot bypass α-1,6 linkages, e.g., β-amylase (exoacting amylases producing maltose as a major end product).

3. Enzymes that hydrolyse α-1,4 and α-1,6 linkages, e.g., amyloglucosidase (glucoamylase) and exoacting amylase.
4. Enzymes that hydrolyse only α-1,6 linkages, e.g., pullulanase and other debranching enzymes.
5. Enzymes that hydrolyse preferentially α-1,4 linkages in short chain oligosaccharides produced by the action of other enzymes on amylose and amylopectin, e.g., α-glucosidases.
6. Enzymes that hydrolyse starch to a series of non-reducing cyclic D-glucosyl polymers called cyclodextrins or Sachardinger dextrins, e.g., *Bacillus macerans* amylase (cyclodextrin producing enzyme).

5.2.1 Starch

Before describing the action pattern and properties of amylolytic enzymes, it is essential to discuss the features of the natural substrate, starch. Starch is a major reserve carbohydrate of all higher plants. In some cases it accounts for as high as 70% of the undried plant material. It occurs in the form of water insoluble granules. The size and shape of the granules are often characteristic of the plant species from which they are extracted. When heated in water the hydrogen bonds holding the granules together begin to weaken and this permits them to swell and gelatinise. Ultimately they form paste or dispersion, depending on the concentration of polysaccharide.

Starches are produced commercially from the seeds of plants, such as corn, wheat, sorghum or rice, from the tubers and roots of plants such as cassava, potato, arrowroot and the pith of sago palm. The major commercial source of starch is corn from which it is extracted by a wet milling process. Starch is a heterogeneous polysaccharide composed of two high molecular weight entities called amylose and amylopectin. These two polymers have different structures and physical properties (Table 5.1).

Table 5.1: Comparison of amylose and amylopectin.

Properties	*Amylose*	*Amylopectin*
Basic structure	Essentially linear	Branched
Stability in aqueous solution	Retrogrades	Stable
Degree of polymerisation	C. 10^3	C. 10^4–10^5
Average chain length	C. 10^3	C. 20–25
β-amylase hydrolysis	87%	54%
β-amylase and debranching enzyme hydrolysis	98%	79%
Iodine complex max	650 nm	550 nm

Starch may be separated into its two components by addition of a polar solvent, e.g., *n*-butanol, to a dispersion of starch. The insoluble amylose complex can then be separated from soluble amylopectin fraction. Amylose is composed of linear chains of α-1,4 linked D-glucose residues. Hence it is extensively degraded by α-amylase. Some amylose is not totally degraded to maltose by this enzyme. Amylose has a degree of polymerisation of several thousands of glucose units. Because of the molecular shape and structure of amylase, it is not stable in aqueous solution and retrogrades (precipitates spontaneously). This is because linear chains align themselves by hydrogen bonding and thus forms aggregates. This process is irreversible. Retrograded amylase will only dissolve in alkaline solution. Amylose has considerable viscosity in alkaline solutions due to its molecular shape. Amylase forms complex with iodine to form intense blue colour and this forms the basis of a method for quantitative determination of amylase.

Amylopectin may account for 75 to 85% of most starches. It has molecular weight in excess on 10^7–10^8 and has a branched structure composed of chains of about 20–25 α-1,4 linked D-glucose residues. Amylopectin which is branched by α-1,6 linkages may contain 4 to 5% α-1,6-D-glucosidic bonds. In aqueous solutions, amylopectins are relatively stable due to branched molecules and are not able to form compact aggregates. There is no apparent relationship between the limiting viscosity number and the degree of polymerisation. Due to the nature of branched structure, the iodine binding power is reduced. The branched components of starch is amylopectin which has different types of chains referred to as A, B and C chains. The hydrolysis of starch may be carried out using either acid or enzyme as catalyst. Enzyme hydrolysis has several advantages: it is more specific, therefore fewer by-products are formed and hence yields are higher. Conditions for enzyme hydrolysis are milder therefore refining stages to remove ash and colour is minimised. The enzymatic hydrolysis of starchy has been practiced on an industrial scale for many years and is gradually replacing the traditional acid hydrolysis process.

5.2.2 Applications of amylases

The history of the industrial production of enzymes dates back to the time when Dr. Jhokichi Takamine began the production of digestive enzyme preparation by wheat bran koji culture of *Aspergillus oryzae* in 1894. Industrial production of dextrose powder and dextrose crystals from starch using α-amylase and glucoamylase began in 1959. Since then, amylases are being used for various purposes. Conversion of starch into sugar, syrups and dextrins forms the major part of the starch processing industry. The hydrolysates are used as carbon sources in fermentation as well as sources of sweetness in a range of manufactured food products and beverages. Hydrolysis of starch to products containing glucose, maltose, etc., is brought about by controlled degradation. Some of the applications of amylase are discussed below.

Liquefaction

Liquefaction is a process of dispersion of insoluble starch granules in aqueous solution followed by partial hydrolysis using thermostable amylases. In industrial processes, the starch suspension for liquefaction is generally in excess of 35% (w/v). Therefore the viscosity is extremely high following gelatinisation. Thermostable α-amylase is used as a thinning agent, which brings about reduction in viscosity and partial hydrolysis of starch. Retrogradation of starch is thus avoided during subsequent cooling.

The traditional thinning agent used in starch technology was acid (hydrochloric or oxalic acids, pH 2 and 140–150°C for 5 min). The introduction of thermostable amylases has meant milder processing conditions. The formation of by-products is reduced and refining and recovery costs are lowed.

In the enzymatic process the hydrolytic action is terminated when the average degree of polymerisation is about 10–12. Two distinct types of thermostable α-amylases are commercially available and used extensively in starch processing technology. The amylase of *Bacillus amyloliquefaciens* was the first liquefying α-amylase used on a large scale. Later, a more heat stable enzyme from *Bacillus licheniformis* was introduced commercially. Liquefaction can be done by two methods:

Single stage enzyme liquefaction: In this process, starch slurry containing 30–40% dry solids is prepared in the feed tank. The pH is adjusted to about 6–6.5 with sodium hydroxide. Calcium salts may be added if the level of free calcium ions is below 50 ppm. The liquefying enzyme is then added. The slurry is then pumped continuously through a jet cooker where the temperature is raised to 105°C by direct injection of live steam. Tremendous shearing forces are exerted on the slurry as it is pumped through the jet cooker. So in addition to the viscosity reduction action of the enzyme, some mechanical thinning also occurs. The slurry is maintained at this high temperature in the pressurised holding cell for about 5 min, after which it is discharged via a spring loaded release valve into a reaction, where enzyme action is allowed to continue for about 2 hr at 95°C. After this treatment the liquefied starch will have dextrose equivalent (DE) of 10–20 depending on the amount of enzyme used. DE is defined as reducing sugars expressed as dextrose and calculated as a percentage of dry substance. This process is simple energy consumption is relatively low because the maximum operating temperature is only 105°C as compared to 140–150°C normally used.

Acid enzyme liquefaction: This is another process which takes advantage of the thermostability of *B. licheniformis* amylase. The enzyme is added after the starch has been cooked and cooled to 100 – 95°C. A starch slurry containing 30–40% dry solids is cooked at a high temperature for about 5 min. A jet cooker is used so that sufficient mechanical thinning, due to shearing takes place. The pH may be in the range 2–5, but if it is too low, by-product formation will be significant. If it is too high there will be no thinning effect from the acid and there will be an increased colour formation. After cooking, the slurry is flash cooled to about 100°C and the pH is set to 6 to 6.5 before the addition of enzyme. By this process the enzyme consumption is slightly reduced. The filtration properties are also improved because better fat/protein separation is achieved. There is an increase in steam consumption and hence fuel costs due to high temperature cooking.

Liquefaction is the first and most important step in starch processing. The purpose is to provide a partially hydrolysed starch suspension of relatively low viscosity which is free from by products, stable to retro gradation and suitable for further processing, i.e., saccharification. If the liquefaction process does not go well, problems like poor filtration and turbidity of the processed solution occurs. The most important factor for ideal liquefaction of starch is that the starch slurry which contains suitable amount of α-amylase is treated at 105 to 107°C as quickly and uniformly as possible.

Thermostable amylase are not sufficiently heat stable to be used during liquefaction process, but they can be used as saccharifying enzymes. The most widely used enzymes in this group are the maltogenic enzymes.

Manufacture of maltose

Maltose is a naturally occurring disaccharide. Its chemical structure has 4-O-α-D-glucopyranosyl-D-glucopyranose. It is the main component of maltosugar syrup. Maltose is widely used as sweetener and also as intravenous sugar supplement. It is used in food industries because of low tendency to be crystallised and is relatively non-hygroscopic.

Corn, potato, sweet potato and cassava starches are used for maltose manufacture. The concentration of starch slurry is adjusted to be 10–20% for production of medical grade maltose and 20–40% for food grade. Thermostable α-amylase from *B. licheniformis* and *B. amyloliquefaciens* are used.

Manufacture of high fructose containing syrups

High fructose containing syrups (HFCS) (Fructose content = 42%) is prepared by enzymic isomerisation of glucose with glucose isomerase. The starch is first converted to glucose by enzymic liquefaction and saccharification.

Manufacture of oligosaccharides mixture

Oligosaccharides mixture (malto-oligomer mix) is obtained by digestion of corn starch with α-amylase, β-amylase and pullulanase. Malto-oligomer mix is a new commercial product. Its composition is usually as follows: glucose, 2.2%, maltose, 37.5%, maltotriose, 46.4% and maltotetraose and larger maltooligosaccharides, 14%. Malto-oligomer mix powder obtained by spray drying is highly hygroscopic. Therefore it serves as a moisture regulator of the food with which it is mixed. Malto-oligomer mix tastes less sweet than sucrose. Its solution shows a lower viscosity than corn syrup because of its low content of glucose. Malto-oligomer mix is mainly used as a substitute for sucrose and other saccharides. It is also used for preventing crystallisation of sucrose in foods and keeping a certain level of hardness of the texture during storage.

Manufacture of maltotetraose syrup

Maltotetraose syrup (G4 syrup) is produced by subjecting starch to the action of maltotetraose forming amylase (G4 amylase). The sweetness of the syrup is as low as 20% of sucrose. Therefore a partial replacement of sucrose with G4 syrup reduces the sweetness of foods without affecting their inherent taste and flavour. It has high moisture retention power which serves to prevent retrogradation of starch ingredient and retains suitable moisture in foods. G4 syrup shows less Millard reaction as it has less glucose and maltose content. It has higher viscosity than sucrose thus improving the food texture. G4 syrups depresses the freezing point of water more moderately than sucrose or high fructose syrup. Therefore, G4 syrup can be used to control the freezing points of frozen foods. G4 syrup imparts gloss and can be used in industry such as a paper sizer. Commercial thermostable α-amylase of *B. licheniformis* or *B. subtilis* is used to make G4 syrups.

Production of anomalously linked oligosaccharides mixture (Alo mixture)

'Alo mixture' is a mixture of isomaltose, panose, isomaltotriose and branched oligosaccharide composed of 4 and 5 glucose residues. The 'Alo mixture' has properties that are favourable to food industry. It is mildly sweet, has low viscosity, high moisture retaining capacity and low water activity convenient in controlling microbial contamination.

For the manufacturing of 'Alo mixture', starch is dextrinised using thermostable bacterial α-amylase. The degree of hydrolysis (DE) of starch is kept between 6 to 10. The simultaneous reaction of sacharification and transglucosidation of dextrin is done by using soyabean β-amylase and *Aspergillus niger* transglucosidase. The reaction mixture is finally purified and concentrated to 25% moisture.

Manufacture of high molecular weight branched dextrins

Branched dextrins of high molecular weight are prepared by hydrolysis of corn starch with α-amylase. The extent of starch degradation to be carried out depends on the type of starch and the physical properties desired. The branched dextrins are obtained as powder after chromatography and spray drying. These are used as extender and a glozing agent for production of powdery foods and rice cakes, respectively.

Removal of starch sizer from textile (desizing)

In textile weaving, starch paste is applied for warping. This gives strength to the textile at weaving. It also prevents the loss of string by friction, cutting and generation of static electricity on the string by giving softness to the surface of string due to laid down warp. After weaving the cloth, the starch is removed and the cloth goes to scouring and dyeing. The starch on cloth is usually removed by application of α-amylase.

Direct fermentation of starch to ethanol

The amylolytic activity rate and amount of starch utilisation and ethanol yields increase in several fold in cocultures. Molds amylases are used in alcohol production and brewing industries. The advantages of such system are uniform enzyme action in mashes, increase rate of saccharification, alcohol yield and yeast growth.

Treatment of starch processing wastewater (SPW)

Starch is also present in waste produced from food processing plants. Starch waste causes pollution problems. Biotechnological treatment of food processing wastewater can produce valuable products such as microbial biomass protein and also purifies the effluent.

Other applications

Amylases, especially alkaline amylases are used in detergents. To some extent amylases are also used as digestive aids to supplement the diastatic activity of flour and to improve digestibility of some of the animal feed ingredients.

5.3 Pectic enzymes

Pectic enzyme, also known as pectinase, is a protein that is used to break down pectin, a jelly like glue that holds plant cells together. In wines pectin can cause troublesome 'pectin haze' that is not easily cleared without the use of pectic enzymes.

While this enzyme does occur naturally in grapes as well as yeast there is not enough of it to overcome the amount of pectin present in the must. Other sources of pectic enzyme include plants, bacteria and fungus.

It turns out that fungus produces a special kind of pectic enzyme that is particularly adept at breaking down pectin even in the harsh environment created during fermentation. Most commercially sold pectic enzymes come from fungus.

Pectic enzymes may be purchased in a liquid form or as a powder at any home brewing supply store. Pectins are polysaccharides ubiquitous in the plant kingdom and constitute the major component of plant cell walls. The pectinases are a group of related enzymes capable of degrading pectin. Therefore, this group of enzymes have been used for decades in the food and wine making industry for the processing of fruit juices.

The pectinases are synthesised by plants and micro-organisms, the latter being used for industrial production. Micro-organisms are used to produce many enzymes of industrial interest in processes relatively inexpensive and environmentally friendly. Moreover, enzymatic catalysis is preferred over other chemical methods since it is more specific, less aggressive and generates less toxicity. Advances in biotechnology, especially in the fields of molecular biology and microbial genetics, have led to major advances in enzyme technology and have allowed, in many cases, the development of new producing strains and microbial enzymes. The production of pectinases may also benefit from these technologies.

This section reviews the characteristics of pectic substances, the types and mode of action of enzymes which degrade them and the main applications of commercial preparations of microbial pectinases in the food and wine making industry, followed by a review of new micro-organisms and pectolytic enzymes, evaluating new approaches to their production, marketing and use.

Pectic substances are polysaccharides of high molecular weight, with a negative charge, appearing mostly in the middle lamella and the primary cell wall of higher plants, found in the form of calcium pectate and magnesium pectate. They are formed by a central chain containing a variable amount although in high proportion of galacturonic acid residues linked through α-(1–4) glycosidic bonds partially esterified with methyl groups.

The generic name of pectic substances is used for referring to four types of molecules: protopectin (pectic substance in intact tissue), pectinic acids (polygalacturonan containing >0–75% methylated galacturonate units), pectic acids (polygalacturonan that contains negligible amount of methoxyl groups) and pectins (pectinic acid with at least 75% methylated galacturonate units). Protopectines are insoluble in water, while the rest are wholly or partially soluble in water.

Pectic substances represent between 0.5–4% of fresh weight plant material. In addition to their role as cementing and lubricating agents in the cell walls of higher plants, they are responsible for the texture of fruits and vegetables during growth, maturation and their storage. Furthermore, pectic substances are involved in the interaction between plant hosts and their pathogens.

Pectins have numerous and important applications in the food and pharmaceutical industries. In the food sector, it is primarily used as a gelling agent, replacing sugars and/or fats in low-calorie food and as nutritional fibre. The pharmaceutical industry offers them as preparations to reduce cholesterol or to act as a lubricant in the intestines thus promoting normal peristaltic movement without causing irritation. In addition, these polysaccharides are used as drug delivery systems, which can also reduce the toxicity of these and make their activity longer lasting without altering their therapeutic effects.

5.3.1 Pectic enzymes in nature: Microbial pectinases

The microbial world has shown to be very heterogeneous in its ability to synthesise different types of pectolytic enzymes with different mechanisms of action and biochemical properties. Pectic enzymes are produced by both prokaryotic micro-organisms, which primarily synthesise alkaline pectinases and by eukaryotic micro-organisms, mostly fungi that synthesise acid pectinases.

There are many studies that have been conducted related to the characterisation of different microbial pectic enzymes concerning their mechanisms of action and biochemical properties. The optimal pHs that these enzymes may act range between 3.5–11, while the optimal temperatures vary between 40–75°C. Given the features of the substrate on which they act and the effect that is required of them, acidic and depolymerising pectinases are of great interest for the food industry although some applications such as the extraction of oils requires the alkaline ones.

5.3.2 Pectinases in the food industry

Extraction of fruit and vegetable juices

The main industrial application of pectinases is the extraction and clarification of fruit and vegetable juices. Most of the microbial pectinases produced by the industry are dedicated to this purpose. Pectins are responsible for the turbidity and consistency of the juice causing an increase in their viscosity which hinders its clarification, filtration and concentration. The degradation of pectic substances in mashed fruit purees is achieved through the addition of pectolytic enzymes resulting in an increase in juice yield and its clarification as well as a decrease in viscosity. Treatment with pectinases also provides filtering of the product.

The fruit and vegetable juice industry uses mainly acidic pectinases of fungal origin, principally from *Aspergillus* spp. Commercial preparations are mixtures of polygalacturonases, pectate lyases and pectin esterases. Pectate lyases can act on the esterified pectin while the polygalacturonases act on the desesterified pectin thus it might require previous action of the pectin esterases. Pectic enzymes treatments vary depending on the type of juice.

Clear juices (i.e., apple, pear, grape): In this type of product, pectolytic enzymes are added to increase the yield in juice during the pressing and for removing matter in suspension. In the case of apple juice, the most commonly used enzymes are those that can depolymerise the highly esterified apple pectin. Apple juice can be obtained through a two-step process consisting of a first treatment of the crushed apple mush with pectinases to obtain the premium juice followed by pomace liquefaction treatment made with a mixture of different pectinases and cellulases for the complete extraction of the juice.

After washing and crushing, the apples are pressed to obtain the juice. Pectic enzymes are used to facilitate the pressing and juice extraction and to assist in the separation of a flocculant precipitate by means of sedimentation, filtration or centrifugation. If a cloudy product is required, the juice is pasteurised after pressing to inactivate residual enzymes. Centrifugation removes the large-size remains leaving small particles in suspension.

However, if a clear juice is required, these suspended particles have to be withdrawn. In order to do this a treatment with mixtures of commercial enzymes is carried out containing pectinases, cellulases and hemicellulases. Finally, the fluid is centrifuged to obtain the clear juice. Although it has been noted that the proper clarification of apple juice requires mixtures of polygalacturonase and pectin methyl esterase, subsequent studies have shown that it can be clarified by only using a pure pectin lyase.

Cloudy juices (i.e., citrus, tomato, nectars): In the case of orange juice, where natural pectin esterases are present, pectin is only partially methylated. Polygalacturonases are the pectic enzymes which are most commonly used and of great interest for this type of fruit juice.

In the process of orange juice extraction, pectinases can be added at the end of the pulp wash extraction to reduce viscosity or, preferably, at the end of the first finisher. This leads to higher yield in juice, a better extraction of soluble solids and to a lower viscosity. The action of these enzymes just reduces the viscosity without attacking the insoluble pectin that maintains the stability of the cloud. Enzyme preparations should lack or have the least possible amount of pectin methyl esterases to avoid the clarification of the product. It has been suggested that the best enzyme might be pure pectin lyase.

Maceration products of plant tissues

The enzymatic maceration of plant tissues allows the transformation of these organised tissues in suspensions of intact cells that constitute the pulpy products that are used as a basis for preparing juices, nectars, baby food and some dairy products such as yoghurts. Enzyme preparations for this purpose contain cellulases, hemicellulases and pectic enzymes which should only act on the middle lamella of the plant tissue.

Extraction of vegetable oil

Vegetable oils of olive, sunflower, coconut, palm or canola are obtained by extraction with organic solvents such as hexane. The use of pectolytic enzymes, in this case preferably alkaline, allows the extraction of vegetable oils in an aqueous process. Enzyme preparations based on cellulases, hemicellulases and pectinases have been used successfully in the extraction of olive oil. The enzyme treatment not only improves oil yield and stability but also increases polyphenols and vitamin E content enhansing its organoleptic quality.

Coffee, cocoa and tea fermentation

Traditionally, fermentation of coffee is made with pectolytic micro-organisms in order to remove the layer of mucilage from the coffee beans. With the same purpose, commercial enzyme preparations containing pectinase is sprayed onto the beans to ferment. A cheaper alternative is to use, with the same purpose, the filtrate of inoculated fermentations. The enzyme treatment significantly reduces the fermentation time. Cocoa fermentation is essential to develop the chocolate flavour. This fermentation is carried out by a succession of different micro-organisms, some of them pectolytic. Pectic enzymes allow the degradation of the cocoa pulp and are indispensable for the fermentation process and the good quality of fermented beans. Similarly, treatment of tea leaves with pectic enzymes of fungal origin (at a dose adjusted to avoid damaging the leaf), facilitates and accelerates the fermentation.

5.3.3 Pectinases in the wine industry

Wine is the result of the fermentation of grape juice. Pectinases are the most important enzymes used by the wine making industry although commercial preparations may contain other enzymatic activities such as hemicellulases, glucanases and glycosidases.

Pectic enzymes are synthesised naturally by the plant and are present in the grape. However, they have low activity during the wine producing process. Microbial pectolytic enzymes especially of fungal origin are resistant to the conditions of fermentation and can be used to facilitate processes, improve quality and diversify products. So far, commercial enzymes are produced all from fungi, mainly of the genus *Aspergillus*. Although all enzymes are produced by *Aspergillus*, pectinase preparations currently available for the wine market are very different. Both the type of activities as well as their concentration in the preparations depend on the strain of *Aspergillus* used, the fermentation conditions for production, the nature of the fermented substrate and the degree of partial purification.

Various research studies have shown that the addition of pectolytic enzymes leads to increased levels of methanol in wine due to the activity of pectin methyl esterase. Methanol is toxic and its maximum concentration in wine is regulated. Therefore, pectin methyl esterase activity should be at low concentrations in commercial mixtures. The functions of pectic enzymes in the wine making process are to support the extraction process, maximise juice yield, facilitate filtration and intensify the flavour and colour.

Extraction, clarification and filtration

The addition of pectinase to the must reduces its viscosity and causes the grouping of suspended particles in larger aggregates that can be removed by sedimentation. If the enzymes are added to the pulp before pressing, must yield increases, facilitating the pressing and enhancing the colour. Macerating enzyme preparations for this purpose contain pectinases as well as cellulases. A high level of polygalacturonase is very effective for clarification but may require the prior action of pectin lyase activity. For this reason the enzyme preparations with a high content of pectin lyase are desirable when very fast racking is required to prevent any problems related to must oxidation, development of endogenous microbiota and nutrient loss. In addition, wines made with pectic-enzyme-treated grape must significantly reduces filtration times.

Intensification and stabilisation of colour

In the case of musts obtained from red grapes, the degradation of cell walls in the skin of the grapes through pectolytic enzyme treatment results in an increased release of phenolic compounds responsible for colour. Early work related to the use of pectinases to enhance the colour of the wines were a bit confusing because Ough and others were able to intensify the colour of wine by the enzyme treatment, but Wightman and others found that some enzyme preparations containing pectic enzymes reduced red wine colour. Subsequent studies performed with two different enzyme preparations confirmed that both allow to produce wines with higher concentrations of anthocyanins and total phenols and have a higher colour intensity.

Boosting aroma and flavour

The aromatic profile of wines consists of two components: the varietal aromas characteristic of the variety of grape used and the aromas originated by the yeast during fermentation. In many cases the grape variety used completely determines the aroma of the wine, especially in young wines. The volatile compounds of grapes include monoterpenes, C13 nor-isoprenoides, benzene derivatives and aliphatic alcohols. The aromatic components of the grape may appear as free forms, which contribute directly to its scent or bound forms, of greater concentration, to sugars and nonvolatile. Nonodorous glycoside flavour precursors accumulate in the grape especially in the skin during the ripening process. The hydrolysis of these precursors via beta-glucosidases releases the olfactory active aglycones. Pectic enzymes help break down the cell walls of the grapes and thus to extract the aromatic precursors. The addition of pectic enzymes during the extraction or fermentation of the must results in an increase in aromatic precursors susceptible to being attacked by beta-glucosidases from the must, those produced by yeasts and bacteria during fermentation or those which are included in commercial enzyme preparations, thereby enhancing the aroma of wines.

5.3.4 Biotechnological aspects for production of microbial pectic enzymes

As discussed until now, there are numerous applications of microbial pectic enzymes in food and wine-making industry. Not surprisingly, the sales volume of these enzymes represents 25% of the enzymes that are commercialised in the food and alcoholic beverages industry. These many applications require one or more types of pectinases that must act in very different condition according to the process in which they are involved. For example, while in the extraction and clarification of fruit juice enzymes can be used at temperatures between 45–95°C, the wine industry employs temperatures below 15–10°C. Although most food applications pectinases have to act in a medium acid, in others, such as oil extraction, they should perform in a medium alkaline.

So far, commercial pectic enzymes are prepared only from cultures of filamentous fungi, mainly of the genus *Aspergillus*. Commercial products contain mainly a mix of polygalacturonase, pectin lyase and pectin methyl esterase. The use of different strains of *Aspergillus* and modification of substrates and culture conditions can lead to mixtures enriched in one type of enzyme.

However, these commercial pectic preparations are not always optimal for each process and its use is not without side effects and controversy. Thus, while pectin methyl esterase activity present in the samples may be necessary for the action of polygalacturonase in the case of pectin with a high degree of esterification, its action may lead to an undesirable increase of methanol in the products. Moreover, although the pectic enzymes are the most abundant in commercial mixtures, they can also contain other undesirable activities, such as in the case of making wine, polyphenoloxidases or cinnamyl esterases.

At present especially in recent years, the accumulated knowledge of new microbial pectolytic enzymes as well as methodological and technological advances can address the production and use of these enzymes with a different approach. The diversity of applications and conditions in which these enzymes must work also demand a large number of different enzymes capable of acting in such conditions. Even more interesting would be having more robust, broad-spectrum enzymes which allow for a more versatile use in different applications.

Considering the variety of enzymes, traditionally the majority of studies refer to the pectinases of *Erwinia* and *Bacillus* within the bacteria and various fungi especially *Aspergillus* although there has been a major advance in the description and characterisation of pectic enzymes produced by yeast in the last 15 years. In recent years there has been also a growing interest in studying pectic enzymes with very interesting properties from the point of view of their application. These include thermostable pectinases or pectinases with optimal activity at low temperatures.

The possibility of producing different types of pectolytic enzymes separately and for later preparation of their mixtures in the proper proportions would allow to provide more suitable commercial preparations for each application and lacking undesirable activities. In this sense, micro-organisms such as some strains of yeast *Saccharomyces cerevisiae* and *Kluyveromyces marxianus* which produce only one type of enzyme or that constitutively synthesise it are of great interest. Similarly, obtaining constitutive mutants from producing strains allow the optimisation of production and contribute to making cost-effective production processes. Furthermore, these strains can be used to produce pectinases that accumulate together with other metabolites in the culture broth, which contributes favourable to the overall economy of the process. Heterologous expression of enzymes in prokaryotic or eukaryotic systems is a technique of great interest for the production of a single type of enzyme. Table 5.2 shows some of the many pectic enzymes, both from bacteria, yeast and fungi, which are rightly expressed in *Escherichia coli* and different species of yeast. Although *E. coli* is unable to carry out post-translational modifications of proteins, fungal pectolytic enzymes expressed in these bacteria are active. However, one of the most interesting strategies seems to be the expression of pectinase genes in yeast, particularly in *Pichia pastoris*, in which very high levels of constitutive expression has been achieved. In some cases changes in glycosylation patterns conducted by yeast did not affect the activity and characteristics of recombinant pectinases, while other changes do occur with respect to the characteristics of the native protein, which even lead to enzymes with interesting properties for certain applications.

Thus, the enzymes that degrade the pectic substances play an essential role in the food and wine making industries because they are used to degrade the pectins that interfere with the extraction and clarification of fruit juices and oils as well as being important in the fermentation of coffee, cocoa and tea. Also, in the wine industry they play an important role by contributing to the release of the molecules responsible for aroma and colour, two of the major components that characterise a wine.

Table 5.2: Microbial pectic enzymes expressed in different host strains.

Micro-organism gene origin	***Type of enzyme***	***Host strain***
Xanthomonas campestris	Pectate lyase Polygalacturonase	*E. coli*
Streptomyces coelicolour	Polygalacturonase	*E. coli*
Pseudoalteromonas haloplanktis ANT/505	Pectate lyase	*E. coli*
Thermotoga maritima	Exopolygalacturonase	*E. coli*
Burkholderia capacia	Endopolygalacturonase	*E. coli*
Phytophthora capsici	Pectate lyase	*E. coli*
Erwinia chrysantemy	Pectate lyase Polygalacturonase	*S. cerevisiae*
Aspergillus niger RH5344	Polygalacturonase	*S. cerevisiae*
Aspergillus aculeatus	Pectin methyl esterase	*S. cerevisiae*
S. cerevisiae IM1-8b	Endopolygalacturonase	*S. cerevisiae*
S. cerevisiae IM1-8b	Endopolygalacturonase	*Schizosaccharomyces pombe*
K. marxianus CECT1043	Endopolygalacturonase	*P. pastoris*
Bispora sp. MEY-1	Endopolygalacturonase	*P. pastoris*

Traditionally, this industry uses different mixtures of pectolytic enzymes derived from fungi cultures, mainly of the genus *Aspergillus*, not always completely adequate for the processes they must carry out because of the type and concentration of different enzyme activities that they contain, not without undesirable effects due to other non-pectic enzymes that may be present in the mixtures.

The exploration of microbial biodiversity has allowed, especially in recent years, to identify and characterise new pectic-enzyme-producing micro-organisms with different biochemical characteristics, some potentially very interesting from the point of view of their application. Also, it has been technically possible, on the one hand, to select wild strains and constitutive mutants that produce a single enzyme, and, on the other hand, the heterologous expression in bacteria and yeast of numerous genes which encode pectic enzymes, obtaining producing strains of interest. All this opens the possibility of producing different pectic enzymes individually and preparing commercial mixtures of these, adapted to each process.

5.4 Lactase enzyme

For those with lactose intolerance, we may be all too familiar with what we picked from our digestive enzymes, lactase. Lactase digestion helps you process the lactose in dairy, helping to prevent bloating, gas, constipation, diarrhea and cramping that happens to many who ingest milk products. Lactose intolerance is not an allergy, it is about a missing enzyme necessary for the proper digestion of dairy. For those who can digest dairy, lactase is produced by cells in the small intestine. For those who are intolerant, it is not—or it is not produced in sufficient quantity.

Lactase's primary function is to break down lactose, a type of sugar found in milk and other dairy products. Our body cannot naturally absorb lactose, so lactase enzymes help digest this milk sugar. This in done in order for our body to metabolise this form of sugar by breaking down the lactose into smaller, more digestible sugars called glucose and galactose.

When our body is unable to produce enough lactase to function properly, a condition known as lactose intolerance develops. Lactase deficiency is the most common and well-known form of carbohydrate intolerance.

Most mammals, including humans, have high intestinal lactase activity at birth (which makes sense when you consider all mammals nurse during their first few months). But, in many cases, this activity declines to low levels during childhood and drops even further (or completely disappears) in adulthood. The low lactase levels cause incomplete digestion of milk and other foods containing lactose. Undigested lactose in the bowel then is subject to fermentation which causes the bloating, gas, constipation, diarrhea and cramping associated with lactose intolerance.

It is estimated that approximately 70% of the world's population is deficient in intestinal lactase with more than one-third of the U.S. population presumed to be lactose intolerant and unable to digest dairy products, causing sometimes severe digestion problems.

According to MedlinePlus, a service of the National Institutes of Health, symptoms of lactase deficiency begin 30 min to 2 hr after ingesting milk or a similar dairy product. Symptoms can include abdominal bloating, abdominal cramps, diarrhea, flatulence and diarrhea. Supplemental lactase enzymes have been found to decrease the symptoms of lactose intolerance associated with the consumption of dairy foods. One way of treating lactose intolerance is by adding lactase enzymes to regular milk, or take digestive enzymes that contain lactase in capsule or chewable tablet form.

5.5 Invertase

Invertase, also called beta-fructofuranosidase cleaving the terminal non-reducing beta-fructofuranoside residues, is a glycoprotein with an optimum pH 4.5 and stability at 50°C. Invertase is used for the inversion of sucrose in the preparation of invert sugar and high fructose syrup (HFS). It is one of the most widely used enzymes in food industry where fructose is preferred than sucrose especially in the preparation of jams and candies, because it is sweeter and does not crystallise easily. The enzymatic activity of Invertase has been characterised mainly in plants and micro-organisms. Among micro-organisms, *Saccharomyces cerevisae*, *Candida utilis*, *Aspergillus niger*, *Thermomyces lanuginosus* and *Penicillium chrisogenum* has been widely studied.

Invertase exhibits marked stability towards temperature, pH changes and denaturants. Temperature of the reaction mixture determines the rate of sucrose inversion by the active enzyme.

Sources of Invertase: The official name for Invertase is beta-fructofuranosidase (EC.3.2.1.26), which implies that the reaction catalysed by the enzyme, is the hydrolysis of the terminal non-reducing beta-fructofuranoside residues in beta-fructofuranosides. Invertaseis widely distributed among the biosphere. It is mainly characterised in plants and micro-organisms. *Saccharomyces cerevisiae* commonly called Baker's yeast is the chief strain used for the production of Invertase commercially. They are found in wild growing, on the skin of grapes and other fruits. Though plants like Japanese Pear fruit (Pyrus pyrifolia), Pea (Pisum sativum), Oat (Avena sativa) can also be used, but generally micro-organisms like *S. cerevisiae*, *Candida utilis*, *A. niger* are considered ideal for their study.

Kinetics of Invertase enzyme: In contrary to most other enzymes, Invertase exhibits relatively high activity over a broad range of pH (3.5–4.5) with the optimum near pH of 4.5. The enzyme activity reaches a maximum at 55°C. The Michaelis–Menten (K_m) value for the free enzyme is typically 30 mM (approximately). The enzyme is a glycoprotein, stable at 50°C. The cations Hg^{2+}, Ag^{+}, Ca^{2+} and Cu^{2+} exhibit a marked inhibition of the enzyme. Competitive inhibition was observed with the fructose analogue 2,5-anhydro-D-mannitol suggesting that the enzyme was inhibited by the furanose form of fructose.

5.5.1 Isozymes of Invertase

Isozymes in baker's yeast

Invertase exists in more than one form in yeasts generally, either extracellular Invertase or intracellular Invertase. The external yeast Invertase is a glycoprotein containing about 50% carbohydrate, 5% mannose, 3% glucosamine, whereas internal Invertase contains no carbohydrate. The former one has a molecular weight of 135 KDa whereas the latter variety has a molecular weight of 270 KDa. It has been established that in depressed cells most of the Invertase is external whereas in fully repressed state all the Invertase is intracellular. Both differ in amino acid sequences particularly the internal Invertase does not contain cysteine. Both the enzymes are inhibited by Iodine and reactivated by mercaptoethanol. Both require an acid with pK_a about 6.8 in its protonated form. Both are inhibited by cyanogen bromide in a biphasic reaction.

Isoforms of Invertase in plants

Several isoforms of Invertase exist with different biochemical properties and subcellular locations in plants. On the basis of solubility, optimum pH, isoelectric point and subcellular localisation, plant Invertase can be classified into three subgroups. Three biochemical subgroups of Invertase in plants: vacuolar (soluble acid), cytoplasmic (soluble alkaline) and cell wall bound Invertase. The presence of multiple isoform of Invertase in nature have functionally beneficial role to the plants.

Insoluble acid/cell wall bound Invertase: Insoluble acid Invertase (INAC-INV) is cell wall bound, glycosylated protein with a variable molecular weight ranging between 28 and 64 KDa. It has an optimum pH of 4.0, temperature optimum of 45°C and an isoelectric point of 9. Its activity is inhibited by 6.2 mM copper sulphate.

Soluble acid/vacuolar Invertase: Vacuolar Invertase has an acidic pI with a pH range between 4.5 and 5.0. The enzyme has a K_m for sucrose in the low-millimoles range. Along with sucrose, it also hydrolyses raffinose or stachiose being as a true member of β-fructofuranoside family. The enzyme loses its activity when reacted by heavy metal ions like mercury or silver. Also, glucose acts as a non-competitive inhibitor for the enzyme and fructose being a competitive inhibitor. The mature polypeptide is *N*-glycosylated and has a molecular mass of approximately 70 KDa.

Soluble acid Invertase plays important biological functions related to sucrose metabolism and predominantly hydrolyses sucrose for growth and developmental processes. Also, sucrose hydrolysis by soluble acid Invertase helps in regulation of osmotic pressure which is controlled by cell expansion which depends on size of vacuole.

Soluble alkaline/cytoplasmic Invertase: Soluble alkaline Invertase is a non-glycosylated polypeptide expressed at low levels. The two isoforms are encoded by the same gene and two transcripts originate from differential splicing of a hetero nuclear mRNA. The native polypeptides are homotetramers with a molecular mass of 54–65 KDa. The enzyme has a K_m of 10 mM for sucrose and is inhibited by glucose, fructose and Tris but not by heavy metal ions. Using cDNA expression, when the amino acid sequence of soluble alkaline Invertase was deduced, it lacks *N*-terminal signal peptide and has no similarity with other forms of Invertases.

Soluble alkaline Invertase is not a member of β-fructofuranosidase family as it hydrolyses sucrose only unlike other acid Invertases. It is found in all plant organs at different developmental stages, especially in the developing tissues implying it has growth related functions.

5.5.2 Role of Invertase in plants

Plant osmoregulation and metabolism

To provide cell, fuel for respiration, carbon and energy for the synthesis of different compounds, Invertase cleave sucrose into corresponding monosaccharide. By generating the necessary sucrose concentration gradient between sites of phloem loading and unloading, Invertase also help in long-distance transport of sucrose. Hydrolysis of sucrose into glucose and fructose influences the osmotic pressure of cells and thus helps in cell elongation and plant growth. Developing roots of carrot or elongating stems of bean are some of the organs of the plant which contain high activity of acid Invertase especially in rapidly growing tissues.

High acid Invertase activity can also be correlated with the accumulation of hexoses in sugar storing sink organs such as fruit. Thus, indicating that a soluble acid Invertase also function as a regulator of sugar composition in the post harvest processes.

Role of Invertase in plant development

The higher levels of Invertase activity can be observed in oat internodes reflecting the increased energy and carbon requirements to sustain the biochemical reactions during growth period. Thus, suggesting that a close relationship exists between growth rate and level of Invertase activity. The degradation of carbohydrate in the tissue is also observed proportional to the enhancements in respiration and protein and cell-wall biosynthesis during the growth period.

Invertase in defence mechanism

Invertase results in a link reaction between carbohydrate degradation and pathogen responses. This includes the phenomenon of high sugar resistance in which key pathogenesis related genes which are sugar inducible get over expressed in the plant apoplast. This results in an increased expression of pathogen related (PR) proteins and thus increased resistance against viral infections. The regulation of extracellular Invertase by phytohormones could also contribute to plant pathogen responses involving in expression of various defences related genes.

In this process the extracellular Invertase induced by sugars provides a mechanism in which the sink strength will elevate increasing the sugar concentration. This induces PR genes and represses photosynthetic genes in addition to signals derived from the pathogen.

5.5.3 Regulation of Invertase in plants

By phytohormones

Gibberellic acid plays a central role in regulating Invertase levels (GA3) promoting cell elongation essential for flower induction. High Invertase activity can be seen in several plant organs such as sugarcane stem, Jerusalem artichoke tubers, beet roots, lentil epicotyls, internodes of beans and oat, etc. Cytokinins promote cell and thus an enhanced demand for carbohydrate is needed for active growth. This phenomenon is bolded by the fact that tissues with higher activity of extracellular Invertase (rapidly growing tissues), also contain elevate concentration of cytokinin phytohormone. Ethylene is the only phytohormone which stimulates the down-regulation the expression of extracellular Invertase by repressing its mRNA transcripts.

Under abiotic stress

Under salt stress, plants produce photo assimilates which support crucial processes such as growth, maintenance and osmotic adjustment. An increase in sucrose in source leaves occurs with a decrease in photosynthesis rate due to feedback inhibition under saline condition. The extracellular Invertase plays a key role in those species in which the step of phloem unloading of sucrose is apoplasmic and also in the control of assimilate allocation.

In the above process, when the extracellular Invertase is impaired or the phloem unloading pathway is symplasmic, the vacuolar acid Invertase and neutral Invertase play the major role.

A decrease in export of assimilates and a decrease in crop production occurs under water stress due to inductions of large alterations in source – link reactions. Under such conditions, the elevated activities of soluble and insoluble Invertase get blocked during pollination and early kernel development in maize. Under drought conditions, in mature maize leaves, cell wall Invertase activity does not get affected but an increase in vacuolar Invertase activity can be seen leading to accumulation of hexoses in the leaves. Low oxygen stress in maize root tips decreases Invertase expression and therefore, decreasing the Invertase/sucrose ratio. Thus, by conserving sucrose and ATP and reduction of the hexose-based sugar signalling system, plants acclimatised to low oxygen condition.

5.5.4 Applications of Invertase enzyme

An equimolar mixture of fructose and glucose (invert syrup) obtained by sucrose hydrolysis is sweeter than sucrose due to high degree of sweetness of fructose, as a result the sugar content can be increased without crystallisation of the material. The production of non-crystallisable sugar syrup from sucrose is one of the major applications of Invertase enzyme. Invert syrup has hygroscopic properties which makes it useful in the manufacturing of soft-centred candies and fondants as ahumectants. Alcoholic beverages, lactic acid, glycerol, etc., produced by fermentation of sucrose containing substrates requires the use of Invertase. It is also associated with insulinase for the hydrolysis of inulin (poly-fructose) to fructose.

Other application of the enzyme is seen in drug and pharmaceutical industries. Also it is used in the manufacture of artificial honey and plasticising agents which are used in cosmetics. Enzyme electrodes are used for the detection of sucrose. Formation of undesirable flavouring agents as well as coloured impurities do not take place on enzymatic hydrolysis of sucrose instead of acid hydrolysis. Immobilised Invertase is used for continuous hydrolysis of sucrose as the resulting shifts in the pH can be used to prevent the formation of oligosaccharides by the transferase activity associated with the soluble enzyme. Invertase being a powerful anti-microbial agent and an anti-oxidant aids in the prevention of bacterial infestations and gut fermentation due to oxidation.

Raw honey was used in ancient India in killing bacteria, reducing intestinal ailments and was given to patients having a weak heart. It can also be used in subsiding bacterial infections because of its ability to extract moisture from the body of the patient. According to a European study on 18000 patients, honey has been proved effective in treating respiratory tract infection such as bronchitis, asthma and allergies. Invertase along with other enzymes has also been shown to help cure colds, flu and other respiratory problems.

5.6 Alpha-galactoside

Alpha-galactosidase is a glycoside hydrolase enzyme that hydrolyses the terminal alpha-galactosyl moieties from glycolipids and glycoproteins. It is encoded by the *GLA* gene. Two recombinant forms of alpha-galactosidase are called agalsidase alfa and agalsidase beta.

5.6.1 Function of alpha-galactosidase

This enzyme is a homodimeric glycoprotein that hydrolyses the terminal alpha-galactosyl moieties from glycolipids and glycoproteins. It predominantly hydrolyses ceramide trihexoside and it can catalyse the hydrolysis of melibiose into galactose and glucose.

5.6.2 Pathology of alpha-galactosidase

A variety of mutations in this gene affect the synthesis, processing and stability of this enzyme, which causes Fabry's disease, a rare lysosomal storage disorder and sphingolipidosis that results from a failure to catabolise alpha-D-galactosyl glycolipid moieties.

Two enzyme replacement therapies are available to functionally compensate for alpha-galactosidase deficiency. Agalsidase alpha and beta are both recombinant forms of the human α-galactosidase A enzyme and both have the same amino acid sequence as the native enzyme. Agalsidase alpha and beta differ in the structures of their oligosaccharide side chains.

5.7 Cellulase

Microbial cellulases find applications in various industries and constitute a major group of the industrial enzymes. Recently, there is resurgence in utilisation of biomass for fuel production employing cellulases and hence forth in obtaining better yields and novel activities. Improving the economics of such processes will involve cost reduction in cellulase production which may be achieved by better bioprocesses and genetic improvement of cellulase producers to yield more of the enzyme.

Cellulose is the most abundant and renewable biopolymer on earth and the dominating waste material from agriculture. A promising strategy for efficient utilisation of this renewable resource is the microbial hydrolysis of lignocellulosic waste and fermentation of the resultant reducing sugars for production of desired metabolites or biofuel. Cellulose is a crystalline polymer, an unusual feature among biopolymers. Cellulose chains in the crystals are stiffened by inter and intra chain hydrogen bonds and the adjacent sheets which overlie one another are held together by weak van der Waals forces. In nature, cellulose is present in a nearly pure state in a few instances whereas in most cases, the cellulose fibres are embedded in a matrix of other structural biopolymers, primarily hemicelluloses and lignin. An important feature of this crystalline array is the relative impermeability of not only large molecules like enzymes but in some cases even small molecules like water. There are crystalline and amorphous regions, in the polymeric structure and in addition there exists several types of surface irregularities.

This heterogeneity makes the fibres capable of swelling when partially hydrated, with the result that the micro-pores and cavities become sufficiently large enough to allow penetration of larger molecules including enzymes. At the molecular level, cellulose is a linear polymer of glucose composed of anhydoglucose units coupled to each other by β-1-4 glycosidic bonds. The number of glucose units in the cellulose molecules varies and degree of polymerisation ranges from 250 to well over 10,000 depending on the source and treatment method. The nature of cellulosic substrate and its physical state are important factors in its enzymatic hydrolysis. Though lignocellulosic biomass is generally recalcitrant to microbial action, suitable pre-treatments resulting in the disruption of lignin structure and increase accessibility of enzymes have been shown to increase the rate of its biodegradation.

Microbial degradation of lignocellulosic waste and the downstream products resulting from it is accomplished by a concerted action of several enzymes, the most prominent of which are the cellulases, which are produced by a number of micro-organisms and comprise several different enzyme classifications. Cellulases hydrolyse cellulose (β-1,4-D-glucan linkages) and produce as primary products glucose, cellobiose and cello-oligosaccharides.

There are three major types of cellulase enzymes [Cellobiohydrolase (CBH or 1,4-β-D-glucan cellobiohydrolase, EC 3.2.1.91), Endo-β-1,4-glucanase (EG or endo-1,4-β-D-glucan 4-glucanohydrolase, EC 3.2.14) and β-glucosidase (BGEC 3.2.1.21)]. Enzymes within these classifications can be separated into individual components, such as microbial cellulase compositions may consist of one or more CBH components, one or more EG components and possibly β-glucosidases. The complete cellulase system comprising CBH, EG and BG components synergistically act to convert crystalline cellulose to glucose. The exocellobiohydrolases and the endoglucanases act togetherto hydrolyse cellulose to small cello-oligosaccharides.

The oligosaccharides (mainly cellobiose) are subsequently hydrolysed to glucose by a major β-glucosidase.

Cellulases are used in the textile industry, in detergents, pulp and paper industry, improving digestibility of animal feeds, in food industry and the enzymes account for a significant share of the world enzyme market. The growing concerns about shortage of fossil fuels, the emission of green house gases and air pollution by incomplete combustion of fossil fuel has also resulted in an increased focus on production of bioethanol from lignocellulosics and especially the possibility to use cellulases and hemicellulases to perform enzymatic hydrolysis of the lignocellulosic material. However, in production of bioethanol, the costs of the enzymes to be used for hydrolysis of the raw material need to be reduced and their efficiency increased in order to make the process economically feasible.

Commercial production of cellulases has been tried by either solid or submerged culture including batch, fed batch and continuous flow processes. Media used in cellulase fermentations contain cellulose in different degrees of purity, or as raw lignocellulosic substrates, which is especially true in the case of solid-state fermentation. Cellulases are inducible enzymes and the most problematic and expensive aspect of industrial cellulase production is providing the appropriate inducer for cellulases. Cellulase production on a commercial scale is induced by growing the fungus on solid cellulose or by culturing the organism in the presence of a disaccharide inducer such as lactose. However, on an industrial scale, both methods of induction result in high costs. Since the enzymes are inducible by cellulose, it is possible to use cellulose containing media for production but here again the process is controlled by the dynamics of induction and repression. At low concentrations of cellulose, glucose production may be too slow to meet the metabolic needs of active cell growth and function. On the other hand, cellulase synthesis can be halted by glucose repression when glucose generation is faster than consumption.

Thus, expensive process control schemes are required to provide slow substrate addition and monitoring of glucose concentration. Moreover, the slow continuous delivery of substrate can be difficult to achieve as a result of the solid nature of the cellulosic materials. The challenges in cellulase production involve developing suitable bioprocesses and media for cellulase fermentation, besides identification of cheaper substrates and inducers. Genetic modification of the cellulase producers to improve cellulase activity has gone a long way to give better producers with high enzyme titers, but still cellulase production economics needs further improvement for commercial production of ethanol from biomass.

5.7.1 Micro-organisms producing cellulases

Cellulolytic microbes are primarily carbohydrate degraders and are generally unable to use proteins or lipids as energy sources for growth. Cellulolytic microbes notably the bacteria *Cellulomonas* and *Cytophaga* and most fungi can utilise a variety of other carbohydrates in addition to cellulose, while the anaerobic celluloytic species have a restricted carbohydrate range, limited to cellulose and or its hydrolytic products.

The ability to secrete large amounts of extracellular protein is characteristic of certain fungi and such strains are most suited for production of higher levels of extracellular cellulases.

One of the most extensively studied fungi is *Trichoderma reesei*, which converts native as well as derived cellulose to glucose. Most commonly studied cellulolytic organisms include: Fungal species-*Trichoderma, Humicola, Penicillium, Aspergillus*, Bacteria-*Bacilli, Pseudomonads, Cellulomonas* and Actinomycetes-*Streptomyces, Actinomucor* and *Streptomyces*.

While several fungi can metabolise cellulose as an energy source, only few strains are capable of secreting a complex of cellulase enzymes, which could have practical application in the enzymatic hydrolysis of cellulose. Besides *T. reesei,* other fungi like *Humicola, Penicillium* and *Aspergillus* have the ability to yield high levels of extracellular cellulases. Aerobic bacteria such as *Cellulomonas, Cellovibrio* and *Cytophaga* are capable of cellulose degradation in pure cultures.

However, the microbes commercially exploited for cellulase preparations are mostly limited to *T. reesei, H. insolens, A. niger, Thermomonospora fusca, Bacillus* sp. and a few other organisms (Table 5.3).

Table 5.3: Major micro-organisms employed in cellulase production.

Major group	*Micro-organism*	
	Genus	*Species*
Fungi	*Aspergillus*	*A. niger* *A. nidulans* *A. oryzae* (recombinant)
	Fusarium	*F. solani* *F. oxysporum*
	Humicola	*H. insolens* *H. grisea*
	Melanocarpus	*M. albomyces*
	Penicillium	*P. brasilianum* *P. occitanis* *P. decumbans*
	Trichoderma	*T. reesei* *T. longibrachiatum* *T. harzianum*
Bacteria	*Acidothermus*	*A. cellulolyticus*
	Bacillus	*Bacillus sp.* *Bacillus subtilis*
	Clostridium	*C. acetobutylicum* *C. thremocellum*
	Pseudomonas	*P. cellulosa*
	Rhodothermus	*R. marinus*
Actinomycetes	*Cellulomonas*	*C. fimi* *C. bioazotea* *C. uda*
	Streptomyces	*S. drozdowiczii* *S. sp.* *S. lividans*
	Thermononospora	*T. fusca* *T. curvata*

5.7.2 Cellulase systems and the control of cellulase

Gene expression

Cellulase systems of microbes can be generally regarded as complexed or non-complexed. Utilisation of insoluble cellulose requires the production of extracellular cellulases by the organism. The cellulase systems consist of either secreted or cell associated enzymes belonging to different classes categorised based on their mode of action and structural properties.

The three major type of cellulase activities recognised are: (i) endoglucanases/1-4-β-D-glucanohydrolases/EG-(EC 3.2.14), (ii) exoglucanases/1-4-β-D-glucan glucanohydrolases/Cellobiohydrolase/CBH-(EC 3.2.1.74) and (iii) β-Glucosidases/BG/BGL/β-glucoside glucohydrolases-(EC 3.2.1.21). Endoglucanases cut at random at internal amorphous sites in the cellulose polysaccharide chain generating oligosaccarides and new chain ends. Exogulcanases act on the reducing and non reducing ends of the cellulose chains liberating glucose, cellobiose or cello-oligosaccharides as major products. β-Glucosidases hydrolyse soluble cellodextrins and cellobiose to glucose.

Cellulases are inducible enzymes and the regulation of cellulase production is finely controlled by activation and repression mechanisms. In *T. reesei*, genes are coordinately regulated. The production of cellulolytic enzymes is induced only in presence of the substrate and is repressed when easily utilizable sugars are available. It is proposed that the inducer is generated by the *trans*-glycosylation activity of basally expressed β-glucosidase. Cellobiose, δ-cellobiose-1-5 lactone and other oxidised products of cellulose hydrolysis can also act as inducers of cellulose. Lactose is another known inducer of cellulases and it is utilised in commercial production of the enzyme owing to economic considerations. Though the mechanism of lactose induction is not fully understood, it is believed that the intracellular galactose-1-phosphate levels might control the signalling. Glucose repression of cellulase system over rides its induction and derepression is believed to occur by an induction mechanism mediated by trans-glycosylation of glucose.

5.7.3 Bioprocesses for cellulase: Fermentation production of cellulolytic enzymes

Majority of the reports on microbial production of cellulases utilises submerged fermentation technology (SmF) and the widely studied organism used in cellulase production is *T. reesei*, which has also been tested mostly in liquid media. However, in nature, the growth and cellulose utilisation of aerobic micro-organisms elaborating cellulases probably resembles solid substrate fermentation than a liquid culture. Nevertheless, the advantages of better monitoring and handling are still associated with the submerged cultures.

Cellulase production in cultures is growth associated and is influenced by various factors and their interactions can affect cellulase productivity. Among known inducers of cellulase genes, lactose is the only economically feasible additive in industrial fermentation media. In *T. reesei*, a basal medium after Mandels and Reese has been most frequently used with or without modifications. Carbon sources in majority of commercial cellulase fermentations are cellulosic biomass including straw, spent hulls of cereals and pulses, rice or wheat bran, bagasse, paper industry waste and various other lignocellulosic residues.

Though majority of the processes are batch processes, there has been attempts to produce cellulase in fed batch or continuous mode, which supposedly helps to over ride the repression caused by accumulation of reducing sugar. The major technical limitation in fermentative production of cellulases remains the increased fermentation times with a low productivity. Solid-state fermentation (SSF) for production of cellulases is rapidly gaining interest as a cost effective technology, not only for production of the enzyme but also for the bioconversion of lignocellulosic biomass employing cellulolytic micro-organisms.

5.7.4 Applications of cellulases

Cellulases were initially investigated several decades back for the bioconversion of biomass which gave way to research in the industrial applications of the enzyme in animal feed, food, textiles and detergents and in the paper industry. With the shortage of fossil fuels and the arising need to find alternative source for renewable energy and fuels, there is a renewal of interest in the bioconversion of lignocellulosic biomass using cellulases and other enzymes. In the other fields, however, the technologies and products using cellulases have reached the stage where these enzymes have become indispensable.

Textile industry

Cellulases have become the third largest group of enzymes used in the industry since their introduction only since a decade. They are used in the biostoning of denim garments for producing softness and the faded look of denim garments replacing the use of pumice stones which were traditionally employed in the industry. They act on the cellulose fibre to release the indigo dye used for colouring the fabric, producing the faded look of denim. *H. insolens* cellulase is most commonly employed in the biostoning, though use of acidic cellulase from *Trichoderma* along with proteases is found to be equally good. Cellulases are utilised for digesting off the small fibre ends protruding from the fabric resulting in a better finish. Cellulases have also been used in softening, defibrillation and in processes for providing localised variation in the colour density of fibres.

Laundry and detergents

Cellulases, in particular EGIII and CBH I, are commonly used in detergents for cleaning textiles. Several reports disclose that EG III variants, in particular from *T. reesei*, are suitable for the use in detergents. *T. viride and T. harzianum* are also industrially utilised natural sources of cellulases, as *A. niger*. Cellulase preparations, mainly from species of *Humicola* (*H. insolens* and *H. grisea* var. *thermoidea*) that are active under mild alkaline conditions and at elevated temperatures, are commonly added in washing powders and in detergents.

Food and animal feed

In food industry, cellulases are used in extraction and clarification of fruit and vegetable juices, production of fruit nectars and purees and in the extraction of olive oil. Glucanases are added to improve the malting of barley in beer manufacturing and in wine industry, better maceration and colour extraction is achieved by use of exogenous hemicellulases and glucanases. Cellulases are also used in carotenoid extraction in the production of food colouring agents. Enzyme preparations containing hemicellulase and pectinase in addition to cellulases are used to improve the nutritive quality of forages. Improvements in feed digestibility and animal performance are reported with the use of cellulases in feed processing. Bedford and others describes the feed additive use of *Trichoderma* cellulases in improving the feed conversion ratio and/or increasing the digestibility of a cereal-based feed.

Pulp and paper industry

In the pulp and paper industry, cellulases and hemicellulases have been employed for biomechanical pulping for modification of the coarse mechanical pulp and hand sheet strength properties, de-inking of recycled fibres and for improving drainage and runnability of paper mills. Cellulases are employed in the removing of inks, coating and toners from paper. Biocharacterisation of pulp fibres is another application where microbial cellulases are employed. Cellulases are also used in preparation of easily biodegradable cardboard. The enzyme is employed in the manufacture of soft paper including paper towels and sanitary paper and preparations containing cellulases are used to remove adhered paper.

Biofuel

Perhaps the most important application currently being investigated actively is in the utilisation of lignocellulosic wastes for the production of biofuel. The lignocellulosic residues represent the most abundant renewable resource available to mankind but their use is limited only due to lack of cost effective technologies. A potential application of cellulase is the conversion of cellulosic materials to glucose and other fermentable sugars, which in turn can be used as microbial substrates for the production of single cell proteins or a variety of fermentation products like ethanol. Organisms with cellulase systems that are capable of converting biomass to alcohol directly are already reported. But none of these systems described are effective alone to yield a commercially viable process. The strategy employed currently in bioethanol production from lignocellulosic residues is a multi-step process involving pre-treatment of the residue to remove lignin and hemicellulase fraction, cellulase treatment at 50°C to hydrolyse the cellulosic residue to generate fermentable sugars and finally use of a fermentative micro-organism to produce alcohol from the hydrolysed cellulosic material. The cellulase preparation needed for the bio-ethanol plant is prepared in the premises using same lignocellulosic residue as substrate and the organism employed is almost always *T. ressei*. To develop efficient technologies for biofuel production, significant research have been directed towards the identification of efficient cellulase systems and process conditions, besides studies directed at the biochemical and genetic improvement of the existing organisms utilised in the process. The use of pure enzymes in the conversion of biomass to ethanol or to fermentation products is currently uneconomical due to the high cost of commercial cellulases. Effective strategies are yet to resolve and active research has to be taken up in this direction. Overall, cellulosic biomass is an attractive resource that can serve as substrate for the production of value added metabolites and cellulases as such.

Apart from these common applications, cellulases are also employed in formulations for removal of industrial slime, in research for generation of protoplast and for generation of antibacterial chitooligosaccharides, which could be used in food preservation, immuno-modulation and as a potent antitumor agent.

5.7.5 Future perspectives – The challenges in cellulase research

Lignocellulose is the potential source of biofuels, biofertilisers, animal feed and chemicals, besides being the raw material for paper industry. Exploitation of this renewable resource needs either chemical or biological treatment of the material and in the latter context cellulases have gained wide popularity over the past several decades. Research has shed light into the mechanisms of microbial cellulase production and has led to the development of technologies for production and applications of cellulose degrading enzymes. However, there is no single process, which is cost effective and efficient in the conversion of the natural lignocellulosic materials for production of useful metabolites or biofuel. Use of the current commercial preparations of cellulase for bioconversion of lignocellulosic waste is economically not feasible.

The major goals for future cellulase research would be: (i) reduction in the cost of cellulase production and (ii) improving the performance of cellulases to make them more effective, so that less enzyme is needed.

The former task may include such measures as optimising growth conditions or processes, whereas the latter require directed efforts in protein engineering and microbial genetics to improve the properties of the enzymes.

Optimisation of growth conditions and processes has been attempted to a large extent in improving cellulase production. The section on fermentation production of cellulases describes many of these works basically dealing with empirical optimisation of process variables to improve productivity. Many of the current commercial production technologies utilise submerged fermentation technology and employ hyper producing mutants. In spite of several efforts directed at generating hyper producers by directed evolution, the cost of enzymes has remained high. Alternative strategies thought of in cellulase production include mainly solid substrate fermentation on lignocellulosic biomass particularly by using host/substrate specific micro-organisms. There are several reports on such use of filamentous fungi in production of optimal enzyme complex for the degradation of host lignocellulose.

Performance of enzyme complexes on lignocellulosic material is best when these complexes are prepared with the same lignocellulosic material as the host/substrate in fermentation. Another strategy is to use mixed culture in the production of enzyme. Several reports have shown that mixed culture gives improved production and enzyme complexes with better hydrolytic activity. Thus, SSF may be considered as a cost effective means for large scale production of cellulases which probably would be several fold cheaper compared to the current commercial preparations.

Cellulases are subject to regulation by various factors and some of the *cis*-acting promoter elements have been characterised. Active research in this field has led to genetic improvement of cellulase production by various methods including over expressing cellulases from the *cbh1* promoter of *T. reesei* and generation of desired variation in the cellulase production profile of organism. The *cbh1* and *cbh2* promoters of *T. reesei* have also been exploited for expression of foreign proteins in *Trichoderma*. Feedback inhibition of cellulase biosynthesis by the end products, glucose and cellobiose, generated by endogenous cellulolytic activity on the substrate is another major problem encountered in cellulase production. Cellobiose is an extremely potent inhibitor of the *CBH* and *EG* biosynthesis. *Trichoderma* and the other cellulase producing microbes make very little β-glucosidase compared to other cellulolytic enzymes. The low amount of β-glucosidase results in a shortage of capacity to hydrolyse the cellobiose to glucose resulting in a feed back inhibition of enzyme production and in the case of biomass conversion applications in the inhibition of cellulases. This issue has been addressed by various means like addition of exogenous β-glucosidases to remove the cellobiose and engineering β-glucosidase genes into the organism so that it is over produced. More and more research is oriented in genetic manipulations of cellulase producers for improving productivity. The developments in process design and medium formulations have come to an age and the future definitely requires controlled genetic interventions into the physiology of cellulase producers to improve production and thereby make the cellulase production process more cost effective. The major tasks ahead include over riding the feedback control by glucose and development of integrated bioprocesses for the production of cellulases. Improvements in cellulase activities or imparting of desired features to enzymes by protein engineering are probably other areas where cellulase research has to advance. Active site modifications can be imparted through site directed mutagenesis and the mutant proteins can be used for understanding the mechanisms of action as well as for altering the substrate specificities or improving the activities. There are several reports of developments made in this direction. Meinke and others has generated a mutant enzyme with endoglucanase like features and improved activity by deleting C-terminal loop of *Clostridium fimi CELB*. Protein engineering has been successfully employed to improve the stability of a *Humicola* cellulase in presence of detergents, to improve the thermostability of an alkaline, mesophilic *endo*-1,4-β-glucanase from alkaliphilic *Bacillus* sp. and for altering pH profile of cellobiohydrolase and more recently endoglucanase from *T. reesei*. Such modifications affecting the enzyme properties may be beneficial in improving the overall performance of cellulases and a better understanding of their mode of action, which will enable better utilisation of enzymes in biomass conversion. More basic research is needed to make designer enzymes suited for specific applications.

5.8 Hemicellulase

'Hemicellulase' is a collective term for a group of enzymes that break down hemicellulose.

1. 'Hemicellulose' is a collective term for various components of cell walls in plants (glucans, galactans, mannans, pentosans, xylans) with the exception of cellulose. They cannot be digested by human beings and are counted as roughage.
2. In cereals, hemicelluloses are found in comparatively high proportions (2 to 12%) and also are referred to as fibrils.

5.8.1 Applications of hemicellulase

Often in combination with amylase, glucanase, cellulase or xylanase, hemicellulase is used foremost:

1. As a baking enzyme (in cake mixes) for the enhancement of dough qualities (mechanical handling, stability) and for product optimisation (volume, consistency, storage life), the enzyme is also used for deep-frozen doughs and baked goods.
2. In the production of fruit juice and other beverages.
3. In the production of spirits and in the alcohol industry in general (hemicellulasen solubilise the cereal fibrils, which then may be used in fermentation).

4. In combination with other enzymes in wine production.
5. As an additive to animal feed (better use of animal feed).

5.8.2 Gene technology of hemicellulase

Hemicellulases are produced by fermentation with a variety of fungi and bacteria.

1. In Europe, one type of hemicellulase may be produced with the aid of genetically modified bacterium cultures (*Bacillus*).
2. In fermentation processes using fungus cultures, predominantly non-genetically-modified cultures are used to date.

The hemicellulase enzyme has the ability to take non-cellulose polysaccharides (long chains of sugars) we eat and convert them into usable constituents. Hemicellulase, which is crucial for the breaking down of fruits, vegetables and many grains is required to break down the 'hard' hemicellulose carbohydrates, which are known to slow digestion and the absorption of various nutrients. If we eat large amounts of plant material but do not have enough hemicellulase, we get a very limited prebiotic intake. These prebiotics are composed of smaller indigestible saccharide units that serve as fuel for various types of probiotics (good bacteria) resident in the digestive tract.

Hemicellulase possesses the distinct ability to boost this prebiotic activity. Specifically, common hemicellulose carbohydrates found in the food we eat every day, (hydrocolloid gums, galactomannoglucans, pentosans, betaglucans) are actually added to things like ice cream and dairy products. For this reason, it is important to provide our body with a prebiotic boost by taking a supplement containing hemicellulase.

Hemicellulase is a fairly complex enzyme and there are different permutations of hemicellulase. Different types of this enzyme have been used for different purposes in food technologies, particularly for its ability to enhance the quality of dough, as well as produce fruit juices and alcoholic beverages. In fact, it is a commonly-added enzyme in the production of wines, as the enzyme helps strip away the unwanted compounds from the skins of the grapes that might change the taste of the wine. Although plants make hemicellulases for growth and development, most of the commercial interest is in the enzymes produced by micro-organisms.

5.8.3 Health benefits of hemicellulase

1. May reduce candida: Some research suggests that an increase in this enzyme can help prevent and reduce yeast infestations such as Candida. This is perhaps related to the fact that the cell wall of candida is composed of hemicellulose. Again, as hemicellulase digests hemicellulose, it may help in reducing Candida.

2. Better digestive capacity with ageing: Taking digestive enzymes such as hemicellulase can counteract the net loss of enzymes that occurs with ageing, whether due to a loss in pancreatic production of enzymes or a change in beneficial flora in the gut. Research supports this. In Japan, scientists have confirmed that intestinal enzyme output showed a gradual decrease with age. Individuals 65 years and older were tested and women, in particular, were found to have the sharpest declines in enzyme health.
3. Improves overall health: Animal studies suggest that supplementing with hemicellulase in feedings helps not only increase nutrient digestibility but also betters performance and boosts food conversion ratio. A study conducted in 2012 and published in the Animal Science Journal found that hemicellulase could boost overall health. Hemicellulase enzyme supplements were given to chickens and their subsequent nutrient utilisation, performance and digestion capacities were analysed. It was found that the animals given enzymes grew faster, had decreased abdominal fat and energy content of the diet was improved by the mixed enzyme. Ash-retention (related to the processing of free radicals and the alkalinity in the body) was increased by adding hemicellulase to the diet. This offers proof that hemicellulase has a synergistic effect on the performance and health in animals. More research on humans is needed within the scientific realm.

5.9 Dextranase

Dextranase is an endo-dextranase enzyme belonging to the family of glucosyl transferases obtained from selected strain of micro-organism. It preferentially cleaves the 1,6-A-glucosidic linkages in dextran and its degradation products.

It enhances sugar mill processing by hydrolysing dextran in sugarcane juice and syrup, thereby reducing viscosity and aiding refining. This results in decreasing the viscosity of cane facilitating sugar processing.

Enzyme stability: Dextranase is stable over a wide pH and temperature range making it most suitable for use in sugar processing. It is active over a pH range of 3.0–7.5 and temperature of 35°C to 70°C.

5.9.1 Properties of dextranase

Specificity: Dextranase preferentially cleaves the α-1,6 linkages of dextran, releasing shorter isomaltosaccharides, with net inversion of the anomeric carbon configuration.

Molecular characteristics: Extracellular dextranase is encoded by the dex gene. Post-translational modification includes cleavage of the signal peptide and *N*-glycosylation. Amino acid sequence analyses have revealed 29% identity between Penicillium minioluteum and *Arthrobacter* sp.

Composition: Dextranase enzymes belong to two glycoside hydrolase families, either 49 or 66, which do not share significant sequence similarity. Dextranases from *Penicillium* and *Arthrobacter* species are classified as family 49, while those of the *Streptococcus* species are found in family 66.

Penicillium dextranase and related enzymes contain two domains. The first domain resembles the immunoglobulin fold and consists of 200 amino acid residues forming 13 beta strands. A beta sandwich is formed by nine of those strands, with all but strands 5 and 13 being antiparallel. The second domain contains a right-handed parallel beta-helical fold containing 3 parallel beta sheets. These two domains are connected by a large interface, which contains 29 amino acids completely conserved in the glycoside hydrolase family 49. Disulphide bridges are formed by four of the six cysteines present in the protein, none of which are conserved within the glycoside hydrolase family 49.

Molecular weight: 64.6 kDa (Theoretical).

Optimal pH range: 5.0–7.0.

Isoelectric point: (i) 4.55 ± 0.05 and (ii) 3.88.

Extinction coefficient: 119,350 cm^{-1} M^{-1} (Theoretical).

Activators: Co^{2+}, Mn^{2+}, Cu^{2+}.

Inhibitors: (i) Hg^{2+}, Ag^{2+}, (ii) *N*-bromosuccinimide and (iii) I_2.

Storage and shelf life: Dextranase should be stored in cool and dry environment. When stored in closed containers at temperatures less than 35°C it maintains its activity for a period of 9–12 months.

Safety and enzyme handing: Dextranase is non carcinogenic and is perfectly safe to handle. However prolonged contact is not advised as this may irritate skin, eyes and other mucous membranes. Inhalation of enzyme must should be avoided. In case of contact with eyes or other sensitive organs, rinse with clean water for at least 15 min.

5.9.2 Advantages of dextranase

1. Promotes hydrolysis of dextran thereby reducing viscosity resulting in improved refined product.
2. Acts upon a sugarcane juice with wide range of sucrose content and pH optimum for use in mixed juice.
3. High specificity for inner bonds of dextran polymer as compared to end sugar unit, thereby making it more effective in cleaving dextran.
4. Free from Invertase activity hence quality of sugarcane juice and final product is not affected.

6

Protease

6.1 Introduction

Proteolytic enzyme, also called protease, proteinase, or peptidase, any of a group of enzymes that break the long chainlike molecules of proteins into shorter fragments (peptides) and eventually into their components, amino acids. Proteolytic enzymes are present in bacteria, archaea, certain types of algae, some viruses and plants, they are most abundant, however, in animals. There are different types of proteolytic enzymes, which are classified according to sites at which they catalyse the cleavage of proteins. The two major groups are the exopeptidases, which target the terminal ends of proteins and the endopeptidases, which target sites within proteins. Endopeptidases employ various catalytic mechanisms, within this group are the aspartic endopeptidases, cysteine endopeptidases, glutamic endopeptidases, metalloendopeptidases, serine endopeptidases and threonine endopeptidases. The term oligopeptidase is reserved for those enzymes that act specifically on peptides.

Among the best-known proteolytic enzymes are those that reside in the digestive tract. In the stomach, protein materials are attacked initially by a gastric endopeptidase known as pepsin. When the protein material is passed to the small intestine, proteins, which are only partially digested in the stomach, are further attacked by proteolytic enzymes secreted by the pancreas. These enzymes are liberated in the small intestine from inactive precursors produced by the acinar cells in the pancreas. The precursors are called trypsinogen, chymotrypsinogen, proelastase and procarboxypeptidase. Trypsinogen is transformed to an endopeptidase called trypsin by an enzyme (enterokinase) secreted from the walls of the small intestine. Trypsin then activates the precursors of chymotrypsin, elastase and carboxypeptidase. When the pancreatic enzymes become activated in the intestine, they convert proteins into free amino acids, which are easily absorbed by the cells of the intestinal wall. The pancreas also produces a protein that inhibits trypsin. It is thought that in this manner the pancreas protects itself from autodigestion.

Proteases occur in all organisms, from prokaryotes to eukaryotes to viruses. These enzymes are involved in a multitude of physiological reactions from simple digestion of food proteins to highly regulated cascades (e.g., the blood-clotting cascade, the complement system, apoptosis pathways and the invertebrate prophenoloxidase-activating cascade).

Proteases constitute one of the most important groups of industrial enzymes, accounting for about 60% of the total enzyme market. A protease is an enzyme that conducts proteolysis, that is, begins protein catabolism by hydrolysis of the peptide bonds that link amino acids together in the polypeptide chain forming the protein. For several physiological processes the action of the proteolytic enzyme is essential, e.g., in digestion of food proteins, protein turnover, cell division, blood clotting cascade, signal transduction, processing of polypeptide hormones, apoptosis and also in the life cycle of disease-causing organisms including the replication of retrovirus. With special reference to their key role in life-cycle of many hosts and pathogens they have great medical, pharmaceutical and academic importance.

Protease is of commercial value and various industrial applications. They are widely used as detergent, in food, pharmaceutical and leather tanning industries. The vast variety of proteases, with their specificity of their action and application have attracted worldwide attention to exploit their physiological as well as biotechnological applications. It has been considered as eco-friendly because the appropriate producers of these enzymes for commercial exploitation are non-toxic and non- pathogenic that are designated a safe.

6.2 Classification of proteases

The physiological function of proteases is essential for all living organism, from viruses to humans and the enzymes can be classified based on their origin: microbial (bacterial, fungal and viral), plant, animal and human enzymes can be disguished. On the basis of the site of action on protein substrates, proteases are broadly classified as endopeptidases or exopeptidases enzymes. Exopeptidases cleave the peptide bond proximal to the amino or carboxy termini of the substrate. Based on the site of action at the nitrogen (N) or carbon (C) terminus, they are classified as aminopeptidases and cabroxypetidases. Endopeptidases cleave peptide bonds distant from the termini of the substrate. Based on the functional group present at the active site, endopeptidases are further classified into four prominent groups, i.e., serine proteases, aspartic proteases, cysteine proteases and metallo-proteases.

Based on the pH optima, they are referred to as acidic, neutral, or alkaline proteases.

6.2.1 Protease from different sources

The proteases are physiologically necessary for living organisms, they are ubiquitous, being constituted in a wide diversity of sources such as plants, animals and micro-organism. The use of plants for the production of proteases is dependent on the availability of land for agriculture and certain climatic conditions.

Papain, bromelain, keratinases are some of the well-known proteases of plant origin. The most familiar proteases of animal origin are pancreatic trypsin, chymotrypsin, pepsin and rennins. These are prepared in pure form in bulk quantities. However, their production depends on the availability of livestock for slaughter, which in turn is governed by political and agricultural policies. The inability of the plant and animal proteases to meet current world demands has led to an increased interest in microbial proteases.

Micro-organisms represent an excellent source of enzymes owing to their broad biochemical diversity and their susceptibility to genetic manipulation. Proteases from microbial sources are preferred to the enzymes from plant and animal sources since they possess almost all the characteristics desired for their biotechnological applications. Major types of proteases and their sources are listed in Table 6.1.

Table 6.1: Types of protease enzymes and the source.

Enzyme	*Source*
Endopeptidases serine proteases	
Trpsin	Animal
Chymotrypsin	Animal
Enterokinase	Animal
Endoproteinase	Microbial
Elastase	Animal
Subtilism	Microbial
Proteinase K	Microbial
Thrombin	Animal
Factor Xa	Animal
WNV protease	Microbial
Cysteine proteases	
Bromelain	Plant
Papain	Plant
Ficin	Plant
Rhino virus 3C	Microbial
TEV protease	Microbial
TVMV protease	Microbial
Metalloproteases	
Endoproteinase	Microbial
Thermolysin	Microbial
Collagenase	Microbial
Dipase	Microbial

(Cont'd...)

Enzyme	*Source*
Aspartic	Proteases
Pepsin	Animal
Cathepsin D	Animal
Exopeptidases	
Serine proteases	
Carboxypeptidases Y	Microbial
Cysteine proteases	
Cathepsin C	Animal
DAPase	Animal
Metalloprotease	
Carboxypeptidase A	Animal
Carboxypeptidase B	Animal

Microbial sources produced enzymes are more advantageous than their equivalents from animal or vegetable sources. The advantages comprise lower production costs, possibility of large-scale production in industrial fermentors, wide range of physical and chemical characteristics, possibility of genetic manipulation, absence of effects brought about by seasonality, rapid culture development and the use of non-burdensome methods.

The above characteristics make microbial enzymes suitable biocatalysts for various industrial applications. The development of new enzymatic systems which cannot be obtained from plants or animals is made possible and important progress in the food industry may be achieved through microbial enzymes.

6.2.2 Plant proteases

Papain, bromelain, keratinases and ficin represent some of the well-known proteases of plant origin. Papain is a cysteine proteinase which depend for their enzymic activity on the thiol group of a cysteine residue. It is extracted from the latex of *Carica papaya* fruits, which are grown in subtropical areas of west and central Africa and India. The performance of the enzyme depends on the plant source, the climatic conditions for growth and the methods used for its extraction and purification. The enzyme is active between pH 5 and 9 and is stable up to 80 or 90°C in the presence of substrates. It is extensively used in industries for the preparation of highly soluble and flavoured protein hydrolysates.

6.2.3 Animal proteases

The main recognisable proteases of animal origin are pancreatic trypsin, chymotrypsin, pepsin and rennins. However, their production depends on the availability of livestock for slaughter, which in turn is governed by political and agricultural policies.

6.2.4 Microbial proteases

The inability of the plant and animal proteases to meet current world demands has led to an increased interest in microbial proteases. Micro-organisms stand for an outstanding source of enzymes due to their broad biochemical diversity and their susceptibility to genetic manipulation. The majority commercial proteases, mainly neutral and alkaline, are produced by organisms belonging to the genus *Bacillus*. Bacterial neutral proteases are active between pH 5 to 8 and have relatively low thermotolerance. Their optimal temperature is around 60°C. Genetic manipulation brings about the over production of the enzymes. The properties of bacterial alkaline proteases make them suitable for use in the detergent industry.

Based on the functional group present at the active site, proteases are further classified into serine proteases and cysteine proteases.

6.2.5 Subtilisins

Subtilisins of *Bacillus* origin represent the second largest family of serine proteases. Two different types of alkaline proteases, subtilisin Carlsberg and subtilisin Novo or bacterial protease Nagase (BPN'), has been identified. Subtilisin Novo or BPN' is produced by *Bacillus amyloliquefaciens*. Subtilisin BPN' is less commercially important. Both subtilisins have a molecular mass of 27.5 kDa but differ from each other by 58 amino acids. They have similar properties such as an optimal temperature of 60°C, an optimal pH of 10 and exhibit broad substrate specificity. The Carlsberg enzyme has broader substrate specificity and does not depend on Ca^{2+} for its stability. Subtilisin Carlsberg was the first serine protease from bacterial source to be discovered. It also possesses three Ca^{2+} binding sites for stabilising the three-dimensional structure. They are produced as preprosubtilisins.

6.2.6 Cysteine/thiol proteases

The activity of all cysteine proteases depends on a catalytic dyad consisting of cysteine and histidine. The order of Cys and His residues differ among the families. Generally, cysteine proteases are active only in the presence of reducing agents such as HCN or cysteine. Based on their side chain specificity, they are broadly divided into four groups: (i) papain-like, (ii) trypsin-like with preference for cleavage at the arginine residue, (iii) specific to glutamic acid and (iv) others. Papain is the best-known cysteine protease.

Proteases play a critical role in many physiological and pathological processes such as protein catabolism, blood coagulation, cell growth and migration, tissue arrangement, morphogenesis in development, inflammation, tumour growth and metastasis, activation of zymogens, release of hormones and pharmacologically active peptides from precursor proteins and transport of secretory proteins across membranes. In general, extracellular proteases catalyse the hydrolysis of large proteins to smaller molecules for subsequent absorption by the cell whereas intracellular proteases play a critical role in the regulation of metabolism. They catalyse important proteolytic steps in tumour invasion or in infection cycle of a number of pathogenic micro-organisms and viruses. This makes proteases a valuable target for new pharmaceuticals. They also participate in protein catabolism in degradative or biosynthetic pathways and in the release of hormones and pharmacologically active peptides from precursor proteins.

6.2.7 Stabilisation strategies for the enzyme

The catalytic stability of the enzymes and proteins *in vitro* conditions remains a critical issue as their operational stability is of paramount importance for many bioprocesses. The inherent Liability of many enzymes in organic solvents or at elevated temperatures limits their roles as potential reaction catalysts even though certain enzymes exhibit marked storage stability in organic solvents. Enhancement of functional stability will clearly broaden the range of enzyme applications. Especially, the improvement of thermal resistance of industrial enzyme has received considerable attention, in order to design more economic and efficient production processes catalysed by these biomolecules. The use of enzymes in these fields requires the easy recovery from the reaction mixture at the end of the procedure, the catalytic activity remaining stable over long time period of time and catalyst applicability to continuous processes.

Several strategies are at hand to increase the operational stability of the enzyme such as: chemical modification, the use of stabilising additives, immobilisation, crystallisation and genetic engineering. First of all, the enzymes can be stabilised by modification of the native structure by protein engineering or chemical modification. However, such modifications are often accompanied by unpredictable changes in the catalytic properties of the enzyme. The second strategy is immobilisation. Many of the problems connected with enzyme stabilisation can be solved by this method.

6.3 Protease functionalised nanoparticles

In recent years, nanotechnology has showed a significant attraction to the preparation of immobilised enzymes. Extensive studies have been carried out on the use of micrometer-sized particles for the immobilisation of enzymes by using nanoporous, nanofibrous and nanoparticles. Overall, nanoparticles provide an ideal remedy to the usually contradictory issues encountered in the optimisation of immobilised enzymes, such as minimum diffusional limitation, maximum surface area per unit mass and high enzyme loading. Protein/peptide oral drug delivery is a complicated process mainly due to poor intrinsic protein permeability. This is due to large molecular weight, degradation by proteolytic enzymes in the stomach and in the small intestine and chemical instability. Hence protein functionalised nanoparticles which are also pH sensitive were made to overcome this problem.

The preparation of biocatalytic ultra micro and nanoparticles of low molecular weight chitosan can be done using nanoprecipitation method. This process involves a non solvent, chemical cross-linking with tripoly phosphate, stabilised by a anionic surfactant and the enzyme was loaded on to the nanoparticle by glutaraldehyde cross linking chemistry. TEM and SEM micrographs confirms the presence of enzyme functionalised nanoparticles. The nanoparticles released the enzyme at the physiological pH of 7.4. The molecular weight of chitosan, concentration of the polymer, chitosan-TPP ratio and the pH of the solution are important parameters for the size, surface charge and aggregation of the nanoparticles.

Compared with conventional solid-phase supports, nanoparticulate matrices have a higher catalyst loading capacity due to their very large surface areas. Important application of enzyme or protein functionalised nanoparticle is in the controlled release of protein and peptide drugs, because they show excellent muco-adhesiveness and permeation enhancing effect across the biological surfaces.

6.4 Chemical modification of papain for use in detergents

In contrast to genetic methods, chemical modification is a generally applicable tool because modified biomolecules are easy to prepare in a large scale from commercial material and to improve enzyme stability. Chemical modification of the surface residues on the enzyme and intermolecular cross-linking has been used by several workers to stabilise enzymes and to enhance their activity in non-aqueous solvents The use of protease enzyme in washing clothes is that it hydrolyses the protein strains in the fabric into peptides. But when used in detergents, it requires stability towards additives at alkaline pH values and also at low temperatures. The serine protease, subtilisin from genetically modified *Bacillus licheniformis* is now widely used in detergents due to its stability in alkaline conditions. Papain can be used as an alternative to subtilisin, provided that it is stable in harsh conditions and is active at low and high temperatures.

Therefore it has got high potential to use in detergents. Lysine residues are not a part of active site in papain, hence it can be chemically modified by different dicarboxylic anhydrides of citraconic, phthalic, maleic and succinic acids. These anhydrides react specifically with the α-amino group of the lysine residues and change its charges from positive to negative, leading to a shift in pH optima of the enzyme from 7 to 9.0. Observed changes in kinetic, thermal and catalytic properties of these enzymes brought about by modification are analysed. By changing the pH optima of Papain and the subsequent immobilisation in carbohydrate matrices to make it stable at alkaline pH, it can be used in detergent formulations, as cheap source of protease. The amino groups of papain can be modified by anhydrides of maleic, phthalic, citraconic and succinic acids giving useful properties to the enzyme. The optimum pH of the modified papain was shifted to pH 9.0 with all the dibasic acid anhydrides. Phthalic papain was found to be the best modified enzyme. The immobilisation of the chemically modified enzymes in a suitable matrix, by adsorption or gel entrapment is a useful strategy for the reuse of the enzyme as well as improving the thermal stability of this biocatalyst. Starch was found to be the best matrix for immobilisation by entrapment. The modified enzyme showed enhanced catalytic properties and was more resistant to thermal and autolytic inactivation. It is possible to enhance the resistance to relatively high washing temperatures, a denaturing condition utilising a simple modification without noticeable effects on activity. This modified enzyme was found to be more robust than the native enzyme when incorporated in domestic cloth washing detergent formulations.

6.5 Multiple industrial applications of protease

Proteases execute a large variety of functions, extending from the cellular level to the organ and organism level, to produce cascade systems such as haemostasis and inflammation, which are responsible for the complex processes involved in the normal physiology of the cell as well as in abnormal pathophysiological conditions. Their involvement in the life cycle of disease- causing organisms has led them to become a potential target for developing therapeutic agents against fatal diseases such as cancer and AIDS. Microbial proteases are increasingly used in treatment of various disorders namely cancer, inflammation, cardiovascular disorders, necrotic wounds, etc. Proteases are used an immune – stimulatory agents. Increased antibiotic concentration at a target site when protease was concomitantly used with an antibiotic. Proteases are used extensively in the pharmaceutical industry for preparation of medicines such as ointments for debridement of wounds. It is also used in denture cleaners and as contact-lens enzyme cleaners.

Proteases have a large variety of applications, mainly in the detergent and food industries. Proteases are envisaged to have extensive applications in leather treatment and in several bioremediation processes. Proteases that are used in the food and detergent industries are prepared in bulk quantities and used as crude preparations, whereas those that are used in medicine are produced in small amounts but require extensive purification before they can be used.

The food industries are the major protease using industries. However, they have also found widespread application in laundry detergents. The thermo stability and their activity at high pH and the alleviation of pollution characteristic have made proteolytic enzymes an ideal candidate for laundry applications. Alkaline proteases are supplemented in different brands of detergents for use in home and commercial establishments. Enzymes have been added to laundry detergents since last 50 years to facilitate the release of proteinaceous material in stains such as those of milk and blood. The proteinaceous dirt coagulates on the fabric in the absence of proteinases as a result of washing condition. The enzyme removes not only the stain, such as blood, but also other materials including proteins from body secretion and food such as milk, egg, fish and meat. An ideal detergent enzyme should be stable and active in the detergent solution and should have adequate temperature stability to be effective in a wide range of washing temperature.

Usually the surgical instruments are washed or cleaned by sterilisation or by using chemical steriliants. However, chemical steriliants cannot remove microbes that usually get trapped behind the bioburden that is encrusted on or within surgical instruments. However, the recent technologies include enzyme-containing formulations and zeolite based detergents. Of these, the enzyme detergents often referred to as 'green chemicals' are proving useful in keeping a check on the environmental pollution and thus improving ecological situation.

In leather industry, removal of hair and unwanted adhering subcutaneous layer by chemicals causes a problem. Hence the need for alternatives to sulphide dehairing is being sought. Tanners are hesitant to use the enzyme because of certain disadvantages in using them at commercial level for reasons of the stability of the enzyme at different environmental conditions such as pH, temperature and duration consistent performance and the cost of production and application. The important factor in choosing an enzyme as a dehairing agent depends on the specificity of the enzyme used, which should not attack the collagenous matter.

Numerous studies carried out from time to time to recover silver from photographic films as well as from X-ray films are patented. The silver recovery methods from these wastes includes: burning the films directly for oxidation of metallic silver followed by electrolysis stripping the silver-gelatin layer using microbial enzymes specifically alkaline proteases and stripping the gelatin silver layer using different chemicals. Recovery of silver by burning the films creates environmental pollution and health hazards. On the other hand, enzyme from microbial source breaks the gelatin layer embedded with silver in films creating pollution free stripping. The amount of silver varies from 5–15 g/kg of film. Enzymatic method although slow is free from pollution and cost-effective too.

The use of proteases in the food industry dates back to antiquity. They have been routinely used for various purposes such as cheese making, baking, preparation of soya hydrolysates and meat tenderisation.

The major application of proteases in the dairy industry is in the manufacture of cheese. The milk-coagulating enzymes fall into three main categories: (i) animal rennets, (ii) microbial milk coagulants and (iii) genetically engineered chymosin. Both animal and microbial milkcoagulating proteases belong to a class of acid aspartate proteases and have molecular weights between 30,000 to 40,000. Rennet extracted from the fourth stomach of unweaned calves contains the highest ratio of chymosin to pepsin activity. A world shortage of calf rennet due to the increased demand for cheese production has intensified the search for alternative microbial milk coagulants.

The microbial enzymes exhibited two major drawbacks, i.e., (i) the presence of high levels of nonspecific and heat-stable proteases, which led to the development of bitterness in cheese after storage and (ii) a poor yield.

Extensive research in this area has resulted in the production of enzymes that are completely inactivated at normal pasteurisation temperatures and contain very low levels of nonspecific proteases. In cheese making, the primary function of proteases is to hydrolyse the specific peptide bond to generate para-k-casein and macro-peptides. Chymosin is preferred due to its high specificity for casein, which is responsible for its excellent performance in cheese making.

Wheat flour is a major component of baking processes. It contains an insoluble protein called gluten, which determines the properties of the bakery doughs. Endo and exo-proteinases from *Aspergillus oryzae* have been used to modify wheat gluten by limited proteolysis. Enzymatic treatment of the dough facilitates its handling and machining and permits the production of a wider range of products. The addition of proteases reduces the mixing time and results in increased loaf volumes. Bacterial proteases are used to improve the extensibility and strength of the dough.

The wide diversity and specificity of proteases are used to great advantage in developing effective therapeutic agents. Oral administration of proteases from *Aspergillus oryzae* has been used as a digestive aid to correct certain lytic enzyme deficiency syndromes. Clostridial collagenase or subtilisin is used in combination with broad-spectrum antibiotics in the treatment of burns and wounds. An asparginase isolated from *E. coli* is used to eliminate aspargine from the bloodstream in the various forms of lymphocytic leukemia. Alkaline protease from *Conidiobolus coronatus* was found to be able to replace trypsin in animal cell cultures.

Besides their industrial and medicinal applications, proteases play an important role in basic research. Their selective peptide bond cleavage is used in the elucidation of structure function relationship, in the synthesis of peptides and in the sequencing of proteins. In essence, the wide specificity of the hydrolytic action of proteases finds an extensive application in the food, detergent, leather and pharmaceutical industries, as well as in the structural elucidation of proteins, whereas their synthetic capacities are used for the synthesis of proteins.

Gene cloning is a rapidly progressing technology that has been instrumental in improving our understanding of the structure function relationship of genetic systems. It provides an excellent method for the manipulation and control of genes. More than 50% of the industrially important enzymes are now produced from genetically engineered micro-organisms.

Many industrial applications of proteases require enzymes with properties that are non physiological. Proteases obtained from natural sources are widely used in molecular biology practice. Their degradative nature make them useful for general protein digestion in tissue dissociation, cell isolation and cell culturing. Protein engineering allows the introduction of predesigned changes into the gene for the synthesis of a protein with an altered function that is desired for the application. Recent advances in recombinant DNA technology and the ability to selectively exchange amino acids by site-directed mutagenesis (SDM) have been responsible for the rapid progress of protein engineering. Identification of the gene and knowledge of the three-dimensional structure of the protein in question are the two main prerequisites for protein engineering. The X-ray crystallographic structures of several proteases have been determined. Proteases from bacteria, fungi and viruses have been engineered to improve their properties to suit their particular applications. The specificity and the predictability of cleavages by proteases enables their use for more specific tasks such as antibody fragmentation production, the removal of affinity tags from recombinant proteins and specific protein digestion in the field of proteomics mainly for protein sequencing.

6.6 Protease inhibitors

6.6.1 Soyabeans and other legumes

Legumes and especially soyabeans, are an abundant source of natural protease inhibitors. Some research suggests that protease inhibitors found in soyabeans may play a role in soya's anti-cancer properties and may help protect against certain cancers such as breast and colon. The particular protease inhibitor found in soya, called Bowman-Birk, appears to help prevent tumour formation, according to 'Pathology for the Health Professions.'

6.6.2 Potato tubers

Another natural source of protease inhibitors is the potato. Protease inhibitors from potatoes are shown to suppress tumour growth in clinical research settings. In addition, potato protease inhibitors suppress a hormone responsible for stimulating appetite. Recently researchers found that potato protease inhibitors may be useful at suppressing appetite. More research is needed to understand how well it works.

6.6.3 Green tea

Antioxidant compounds in green tea called catechins produce protease-inhibiting activity. The major catechin in green tea is epigallocatechin gallate (EGCG). The protease-inhibiting activity of green tea catechins is also reportedly involved in green tea's anti-cancer benefits. The natural protease inhibitors in green tea may suppress infection by other pathogens, such as the flu virus and the adenovirus. These pathogens commonly cause respiratory infection in children.

6.6.4 Blue-green algae

Though more common in Asian cuisine than American, blue-green algae is a nutrient-dense food and supplement. In addition to its vitamin and mineral content, blue-green algae contains protease inhibitors. In fact, blue-green algae is a rich source of highly potent protease inhibitors. Researchers found that protease inhibitors have potential anti-cancer benefits, as cancer cells are more sensitive to the suppression of protein than normal cells.

6.7 Health benefits of protease

Proteolytic enzymes are extremely important for the digestion of many foods. But their intestinal duties are not solely limited to digesting food. They also digest the cell walls of unwanted harmful organisms in the body and break down unwanted wastes such as toxins, cellular debris and undigested proteins. In this way, protease helps digest the small stuff, so that our immune system can work hard to avoid toxin overload.

In this way digestion plays a huge role in overall health and enzymes are a big part of digestive health. With the distinct ability to breakdown peptide bonds and liberate amino acids, proteolytic enzymes are now being studied by modern science and medicine for their clinical and therapeutic use in the realms of general oncology and overall immune function.

The following list offers some health benefits of supplementing with protease, as well some of the exciting research being done on protease and its applications to human health and disease prevention/management.

Inflammatory bowel disease: A 2010 U.S. study on Inflammatory Bowel Diseases found that the proteolytic enzyme bromelain could help reduce indications and problems in the colon.

Skin burns and stomach ulcers: A 2010 Brazilian study in 'Burns' journal found that intake of protease can help cellular repair in cases of skin burns and stomach ulcers in laboratory mice.

Sprains: One double-blind, placebo-controlled trial involving 30 individuals with chronic neck discomfort found that those patients using a protease enzyme experienced a moderate reduction in discomfort when compared to a placebo. Clinical trials show that protease enzymes can speed the healing time of sprains, bruises, fractures and tissue injuries.

Slow or stop irritation: Protease enzymes slow or stop irritation by neutralising the biochemicals associated with the response (i.e., bradykinins and eicosanoids). This may support cardiovascular and brain health.

Osteoarthritis discomfort: Although only preliminary and not to be interpreted as a new therapy, one study done on over 400 people found that protease enzymes could reduce osteoarthritis discomfort.

Sports-related injuries: Research suggests that protease enzyme combinations may aid in the recovery of sports injuries. A double-blind, placebo-controlled study of over 40 people with sports-related injuries found that taking protease enzymes could offer faster recovery, as well as reducing the time away from practicing sports again by up to 50%. Another research has found that taking protease enzymes could significantly increase recovery time after injury, particularly in cases involving mild fractures.

Beneficial gut bacteria: Research shows that protease activities give our cells critical amino acids and are essential for the growth of the good forms of gut flora that break compounds down.

Circulatory and lymph system: Protease enzymes help to cleanse organic debris out of our circulatory and lymph system.

Clots: Some forms of protease can boost the 'quality' of blood cells, allowing a better circulatory response and reducing the risk of clots.

6.8 Future scope of protease

Proteases are a unique class of enzymes, since they are of immense physiological as well as commercial importance. They possess both degradative and synthetic properties. Since proteases are physiologically necessary, they occur ubiquitously in animals, plants and microbes. However, microbes are a goldmine of proteases and represent the preferred source of enzymes in view of their rapid growth, limited space required for cultivation and ready accessibility to genetic manipulation. Microbial proteases have been extensively used in the food, dairy and detergent industries since ancient times. There is a renewed interest in proteases as targets for developing therapeutic agents against relentlessly spreading fatal diseases such as cancer, malaria and AIDS. The development of recombinant rennin and its commercialisation by Pfizer and Genencor is an excellent example of the successful application of modern biology to biotechnology. Analysis of sequences for acidic, alkaline and neutral proteases has provided new insights into the evolutionary relationships of proteases. Despite the systematic application of recombinant technology and protein engineering to alter the properties of proteases, it has not been possible to obtain microbial proteases that are ideal for their biotechnological applications. Industrial applications of proteases have posed several problems and challenges for their further improvements. The biodiversity represents an invaluable resource for biotechnological innovations and plays an important role in the search for improved strains of micro-organisms used in the industry. A recent trend has involved conducting industrial reactions with enzymes reaped from exotic micro-organisms that inhabit hot waters, freezing Arctic waters, saline waters, or extremely acidic or alkaline habitats.

The proteases isolated from extremophilic organisms are likely to mimic some of the unnatural properties of the enzymes that are desirable for their commercial applications. The existing knowledge about the structure-function relationship of proteases, coupled with gene-shuffling techniques, promises a fair chance of success, in the near future, in evolving proteases that were never made in nature and that would meet the requirements of the multitude of protease applications.

Thus, proteases play a decisive role in detergent, pharmaceutical, leather, food and agricultural industries. Currently, the estimated value of the global sales of industrial enzymes is over 3 billion USD, of which proteases account for about 60% of the total sales. Microbial alkaline proteases already play a pivotal role in several industries, mainly in the detergents, leather processing, silver recovery, medical purposes, food processing, feeds and chemical industries, as well as in waste treatment, their potential is much greater and their applications in novel processes are likely to increase in the near future. Advancement in biotechnology offers a constructive position for the development of proteases and will continue to facilitate their applications to provide a sustainable environment for improving the quality of human life.

7

Lipase

7.1 Introduction

Lipase is a physiologically necessary enzyme. It occurs in many plants and animal organisms, as well as in micro-organisms. However, its richest source are bacteria, fungi and yeast.

Lipolytic enzymes are widespread in the plant world, however, the knowledge and experience pertaining to plant-originating lipases is still limited in comparison to the information on lipases from mammals and micro-organisms and this is especially connected to the difficulties with their isolation.

Among mammals, three groups of lipolytic enzymes are distinguished: lipases isolated to the digestive tract, tissue lipases and milk lipases. Mammal tissues and organs which contain lipases include the heart, brain, muscles, veins, kidneys, spleen, lungs, liver, fatty tissue and plasma.

Microbiological lipases, especially those originating from bacteria, are more stable than those from plants or animals. They possess unique qualities and are used more often for industrial purposes. The knowledge about lipolytic bacterial enzymes is developing with surprising rate. The classification of bacterial lipases has been based mainly on the sequence of amino acids and some basic biological properties. Bacterial lipases from the *Pseudomonas* kind constitute the main group from which two types are distinguished – one is connected with the auxiliary protein in order to show proper activity and the other one is not.

Lipases (triacylglycerol acylhydrolases, EC 3.1.1.3) catalyse the hydrolysis and the synthesis of esters formed from glycerol and long-chain fatty acids. Lipases occur widely in nature, but only microbial lipases are commercially significant. The many applications of lipases include speciality organic synthesis, hydrolysis of fats and oils, modification of fats, flavour enhancement in food processing, resolution of racemic mixtures and chemical analyses. This chapter discusses the production, recovery and use of microbial lipases. Immobilised preparations of lipases are also discussed. In view of the increasing understanding of lipases and their many applications in high-value synthesis and as bulk enzymes, these enzymes are having an increasing impact on bioprocessing.

Lipases [triacylglycerol acylhydrolases, (E.C. 3.1.1.3)] are ubiquitous enzymes of considerable physiological significance and industrial potential.

Lipases catalyse the hydrolysis of triacylglycerols to glycerol and free fatty acids. In contrast to esterases, lipases are activated only when adsorbed to an oil-water interface and do not hydrolyse dissolved substrates in the bulk fluid. A true lipase will split emulsified esters of glycerine and long-chain fatty acids such as triolein and tripalmitin. Lipases are serine hydrolases. Lipases display little activity in aqueous solutions containing soluble substrates. In contrast, esterases show normal Michaelis–Menten kinetics in solution. In eukaryotes, lipases are involved in various stages of lipid metabolism including fat digestion, absorption, reconstitution and lipoprotein metabolism. In plants, lipases are found in energy reserve tissues. How lipases and lipids interact at the interface is still not entirely clear and is a subject of intense investigation.

Because of their wide-ranging significance, lipases remain a subject of intensive study. Research on lipases is focussed particularly on structural characterisation, elucidation of mechanism of action, kinetics, sequencing and cloning of lipase genes and general characterisation of performance. In comparison with this effort, relatively little work has been done on development of robust lipase bioreactor systems for commercial use.

Lipases are unique in terms of catalysing fat hydrolysis to fatty acids and glycerol, which takes place on the water-lipid border and in terms of a reverse reaction in a non-water environment. After contact with an insoluble substrate in water, a change in the enzyme's conformation occurs.

The properties of lipases, just like lipases themselves vary. A definitely desirable feature is the thermal stability of this enzyme. The source of thermally resistant lipases are first and foremost bacteria and the thermal durability is linked to their structure. The thermal stability is influenced by such factors as pH and the presence of heavy metals. All activities at the level of protein engineering are meant to improve their stability. A considerable improvement of the resistance of many lipases may be obtained by their immobilisation on substrates and in many types of carriers.

7.2 Isolation and purification of lipase

Lipases available on the market are usually isolated from higher eukaryotic organisms and micro-organisms which include bacteria, fungi, yeast and actinomycetes. Lipase-producing micro-organisms occur in various environments such as industrial waste-collection plants, plants producing vegetable oils, dairies, oil-polluted soils, decomposing food, composting plants, coal heaps or hot springs.

A reliable method discovering the activity of lipase in micro-organisms is the one used in an unchanged environment with the Tween 80 surfactant for the identification of the lipolytic activity. The occurrence of non-transparent areas around the inspected colony means that the given organism produces

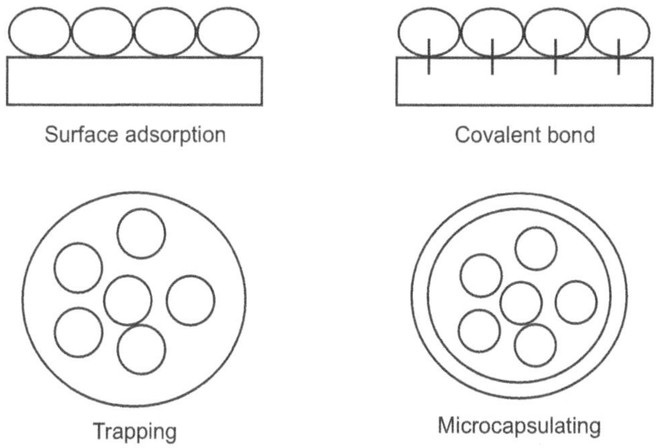

Figure 7.1: Ways of immobilising the lipase on the surface and inside of the carrier.

lipase. This method may be modified by using various tween surfactants in connection with Nile blue (oxazine dye).

Often, also an inoculation is used of micro-organisms producing lipase on agar plates, with the use of tributyrin as the substrate. The transparent areas around the colony indicate the production of lipase.

Microbiological lipases are obtained mainly by a submerged fermentation. Its production is influenced by: the concentration of coal and nitrogen source, environmental pH, temperature and concentration of dissolved oxygen. The majority of microbiological lipases are active in a basic habitat (pH 7.0–9.0). For this reason, the activity of these enzymes is so strongly dependent on pH changes in the reaction mixture which influence the catalytic potential of lipases. Heavy metal ions are also effective lipase antagonists and modulators. On the other hand, chelating agents, e.g., EDTA, inhibit only the activity of metal lipases.

Because of the possibility of intensive production, good stability and numerous stereospecific properties, microbiological lipases still draw scientists' attention. Lipases have been isolated and studied from such micro-organisms as *Pseudomonas*, *Geotrichum*, *Candida rugosa*, *Aspergillus*, *Rhizophus*, *Mucor hirmalis f. hiemalis*, *Streptomyces rimosus*, *Penicillinum*. These lipases not only differ from each other molecularly, but they also possess different catalytic properties.

Apart from the catalyst function, microbiological lipases used in waterless solvents offer also new possibilities, e.g., the improvement of the enzyme's thermal stability or the possibility of using hydrophobic substrates.

The majority of industrial products do not require homogenous lipase agents, however, the specified degree of purification allows for an effective and successful application. Moreover, the purification of enzymes enables a proper determination of the sequence of their main amino acids in the 3-dimensional structure. X-ray examinations of pure lipase allowed to determine the relation between the structure and the function and they have contributed to a better understanding of the kinetic mechanism of lipase's functioning in hydrolysis or synthesis reactions. The purity of lipase products is very important in an industry using enzymes for the biocatalytic production of pharmaceuticals or cosmetics.

Many lipases have been carefully purified and characterised, it was shown that their activity and stability depend on the pH, temperature, influence of metal ions and chelating factors. The purification methods used up to date involved mainly the use of such techniques as precipitation, chromatography, gel filtration. In some cases, an affinity chromatography has been used in order to reduce the number of purification levels.

The main limitations in traditional methods include low efficiency and time consumption. Alternative new technologies, such as membrane processes, biphasic water systems and immuno-cleansing gradually stand against the outdated methods of lipase purification. Currently, the industry is searching for new methods which are relatively cheap, fast and highly efficient.

Many lipases are active in organic solvents during the catalysis of a number of chemical reactions including esterification, trans-esterification, regioselective acylation of glycols and methanol, synthesis of peptides and other substances. It is expected that lipase will become an equally important industrial enzyme as protease or carbohydrase.

7.3 Lipase immobilisation

Lipases are readily and often used in biotechnological processes. Their activity may be improved by immobilisation. Because of this, new studies are conducted which pertain to the activity and application of immobilised lipase enzymes.

There exist many methods of immobilising lipase, starting from adsorption or precipitation on hydrophobic materials, covalent bonding to functional groups, trapping in polymer gels, adsorption on macroporous, anionic ion-exchange resins, or microencapsulation in lipid membranes. Among the mentioned methods, covalent bonding of lipase is dominating.

On non-organic substrates (aluminium oxide, silica and porous glass) Moreno and others covalently immobilised the Candida cylindracea lipase, previously activating the carriers with cyanuric acid. Derivative immobilising on silica and aluminium oxide were characterised by a higher activity and thermal resistance than pure enzymes or even those immobilised on porous glass.

Lipase (Candida rugosa) has been covalently immobilised also by Erdemir and others using aminofunctional calixarenes on substrates of silica polymer. This way, a stable biocatalyst has been obtained with the possibility of repeated use and a lot better hydrolytic activity compared to the free enzyme.

Lipase has also been covalently immobilised on silica gel by Ozyilmaza who used two bonding factors – glutarate aldehyde and 1,6-diaminohexane. The Candida lipolitica lipase has been catalysed on modified palygorskite substrates by Huang and others.

Ozmen and others covalently immobilised the Candida rugosa lipase on β-cyclodextrin polymers cross-linked with glutarate aldehyde and hexamethylenediizocyjanian. Aldehyde groups of the substrate and amino groups of the enzyme provide multipoint connections in the compound with which a stable enzyme derivative is obtained. Lipase has been covalently bound on organic carriers also by Mieletic and others, Huang and others, using various copolymers and resins.

Often, lipases are immobilised also by means of adsorption. Lee and others used the adsorption method to immobilise pig pancreatic lipase on magnetite nanoparticles. Based on the discovery that lipase is strongly adsorbed on hydrophobic substrates, magnetite nanoparticles have been modified with sodium dodecyl sulphate. This method has proved to be interesting because the bound enzyme may be recovered by magnetic separation. On magnetic nanoparticles protected by ionic fluid, with a different length of the cationic chain (C1, C4, C8) and various anions (Cl^-, BF_4^-, PF_6^-), also the Candida rugosa lipase was immobilised with the use of the ionic adsorption method.

Salis and others conducted studies on the influence of the substrate on the enzyme load and activity of the immobilised lipase (*Pseudomonas fluorescens*). Various functional groups occurring on the porous substrates used by them were responsible for the physical (polypropylene, organosilicate, SBA-15) or chemical adsorption (polymethacrylate and modified SBA-15).

With regard to lipase trapping, alginate substrates are used relatively often for this purpose because of their delicately gelating properties and nontoxicity. Cheirsilp and others used *Pseudomonas* lipase immobilised with calcium alginate to obtain monoacylglycerols from palm oil.

For trapping enzymes the sol-gel technique is used which is proven in case of bioparticles susceptible to high temperatures. Up till now the immobilisation of lipases via this technique has been carried out with the use of silica materials. Pirozzi and others have trapped lipase in their studies in zirconium oxide obtained with the sol-gel method. This technique has allowed to create a substrate and to trap the enzyme in it in one stage.

Trapping and encapsulation of enzymes in natural polymers is becoming increasingly popular because of the low cost, repeatability of the process and nontoxicity. Smith and others have presented a fairly innovative way of encapsulating the Burkholderia cepacia lipase in the naturally occurring κ-carrageenin polysaccharide with the use of coextrusion to create a capsule with a liquid nucleus.

Commercially useful lipases are usually obtained from micro-organisms that produce a wide variety of extracellular lipases. Many lipases are active in organic solvents where they catalyse a number of useful reactions including esterification, trans-esterification, regioselective acylation of glycols and menthols and synthesis of peptides and other chemicals.

Lipases find promising applications in organic chemical processing, detergent formulations, synthesis of biosurfactants, the oleochemical industry, the dairy industry, the agrochemical industry, paper manufacture, nutrition, cosmetics and pharmaceutical processing. Development of lipase-based technologies for the synthesis of novel compounds is rapidly expanding the uses of these enzymes.

One limiting factor is a shortage of lipases having the specific required processing characteristics. An increasing number of lipases with suitable properties are becoming available and efforts are underway to commercialise biotransformation and synthesis based on lipases. The major commercial application for hydrolytic lipases is their use in laundry detergents. Detergent enzymes make up nearly 32% of the total lipase sales. Lipase for use in detergents needs to be thermostable and remain active in the alkaline environment of a typical machine wash.

Lesser amounts of lipases are used in oleochemical transformations. Lipases can play an important role in the processing of γ-linolenic acid, a polyunsaturated fatty acid (PUFA), astaxanthine, a food colourant, methyl ketones, flavour molecules characteristic of blue cheese, 4-hydroxydecanoic acid used as a precursor of γ-decalactone, a fruit flavour, dicarboxylic acids for use as prepolymers, interesterification of cheaper glycerides to more valuable forms (e.g., cocoa butter replacements for use in chocolate manufacture), modification of vegetable oils to obtain fats similar to human milkfat for use in baby feeds, lipid esters including isopropyl myristate, for use in cosmetics and monoglycerides for use as emulsifiers in food and pharmaceutical applications.

The increasing awareness of the importance of chirality in the context of biological activity has stimulated a growing demand for efficient methods for industrial synthesis of pure enantiomers including chiral anti-inflammatory drugs such as naproxen and ibuprofen, antihypertensive agents such as angiotensin-converting enzyme (ACE) inhibitors (e.g., captopril, enalapril, ceranopril, zofenapril and lisinopril) and the calcium channel blocking drugs such as diltiazem, Lipases are used in synthesis of these drugs.

7.4 Applications of lipases

Because of their significance, for a long time lipases have been the subject of numerous studies which concentrate mainly on the structure of this enzyme, the mechanism of functioning, kinetics, sequencing and cloning genes, as well as on the general characteristics of their occurring. Lipases are commonly used in processing fats and oils, as additions to detergents and degreasing agents, in food processing, chemical and pharmaceutical synthesis, in paper production and the cosmetic industry. They may also be used in increasing the speed of fatty and polyurethane waste decomposition.

Lipases used as additions to detergents must fulfill the following criteria: low substrate selectivity, resistance to fairly raw washing conditions (pH 10–11 at 30–60°C), resistance to destructive surfactants and enzymes (e.g., linear AAS and proteases) which are very important components of detergents.

Fats constitute an important component of food. The nutrition and taste values, as well as the physical properties of triglycerides are largely dependent on such factors as the position of the fatty acid in the glycerol chain, the length of the fatty acid chain and the degree of its saturation. Lipases enable the modification of lipid properties by a change of the fatty acid chains position in glycerides and the substitution of one or more fatty acids with others. This way is relatively cheap and less desired lipids may be modified in order to obtain more valuable fats.

Lipases are widely used in the processing of fats and oils, detergents and degreasing formulations, food processing, the synthesis of fine chemicals and pharmaceuticals, paper manufacture and production of cosmetics and pharmaceuticals. Lipase can be used to accelerate the degradation of fatty waste and polyurethane.

Major applications of lipases are summarised in Table 7.1. Most of the industrial microbial lipases are derived from fungi and bacteria (Table 7.2).

Table 7.1: Industrial applications of microbial lipases.

Industry	*Action*	*Product or application*
Detergents	Hydrolysis of fats	Removal of oil stains from fabrics
Dairy foods	Hydrolysis of milk fat, cheese ripening, modification of butter fat	Development of flavouring agents in milk, cheese and butter
Bakery foods	Flavour improvement	Shelf-life prolongation
Beverages	Improved aroma	Beverages
Food dressings	Quality improvement	Mayonnaise, dressings and whippings
Health foods	Trans-esterification	Health foods
Meat and fish	Flavour development	Meat and fish products, fat removal

(Cont'd...)

Industry	*Action*	*Product or application*
Fats and oils	Trans-esterification, hydrolysis	Cocoa butter, margarine, fatty acids, glycerol, mono- and diglycerides
Chemicals	Enantioselectivity, synthesis	Chiral building blocks, chemicals
Pharmaceuticals	Trans-esterification, hydrolysis	Specialty lipids, digestive aids
Cosmetics	Synthesis	Emulsifiers, moisturisers
Leather	Hydrolysis	Leather products
Paper	Hydrolysis	Paper with improved quality
Cleaning	Hydrolysis	Removal of fats

Table 7.2: Some commercially available microbial lipases.

Type	*Source*	*Application*	*Producing company*
Fungal	*C. rugosa*	Organic synthesis	Amano, Biocatalysts, Boehringer Mannheim, Fluka, Genzyme, Sigma
	C. antarctica	Organic synthesis	Boehringer Mannheim, Novo Nordisk
	T. lanuginosus	Detergent additive	Boehringer Mannheim, Novo Nordisk
	R. miehei	Food processing	Novo Nordisk, Biocatalysts, Amano
Bacterial	*Burkholderia cepacia*	Organic synthesis	Amano, Fluka, Boehringer Mannheim
	P. alcaligenes	Detergent additive	Genencor
	P. mendocina	Detergent additive	Genencor
	Ch. viscosum	Organic synthesis	Asahi, Biocatalysts

7.4.1 Lipases in the detergent industry

Because of their ability to hydrolyses fats, lipases find a major use as additives in industrial laundry and household detergents. Detergent lipases are especially selected to meet the following requirements: (i) a low substrate specificity, i.e., an ability to hydrolyse fats of various compositions, (ii) ability to withstand relatively harsh washing conditions (pH 10–11, 30–60°C) and (iii) ability to withstand damaging surfactants and enzymes [e.g., linear alkyl benzene sulphonates (LAS) and proteases], which are important ingredients of many detergent formulations. Lipases with the desired properties are obtained through a combination of continuous screening and protein engineering.

7.4.2 Lipases in food industry

Fats and oils are important constituents of foods. The nutritional and sensory value and the physical properties of a triglyceride are greatly influenced by factors such as the position of the fatty acid in the glycerol backbone, the chain length of the fatty acid and its degree of unsaturation. Lipases allow us to modify the properties of lipids by altering the location of fatty acid chains in the glyceride and replacing one or more of the fatty acids with new ones. This way, a relatively inexpensive and less desirable lipid can be modified to a higher value fat.

Cocoa butter, a high-value fat, contains palmitic and stearic acids and has a melting point of approximately 37°C. Melting of cocoa butter in the mouth produces a desirable cooling sensation in products such as chocolate. Lipase-based technology involving mixed hydrolysis and synthesis reactions is used commercially to upgrade some of the less desirable fats to cocoa butter substitutes. One version of this process uses immobilised *Rhizomucor miehei* lipase for the trans-esterification reaction that replaces the palmitic acid in palm oil with stearic acid. Similarly, Pabai and others described a lipase-catalysed interesterification of butter fat that resulted in a considerable decrease in the long-chain saturated fatty acids and a corresponding increase in C18:0 and C18:1 acids at position 2 of the selected triacylglycerol.

Because of their metabolic effects, PUFAs are increasingly used as pharmaceuticals, neutraceuticals and food additives. Many of the PUFAs are essential for normal synthesis of lipid membranes and prostaglandins. Microbial lipases are used to obtain PUFAs from animal and plant lipids such as menhaden oil, tuna oil and borage oil. Free PUFAs and their mono- and diglycerides are subsequently used to produce a variety of pharmaceuticals including anticholesterolemics, anti-inflammatories and thrombolytics. In addition, lipases have been used for development of flavours in cheese ripening, bakery products and beverages. Also, lipases are used to aid removal of fat from meat and fish products.

7.4.3 Lipases in pulp and paper industry

'Pitch,' or the hydrophobic components of wood (mainly triglycerides and waxes), causes severe problems in pulp and paper manufacture. Lipases are used to remove the pitch from the pulp produced for paper making. Nippon Paper Industries, Japan, have developed a pitch control method that uses the Candida rugosa fungal lipase to hydrolyse up to 90% of the wood triglycerides.

7.4.4 Lipases in organic synthesis

Use of lipases in organic chemical synthesis is becoming increasingly important. Lipases are used to catalyse a wide variety of chemo-, regio- and stereoselective transformations. Majority of lipases used as catalysts in organic chemistry are of microbial origin. These enzymes work at hydrophilic–lipophilic interface and tolerate organic solvents in the reaction mixtures.

The enzymes catalyse the hydrolysis of water-immiscible triglycerides at water–liquid interface. Under given conditions, the amount of water in the reaction mixture will determine the direction of lipase-catalysed reaction. When there is little or no water, only esterification and trans-esterification are favoured. Hydrolysis is the favoured reaction when there is excess water.

7.4.5 Lipases in bioconversion in aqueous media

Hydrolysis of esters is commonly carried out using lipase in two-phase aqueous media. Penreac'h and Baratti reported on the hydrolysis of *p*-nitrophenyl palmitate (*p*NPP) in *n*-heptane by a lipase preparation of *P. cepacia*. Collins and others used lipase entrapped in a hydrophobic sol–gel matrix for a variety of transformations.

Mutagenesis has been used to greatly enhance the enantioselectivity of lipases. The lipase-acyl transferase from *C. parapsilosis* has been shown to catalyse fatty hydroxamic acid biosynthesis in a biphasic liquid/aqueous medium. The substrates of the reaction were acyl donors (fatty acid or fatty acid methyl ester) and a hydroxylamine. The transfer of acyl group from a donor ester to hydroxylamine (aminolysis) was catalysed preferentially compared to the reaction of free fatty acids. This feature made the *C. parapsilosis* enzyme the catalyst of choice for the direct bioconversion of oils in aqueous medium. Yeo and others reported a novel lipase produced by *Burkholderia* sp., which could preferentially hydrolyse a bulky ester, *t*-butyl octanoate (TBO). This lipase was confirmed to be 100-fold superior to commercial lipases in terms of its TBO-hydrolysing activity.

7.4.6 Lipases in bioconversions in organic media

Enzymes in organic media without a free aqueous phase are known to display useful unusual properties and this has firmly established nonaqueous enzyme systems for synthesis and biotransformations. Lipases have been widely investigated for various nonaqueous biotransformations.

7.4.7 Lipases in resolution of racemic acids and alcohols

Stereoselectivity of lipases has been used to resolve various racemic organic acid mixtures in immiscible biphasic systems. Racemic alcohols can also be resolved into enantiomerically pure forms by lipase-catalysed trans-esterification. Arroyo and Sinisterra reported that esterification reaction in nonaqueous media using lipase-B from *C. antarctica* was stereoselective towards the R-isomer of ketoprofen in an achiral solvent such as isobutyl methyl ketone and (S+)-carvone.

7.4.8 Lipases in regioselective acylations

Lipases acylate certain steroids, sugars and sugar derivatives with a high regioselectivity. Monoacylated sugars have been produced in anhydrous pyridine from triethyl carboxylates and various monosaccharides used a lipase from *A. niger* to catalyse the regioselective deacylation of preacylated methyl β-D-glucopyranoside. Similarly, Kodera and others reported regioselective deacetylation of preacetylated monosaccharide derivatives in 1,1,1-trichloroethane using a lipase modified with polyethylene glycol.

7.4.9 Lipases in ester synthesis

Lipases have been successfully used as catalyst for synthesis of esters. The esters produced from short-chain fatty acids have applications as favouring agents in food industry. Methyl and ethyl esters of long-chain acids have been used to enrich diesel fuels.

7.4.10 Lipases in oleochemical industry

Use of lipases in oleochemical processing saves energy and minimises thermal degradation during alcoholysis, acidolysis, hydrolysis and glycerolysis. Although lipases are designed by nature for the hydrolytic cleavage of the ester bonds of triacylglycerol, lipases can catalyse the reverse reaction (ester synthesis) in a lowwater environment. Hydrolysis and esterification can occur simultaneously in a process known as interesterification. Depending on the substrates, lipases can catalyse acidolysis (where an acyl moiety is displaced between an acyl glycerol and a carboxylic acid), alcoholysis (where an acyl moiety is displaced between an acyl glycerol and an alcohol) and trans-esterification (where two acyl moieties are exchanged between two acylglycerols).

7.5 Micro-organisms producing lipases

Lipases are produced by many micro-organisms and higher eukaryotes. Most commercially useful lipases are of microbial origin. Some of the lipase-producing micro-organisms are listed in Table 7.3.

Table 7.3: Some lipase-producing micro-organisms.

Source	*Genus*	*Species*
Bacteria	*Bacillus*	*B. megaterium*
(Gram-positive)		*B. cereus*
		B. stearothermophilus
		B. subtilis
		Recombinant B. subtilis 168
		B. brevis
		B. thermocatenulatus

(Cont'd...)

Source	***Genus***	***Species***
		Bacillus sp. IHI-91
		Bacillus strain WAI 28A5
		Bacillus sp.
		B. coagulans
		B. acidocaldarius
		Bacillus sp. RS-12
		B. thermoleovorans ID-1
		Bacillus sp. J 33
	Staphylococcus	*S. canosus*
		S. aureus
		S. hyicus
		S. epidermidis
		S. warneri
	Lactobacillus	*Lactobacillus delbruckii sub sp. bulgaricus*
		Lactobacillus sp.
	Streptococcus	*Streptococcus lactis*
	Micrococcus	*Micrococcus freudenreichii*
	Propionibacterium	*Propionibacterium acne*
		Pr. granulosum
	Burkholderia	*Burkholderia sp.*
		Bu. glumae
Bacteria (Gram-negative)	*Pseudomonas*	*P. aeruginosa*
		P. fragi
		P. mendocina
		P. putida 3SK
		P. glumae
		P. cepacia
		P. fluorescens
		P. aeruginosa KKA-5
		P. pseudoalcaligenes F-111
		Pseudomonas sp.
		P. fluorescens MF0
		Pseudomonas sp. KWI56
	Chromobacterium	*Ch. viscosum*
	Acinetobacter	*Aci. pseudoalcaligenes*
		Aci. radioresistens

(Cont'd...)

Source	*Genus*	*Species*
	Aeromonas	*Ae. hydrophila*
		Ae. sorbia LP004
Fungi	*Rhizopus*	*Rhizop. delemar*
		Rhizop. oryzae
		Rhizop. arrhizus
		Rhizop. nigricans
		Rhizop. nodosus
		Rhizop. microsporous
		Rhizop. chinensis
		Rhizop. japonicus
		Rhizop. niveus
		Aspergillus A. flavus
		A. niger
		A. japonicus
		A. awamori
		A. fumigatus
		A. oryzae
		A. carneus
		A. repens
		A. nidulans
	Penicillium	*Pe. cyclopium*
		Pe. citrinum
		Pe. roqueforti
		Pe. fumiculosum
		Penicillium sp.
		Pe. camambertii
		Pe. wortmanii
		Mucor Mucor miehei
		Mu. javanicus
		Mu. circinelloides
		Mu. hiemalis
		Mu. racemosus
	Ashbya	*Ashbya gossypii*
	Geotrichum	*G. candidum*
		Geotrichum sp.
	Beauveria	*Beauveria bassiana*
	Humicola	*H. lanuginosa*
	Rhizomucor	*R. miehei*

(Cont'd...)

Source	*Genus*	*Species*
	Fusarium	*Fusarium oxysporum*
		F. heterosporum
	Acremonium	*Ac. strictum*
	Alternaria	*Alternaria brassicicola*
	Eurotrium	*Eu. herbanorium*
	Ophiostoma	*O. piliferum*
Yeasts	*Candida*	*C. rugosa*
		C. tropicalis
		C. antarctica
		C. cylindracea
		C. parapsilosis
		C. deformans
		C. curvata
		C. valida
	Yarrowia	*Y. lipolytica*
	Rhodotorula	*Rho. glutinis*
		Rho. pilimornae
	Pichia	*Pi. bispora*
		Pi. maxicana
		Pi. sivicola
		Pi. xylosa
		Pi. burtonii
	Saccharomyces	*Sa. lipolytica*
		Sa. crataegenesis
	Torulospora	*Torulospora globora*
	Trichosporon	*Trichosporon asteroides*
Actinomycetes	*Streptomyces*	*Streptomyces fradiae NCIB 8233*
		Streptomyces sp. PCB27
		Streptomyces sp. CCM 33
		Str. coelicolour
		Str. cinnamomeus

7.5.1 Isolation and screening of lipase-producing micro-organisms

Lipase-producing micro-organisms have been found in diverse habitats such as industrial wastes, vegetable oil processing factories, dairies, soil contaminated with oil, oilseeds and decaying food, compost heaps, coal tips and hot springs.

Lipase-producing micro-organisms include bacteria, fungi, yeasts and actinomyces. A simple and reliable method for detecting lipase activity in micro-organisms has been described by Sierra. This method uses the surfactant Tween 80 in a solid medium to identify a lipolytic activity. The formation of opaque zones around the colonies is an indication of lipase production by the organisms. Modifications of this assay use various Tween surfactants in combination with Nile blue or neet's foot oil and Cu^{2+} salts. Also, screening of lipase producers on agar plates is frequently done by using tributyrin as a substrate and clear zones around the colonies indicate production of lipase. Screening systems making use of chromogenic substrates have also been described. Wang and others used plates of a modified Rhodamine B agar to screen lipase activity in a large number of micro-organisms.

7.6 Production and media development for lipase

Microbial lipases are produced mostly by submerged culture, but solid state fermentation methods can be used also. Immobilised cell culture has been used in a few cases. Many studies have been undertaken to define the optimal culture and nutritional requirements for lipase production by submerged culture. Lipase production is influenced by the type and concentration of carbon and nitrogen sources, the culture pH, the growth temperature and the dissolved oxygen concentration. Lipidic carbon sources seem to be generally essential for obtaining a high lipase yield.

7.6.1 Effect of carbon sources on lipase

Sugihara and others reported lipase production from *Bacillus* sp. in the presence of 1% olive oil in the culture medium. Little enzyme activity was observed in the absence of olive oil even after prolonged cultivation. Fructose and palm oil were reported to be the best carbohydrate and lipid sources, respectively, for the production of an extracellular lipase by Rhodotorula glutinis. When the two carbon sources were compared, palm oil at a concentration of 2% was found to yield 12-fold more lipase than the fructose medium.

Because of their use in alkaline detergents, alkalostable lipases are especially sought after. An alkaline lipase was produced by *P. alcaligenes* M-1 in a medium with citric acid and soyabean oil as substrates in the batch and fed-batch phases, respectively. This lipase had excellent capability for removing fatty stains in an alkaline environment. The gene encoding the alkaline lipase was isolated and characterised. Kim and others reported production of a highly alkaline thermostable lipase by *Bacillus stearothermophilus* L1 in a medium that contained beef tallow and palm oil. This lipase was most active at 60–65°C and pH 9–10. Activity assessments with synthetic substrates showed this enzyme to be especially active towards *p*-nitrophenyl caprylate.

The yeast *C. rugosa* has been shown to secrete an extracellular lipase whose production can be induced by adding fatty acids to the culture broth. This lipase is composed of several isoforms with slightly differing catalytic properties. Lipase production could be induced by adding oleic acid as the carbon source. In the same yeast, the production of a constitutive lipase was induced by using glucose as the carbon source

P. aeruginosa KKA-5 produced the maximal lipase activity when castor oil (2%) was used as the carbon source at pH 6.9. This enzyme could cause up to 90% hydrolysis of castor oil and it was stable in alkaline conditions (pH 7–10). The maximum activity was obtained at pH 8.5.

One study explored 56 strains of molds for the ability to produce lipase. A strain identified as *Pe. wortmanii* was determined to be the best lipase producer. Maximum lipase production (12.5 U/mL) was obtained in a 7-day culture using olive oil (5% wt/vol) as the carbon source. The optimal pH and temperature for the crude lipase activity were 7.0 at 45°C, respectively.

A thermophilic bacterium, *B. thermoleovorans* ID-1, isolated from hot springs in Indonesia, showed extracellular lipase activity and high growth rates on lipid substrates at elevated temperatures. Using olive oil (1.5% vol/vol) as the sole carbon source, the isolate ID-1 grew rapidly at 65°C (specific growth rate of 2.5 h^{-1}) and its lipase activity attained a maximum value of 520 U/L during the late exponential growth phase. The isolate ID-1 could grow on a variety of lipidic substrates such as oils (olive, soyabean and mineral oils), triglycerides (triolein, tributyrin) and synthetic surfactants.

7.6.2 Effect of nitrogen sources on lipase

For an extracellular lipase of *Pe. citrinum*, Sztajer and Maliszewska obtained maximal production in a medium that contained 5% (wt/vol) peptone (pH 7.2). Nitrogen sources such as corn steep liquor and soyabean meal stimulated lipase production but to a lesser extent than peptone. Urea and ammonium sulphate inhibited lipase synthesis. Lipolytic activity (1120 U/L) was determined by titration of the free fatty acids released from olive oil incubated with the cell-free broth.

Thermostable lipase of *Pseudomonas* sp. KW1-56 was produced in a medium that contained peptone (2% wt/vol) and yeast extract (0.1% wt/vol) as nitrogen sources. The lipase was purified by acetone precipitation and gel filtration. The purification factor was 13.9, but the overall recovery was only 2.9%. The enzyme produced a single band on sodium dodecyl sulphate polyacrylamide gel electrophoresis (SDS-PAGE) and its molecular mass was estimated at 33 kDa. The temperature optimum for the enzyme was 60°C and more than 96% of the original activity remained after 24 hr at 60°C.

Acremonium structum produced a large amount of lipase under stationary conditions in a medium containing 35% (wt/vol) soyabean meal as the nitrogen source. Generally, micro-organisms provide high yields of lipase when organic nitrogen sources are used. One exception reported is *Rho. glutinis*. Although good growth of *Rho. glutinis* seems to require organic nitrogen sources (e.g., yeast extract and tryptone), an inorganic nitrogen source such as ammonium phosphate appears to favour lipase production. The enzyme produced had an optimal activity at pH 7.5. The half-life of the enzyme was 45 and 11.8 min at 45 and 55°C, respectively.

A. oryzae produced maximal alkaline lipase in a medium that contained yeast extract (1%), polypeptone (2%) and soyabean meal (3%) as nitrogen sources. The enzyme produced had an activity optimum at pH 7.5 and 10.0, respectively, with olive oil and tributyrin as substrates. A Brazilian strain of *Pe. citrinum* produced a maximal lipase activity of 409 IU/mL in a medium that contained yeast extract (0.5%) as the nitrogen source. A decrease in yeast extract concentration reduced the attainable lipase activity. Replacement of yeast extract with ammonium sulphate diminished lipase production. *A. niger* produced lipase in a lipid free medium but required an inducer for improved production. Lipase production increased when the medium was supplemented with an inorganic nitrogen source (ammonium nitrate). Similarly, the addition of ammonium sulphate and peptone to the medium enhanced lipase production by the fungus *O. piceae*. The enzyme had optimal activity at 60°C and pH 9.5.

Wang and others reported production of a highly thermostable alkaline lipase by *Bacillus* strain A 30-1 (ATCC 53841) in a medium that contained yeast extract (0.1%) and ammonium chloride (1%) as nitrogen sources. The partially purified lipase preparation had an optimal activity temperature of 60°C and the optimum pH was 9.5. This enzyme was stable to both hydrogen peroxide and alkaline protease. Cordenons and others examined various nitrogen sources for producing extracellular lipase from *Acinetobacter calcoaceticus.* Use of amino acids and tryptone improved the lipase yield by a factor of 2 or 3 when compared to the use of ammonium, yeast extract and protease peptone. However, lipase yield and stability could be improved by supplementing the preferred organic nitrogen source with ammonium. The extracellular lipase was measured using *p*NPP as the substrate.

Lin and others reported an extracellular alkaline lipase produced by *P. alcaligenes* F-111 in a medium that contained soyabean meal (1%), peptone (1.5%) and yeast extract (0.5%). The lipase produced was unaffected by various detergents. The cationic surface active agents such as sodium tripoly-phosphate, sodium dodecyl benzene sulphonate and sodium alkyl benzene sulphonate did not affect the enzyme activity, suggesting that this enzyme is a good candidate for detergent applications.

For intracellular lipase production by the fungus *Rhizop. oryzae*, corn steep liquor (7%) was an optimal nitrogen source. At concentrations greater than 7%, corn steep liquor caused a rapid decline in cell growth and lipase production. *P. aeruginosa* KKA-5 produced an extracellular lipase in a medium composed of polypeptone (4%) and yeast extract (0.05%). This enzyme was stable up to 45°C. The lipase was highly stable in aqueous solutions of solvents such as methanol and ethanol, but was weakly inhibited in the presence of acetone.

7.7 Purification and kinetic characterisation of lipases

Many lipases have been extensively purified and characterised in terms of their activity and stability profiles relative to pH, temperature and effects of metal ions and chelating agents. In many cases, lipases have been purified to homogeneity and crystallised. Purification methods used have generally depended on nonspecific techniques such as precipitation, hydrophobic interaction chromatography, gel filtration and ion exchange chromatography. Affinity chromatography has been used in some cases to reduce the number of individual purification steps needed.

7.8 Thermostability of lipase

The rate of a reaction approximately doubles for each 10°C increase in temperature. Assuming the enzyme is stable at elevated temperatures, the productivity of the reaction can be enhanced greatly by operating at a relatively high temperature. Consequently, thermal stability is a desirable characteristic of lipases.

Thermostable lipases have been isolated from many sources, including *P. fluorescens*, *Bacillus* sp., *B. coagulans* and *B. cereus*, *B. stearothermophilus*, *Geotrichum* sp. and *Aeromonas sobria* and *P. aeruginosa*. The latter enzyme was significantly stabilised by Ca^{2+} and was inactivated by EDTA. This inactivation could be overcome by adding $CaCl_2$, suggesting the existence of a calcium-binding site in *P. aeruginosa* lipase.

One of the more notable thermostable enzyme was isolated by Wang and others from a *Bacillus* strain. This enzyme had maximum activity at 60°C and retained 100% of the original activity after being held at 75°C for 30 min. The half-life of the enzyme was 8 hr at 75°C. The enzyme retained at least 90% of the original activity after being incubated at 60°C for 15 hr. Other highly thermostable lipases have been reported.

Thermal stability of a lipase is obviously related with its structure. Thermostability is influenced by environmental factors such as pH and the presence of metal ions. At least in some cases, thermal denaturation appears to occur through intermediate states of unfolding of the polypeptide. Mutations in the 'lid' region of the enzyme can significantly affect heat stability. Attempts are being made to protein engineer lipases for improved thermal stability.

Compared to the native enzyme, thermal and operational stability of many lipases can be significantly enhanced by immobilisation. *C. antarctica* lipase-B could be thermally stabilised by immobilisation. The native enzyme and the covalently immobilised preparation appeared to follow different modes of thermal deactivation.

7.8.1 Effect of metal ions and chelating agents on lipase activity

Chartrain and others observed that an extracellular lipase of *P. aeruginosa* MB5001 was strongly inhibited by 1 mM $ZnSO_4$ (94% inhibition) but was stimulated by adding 10 mM $CaCl_2$ (1.24-fold stimulation) and 200 mM taurocholic acid (1.6-fold stimulation). Mase and others studied the effect of metal ions (1 mM concentration) on a purified lipase of *Pe. roqueforti* IAM7268. The lipase activity was not affected by Ca^{2+}, Mg^{2+}, Mn^{2+}, Na^{+}, K^{+}, Cu^{2+}, EDTA, *p*-chloro mercuribenzoic acid and iodoacetate. In contrast, the enzyme was inhibited by Ag^{+}, Fe^{2+}, Hg^{2+} and isopropyl fluorophosphate.

7.9 Shear tolerance of lipases

Lipase-catalysed reactions occur at interfaces and, therefore, the interfacial area and the quality of the interface are important influences on the observed reaction rate. Use of intense mechanical agitation and emulsifiers in bioreactors helps in providing the requisite interfacial area. A combination of interface and agitation can be damaging to lipases. Intense agitation and liquid–liquid interfaces are especially frequent during lipase-catalysed hydrolysis. Gas–liquid interfaces occur in the production of lipases by aerobic submerged culture.

Shear-associated inactivation at gas–liquid and liquid–liquid interfaces has been reported for numerous enzymes including lipases. The rate of interfacial denaturation increases with increasing temperature and turbulence in the fluid. Denaturation generally obeys the first-order kinetics. The denaturation rate constant depends on the specific power input in the reactor and the amount of gas–liquid interface present. Additives such as polypropylene glycol can greatly reduce the rate of denaturation. Interfacial denaturation of lipases by unfolding apparently occurs without the molecule breaking into multiple peptides.

7.10 Multiple forms of lipases

Certain microbial and other lipases exist in multiple forms. Chang and others identified multiple forms of CRL. The presence of Tween 80 and Tween 20 in the culture medium altered the relative abundance of the various forms of lipase in the medium, relative to when no additives were used. Two types of lipases, Lipases I and II, are known to be produced by *Rhizop. niveus*. Lipases I and II differ in molecular weight and Lipase I appears to be converted to Lipase II by limited proteolysis. *Geotrichum candidum* ATCC 34614 has been found to produce four different lipases. The main lipase (Lipase I) produced is nonspecific in positional specificity, whereas Lipase IV has unusual positional specificity.

Multiple forms of *C. antarctica* lipase have been reported. Of these forms, lipase B is stereoselective towards the R-isomer of ketoprofen in an achiral solvent such as isopentyl methyl ketone and also in S(+)-carvone. Martinelle and others studied interfacial activation of *C. antarctica* lipases A and B (CALB) and compared them with the *Humicola lanuginosa* lipase. CALB displayed no interfacial activation, which indicated an absence of the lid structure that regulates the access to the active site. The hydrolysis of the lipid *p*-nitrophenyl ester by lipases A and B of *C. rugosa* was characterised by Rodendo and others. Lipase A was maximally active on caprylate, whereas lipase B had maximal activity on laurate. The two enzymes were identical in other respects. Similarly, a commercial lipolytic preparation of *Ch. viscosum* was reported to contain two different lipases.

7.11 Immobilisation of lipases

Both native and immobilised lipases are available commercially. Lipases used in laundry detergents and many other applications are not immobilised, however, an increasing number of speciality applications of lipases in synthesis and biotransformation demand an immobilised biocatalyst for efficiency of use. Immobilisation improves recyclability of expensive lipases. Also, immobilisation can enhance enzyme stability and activity.

Many methods have been used to immobilise lipases, including adsorption or precipitation onto hydrophobic materials, covalent attachment to functional groups, entrapment in polymer gels, adsorption in macroporous anion exchange resins, microencapsulation in lipid vesicles and sol–gel entrapment. *G. candidum* lipases A and B were immobilised on Accurel EP 100 porous polypropylene supports, precoated with ovalbumin to increase stability in organic solvents and at elevated temperatures. Bosley and Clayton used hydrophobic controlled pore glasses to immobilise *R. miehei* lipase. Reetz and others employed sol–gel entrapment in silica gel to immobilise various lipases.

C. cylindracea lipase was immobilised on methyl acrylate divinyl benzene copolymer and its derivatives. The immobilised lipase had improved resistance to thermal denaturation than the native enzyme. Reetz and others reported an immobilisation procedure using alkyl silane precursors of the type R Si $(OCH_3)_3$

and mixtures of R Si $(OCH_3)_3$ and Si $(OCH_3)_4$ to immobilise *C. antarctica* lipase. This immobilisation process provided highly active, chemically and thermally stable, heterogeneous biocatalysts.

For immobilisation by adsorption on polymer membranes, hydrophobic membranes tend to load much more lipase than the hydrophilic membranes, however, at least for the hydrolytic reaction, the lipase immobilised on hydrophilic membranes generally appears to be much more active than the enzyme adsorbed on hydrophobic membranes. Use of hollow fibre and flat membrane reactors for biotransformations with immobilised lipases has been reported extensively. Also, packed bed lipase bioreactors have been used.

7.12 Sequencing and cloning of lipase gene

Lipase genes from many micro-organisms and a higher animals have been cloned. The structural lipase gene from a gene library of *Aci. calcoaceticus* BD413 DNA was cloned in *Escherichia coli* phage M13 by Kok and others. The sequence analysis of 2.1-kb chromosomal DNA fragment revealed one complete open reading frame, lip A, encoding a mature protein with a predicted molecular mass of 32.1 kDa. A recombinant plasmid expressing the alkaline lipase of *P. aeruginosa* IGB83 under the tac promoter was constructed by Leza and others. The plasmid was then cloned in *Xanthomonas campestris*, which efficiently produced and secreted the alkaline lipase. An optimisation of culture conditions of recombinant *X. campestris* led to a 12-fold increase in lipase production relative to initial results in shake flasks.

The dimorphic yeast *C. rugosa* has an unusual codon usage that interferes with the functional expression of genes derived from this yeast in conventional heterologous hosts. CRL occurs in several different isoforms encoded by the lip gene family. Of these lipases, the isoforms encoded by the gene lip 1 is the most abundant. The lip 1 gene (1647 bp) was completely synthesised with an optimised nucleotide sequence to simplify genetic manipulation and allow heterologous expression in yeast. The synthetic gene was functionally overexpressed in *Pi. pastoris*, allowing for the production of the specific isoform recombinant lipase at a level of 150 U/mL in the culture medium. The physiochemical and catalytic properties of the recombinant lipase were compared with those of a commercial, nonrecombinant, CRL preparation containing the various isoforms.

Production of *Pseudomonas* lipases requires correct folding and secretion through the membrane. A controllable expression of the gene lip H, encoding a lipase-specific foldase, is important for overexpression of lipase in the homologous host *E. coli*. Construction of appropriate His-tagged fusion proteins permitted overexpression, secretion and one-step purification of lipase from culture supernatants of the homologous host *P. aeruginosa*.

A thermophilic lipase of *B. thermoleovorans* ID-1 was cloned and sequenced by Cho and others. The lipase gene coded 416 amino acid residues and contained the conserved pentapeptide Ala–X–Ser–X–Gly, as do other *Bacillus* lipases. For expression in *E. coli*, the lipase gene was cloned in pET-22b(+) vector with a strong T7 promoter. The lipase activity was approximately 1.4-fold greater than the activity with the native promoter. Pignede and others isolated the lip 2 gene from the lipolytic yeast *Y. lipolytica.* The gene encoded a 334-amino acid precursor protein. The secreted lipase was a 301-amino acid glycosylated polypeptide.

The lip 2p protein is processed by the KEX 2-like endoprotease encoded by XPR6. Deletion of the XPR6 gene resulted in the secretion of an active but less stable pro-enzyme. The proregion did not inhibit lipase secretion and activity and played an essential role in the production of a stable enzyme. The overexpressing strains correctly processed the gene, secreting 100-fold more activity than the wild type.

To sum up, the lipases are versatile enzymes that are used widely. Lipases are becoming increasingly important in high-value applications in the oleochemical industry and the production of fine chemicals. Lipases are capable of regioselective and stereoselective biotransformations and allow resolution of racemic mixtures. Lipases with improved properties are being produced by natural selection and protein engineering to further enhance usefulness of these enzymes. Simultaneously, advances are being made in bioreactor and reaction technologies for effectively using the lipases. Various kinds of immobilised enzyme reactors and multiphase reaction systems have greatly influenced the processes that require catalysis by lipases. Lipase-based processing has a promising future, however, the rate of progress is slow. Factors posing limitations include a relatively high cost of lipases and a lack of enzymes with the optimal range of catalytic specificities and properties required in the various applications.

8

Oxidoreductase

8.1 Introduction

Oxidoreductases are a class of enzymes that catalyse oxidoreduction reactions. Oxidoreductases catalyse the transfer of electrons from one molecule (the oxidant) to another molecule (the reductant). Oxidoreductases catalyse reactions similar to the following:

$$A^- + B \rightarrow A + B^-$$

where, A is the oxidant and B is the reductant. Oxidorecuctases can be oxidases or dehydrogenases. Oxidases are enzymes involved when molecular oxygen acts as an acceptor of hydrogen or electrons. Whereas, dehydrogenases are enzymes that oxidise a substrate by transferring hydrogen to an acceptor that is either NAD^+/$NADP^+$ or a flavin enzyme.

Other oxidoreductases include peroxidases, hydroxylases, oxygenases and reductases. Peroxidases are localised in peroxisomes and catalyses the reduction of hydrogen peroxide. Hydroxylases add hydroxyl groups to its substrates. Oxygenases incorporate oxygen from molecular oxygen into organic substrates. Reductases catalyse reductions, in most cases reductases can act like an oxidases.

Oxidoreductase enzymes play an important role in both aerobic and anaerobic metabolism. They can be found in glycolysis, TCA cycle, oxidative phosphorylation and in amino acid metabolism. In glycolysis, the enzyme glyceraldehydes-3-phosphate dehydrogenase catalyses the reduction of NAD^+ to NADH. In order to maintain the redox state of the cell, this NADH must be reoxidised to NAD^+, which occurs in the oxidative phosphorylation pathway. Additional NADH molecules are generated in the TCA cycle. The product of glycolysis, pyruvate enters the TCA cycle in the form of acetyl-CoA. During anaerobic glycolysis, the oxidation of NADH occurs through the reduction of pyruvate to lactate.

The lactate is then oxidised to pyruvate in muscle and liver cells and the pyruvate is further oxidised in the TCA cycle. All twenty of the amino acids, except leucine and lysine, can be degraded to TCA cycle intermediates. This allows the carbon skeletons of the amino acids to be converted into oxaloacetate and subsequently into pyruvate. The gluconeogenic pathway can then utilise the pyruvate formed.

8.2 Classification of oxidoreductase

Oxidoreductases are classified as EC 1 in the EC number classification of enzymes. Oxidoreductases can be further classified into 22 subclasses:

1. EC 1.1 includes oxidoreductases that act on the CH–OH group of donors (alcohol oxidoreductases).
2. EC 1.2 includes oxidoreductases that act on the aldehyde or oxo group of donors.
3. EC 1.3 includes oxidoreductases that act on the CH–CH group of donors (CH–CH oxidoreductases).
4. EC 1.4 includes oxidoreductases that act on the CH–NH_2 group of donors (amino acid oxidoreductases, monoamine oxidase).
5. EC 1.5 includes oxidoreductases that act on CH–NH group of donors.
6. EC 1.6 includes oxidoreductases that act on NADH or NADPH.
7. EC 1.7 includes oxidoreductases that act on other nitrogenous compounds as donors.
8. EC 1.8 includes oxidoreductases that act on a sulphur group of donors.
9. EC 1.9 includes oxidoreductases that act on a heme group of donors.
10. EC 1.10 includes oxidoreductases that act on diphenols and related substances as donors.
11. EC 1.11 includes oxidoreductases that act on peroxide as an acceptor (peroxidases).
12. EC 1.12 includes oxidoreductases that act on hydrogen as donors.
13. EC 1.13 includes oxidoreductases that act on single donors with incorporation of molecular oxygen (oxygenases).
14. EC 1.14 includes oxidoreductases that act on paired donors with incorporation of molecular oxygen.
15. EC 1.15 includes oxidoreductases that act on superoxide radicals as acceptors.
16. EC 1.16 includes oxidoreductases that oxidise metal ions.
17. EC 1.17 includes oxidoreductases that act on CH or CH_2 groups.
18. EC 1.18 includes oxidoreductases that act on iron-sulphur proteins as donors.
19. EC 1.19 includes oxidoreductases that act on reduced flavodoxin as a donor.
20. EC 1.20 includes oxidoreductases that act on phosphorus or arsenic in donors.

21. EC 1.21 includes oxidoreductases that act on X-H and Y-H to form an X-Y bond.
22. EC 1.97 includes other oxidoreductases.

8.3 Polyphenol oxidase

Polyphenol oxidase (PPO or monophenol mono-oxygenase or polyphenol oxidase I, chloroplastic) is a tetramer that contains four atoms of copper per molecule and binding sites for two aromatic compounds and oxygen.

The enzyme catalyses the *o*-hydroxylation of monophenols (phenol molecules in which the benzene ring contains a single hydroxyl substituent) to *o*-diphenols (phenol molecules containing two hydroxyl substituents). They can also further catalyse the oxidation of *o*-diphenols to produce *o*-quinones. It is the rapid polymerisation of *o*-quinones to produce black, brown or red pigments (polyphenols) that is the cause of fruit browning. The amino acid tyrosine contains a single phenolic ring that may be oxidised by the action of PPOs to form *o*-quinone. Hence, PPOs may also be referred to as tyrosinases.

Common examples of organisms producing the enzyme are *Agaricus bisporus*, *Malus domestica*, *Lactuca sativa*.

Polyphenol oxidase is listed as a morpheein, a protein that can form two or more different homo-oligomers (morpheein forms), but must come apart and change shape to convert between forms. It exists as monomer, trimer, tetramer, octamer, dodecamer. Evidences for this are multiple (protein moonlighting functions, substrate binding/turnover impacts multimerisation, different assemblies have different activities, kinetic hysteresis).

In plants, PPO is a plastidic enzyme with unclear synthesis and function. In functional chloroplasts, it may be involved in some aspect of oxygen chemistry like mediation of pseudocyclic photophosphorylation.

Enzyme nomenclature differentiates between monophenol oxidase enzymes (tyrosinases) and *o*-diphenol: oxygen oxidoreductase enzymes (catechol oxidases).

A mixture of monophenol oxidase and catechol oxidase enzymes is present in nearly all plant tissues and can also be found in bacteria, animals and fungi. In insects, cuticular polyphenol oxidases are present and their products are responsible for desiccation tolerance.

In fact, browning by PPO is not always an undesirable reaction, the familiar brown colour of tea (especially black tea) and cocoa is developed by PPO enzymatic browning during product processing.

Grape reaction product (2-S glutathionyl caftaric acid) is an oxidation compound produced by action of PPO on caftaric acid and found in wine. This compound production is responsible for the lower level of browning in certain white wines.

Arctic Apples are a suite of trademarked apples that contain a nonbrowning trait. Specifically, gene silencing is used to turn down the expression of polyphenol oxidase (PPO), thus preventing the fruit from browning. They are therefore genetically modified food. Polyphenol oxidase has been given two entries in the International Union of Biochemistry (IUB) classification, as EC 1.14.18.1, monophenol monoxygenase and EC 1.10.3.1, catechol oxidase. Common trivial names for monophenol monoxygenase are tyrosinase, phenolase, monophenol oxidase and cresolase. The enzyme complexes from higher plants and fungi are virtually non-specific and oxidise a wide variety of monophenolic and *o*-diphenolic compounds.

Polyphenol oxidases can catalyse many reactions that involve phenolic compounds. When the monophenol *p*-cresol is the substrate, it is oxidised to the diphenol 4-methyl catechol. The oxygen for the hydroxylation comes from O_2, not from water. Thus, the enzyme acts as a monoxygenase in this reaction. Hydroxylation of monophenols is usually referred to as cresolase activity because *p*-cresol is frequently the substrate. When the *o*-diphenol catechol is the substrate, it is dehydrogenated to *o*-benzoquinone.

Polyphenol oxidase is a copper-containing enzyme widely distributed in plants, animals and humans. In this section, we will discuss the enzyme as it occurs in plants. Many reasons have been put forward for the lack of pigment formation in the intact cell, despite the coexistence of polyphenol oxidase and its substrates in the same organ and the same cell. Some of these reasons are separation of substrate and enzyme, substrate presence as precursor, enzyme inactivity, enzyme complexing with an inhibitor and reduction of quinones formed.

Polyphenol oxidase is little understood in its role in higher plants. It is a plastidic enzyme that is nuclear encoded, but is not active until incorporated into the plastid. Apparently, it also exists free in the cytoplasm in degenerating or senescent tissues such as ripe or ripening fruit. Cytochemical techniques to determine the localisation of polyphenol oxidase established that the enzyme is found in root plastids, potato tuber amyloplasts, hypocotyl plastids, epidermal plastids, carrot tissue culture plastids, leukoplasts of *Aegopodium*, apical plastids, etioplasts, chiomoplasts and chloroplasts of many different species. The function of polyphenol oxidase in plants is an enigma because, although it is localised in the plastid, most of the phenolic compounds are in the vacuole, a cellular location not juxtaposed to the plastid. The plastidic location of the enzyme might be necessary to provide a strong reducing environment to prevent the oxidation of *o*-diphenols to *o*-diquinones. The plastidic polyphenol oxidase also may be necessary to provide an electron donor for monophenol oxygenase activity. Polyphenol oxidase and chlorogenic acid have been implicated in resistance of the tomato plant. These enzymes are known to form orthoquinones in damaged plant tissue. Orthoquinones are known to alkylate amine and sulphydryl groups of proteins and amino acids. This alkylation alters solubility, digestibility and infectivity for some pathogenic viruses.

Peroxidase activity generally can be eliminated or minimised by adding catalase or alcohols to the assay solution. Catalase (EC 1.11.1.6) will remove the hydrogen peroxide and the alcohol will be oxidised preferentially to its aldehyde.

A commonly used polyphenol oxidase assay is the measurement of the formation of quinones and quinone polymers spectrophotometrically at 470–500 nm. Another method is the measurement of the decrease of O_2 by polarographic methods. A common dilemma for all polyphenol oxidase assay methods is the low affinity of the enzyme for oxygen. Consequently aeration of the assay solution is necessary for reliable results.

8.3.1 Inhibitors of polyphenol oxidase

There are two types of inhibitor of polyphenol oxidase. Those competitive to oxygen in the copper site of the enzyme and those competitive to phenolics. Tentoxin has also been used in recent research to eliminate the polyphenol oxidase activity from seedlings of higher plants. Tropolone is a grape polyphenol oxidase inhibitor. Another inhibitor of this enzyme is potassium pyrosulphite ($K_2S_2O_5$). Banana root PPO is strongly inhibited by dithiothreitol and sodium metabisulphite.

Potassium dithionite (or potassium hydrosulphite) is also an inhibitor of the polyphenol oxidase.

Related enzymes polyphenol oxidase

Prophenoloxidase is a modified form of the complement response found in some invertebrates, including insects, crabs and worms.

Hemocyanin is homologous to the phenol oxidases (e.g., tyrosinase) since both enzymes sharing type 3 Cu active site coordination. Hemocyanin also exhibits phenol oxidase activity, but with slowed kinetics from greater steric bulk at the active site. Partial denaturation actually improves hemocyanin's phenol oxidase activity by providing greater access to the active site.

Aureusidin synthase is homologous to plant polyphenol oxidase, but contains certain significant modifications.

8.4 Peroxidase

Peroxidases are distributed ubiquitously in both the plant and animal kingdoms. Perhaps the one most widely known to enzymologists is horseradish peroxidase (EC 1.11.1.7), because of its wide use as an indicator enzyme in standard spectrophotometric and immunoassay techniques. All peroxidases act on hydrogen peroxide as an electron acceptor and oxidise a multitude of donor compounds, many times to coloured end products. Horseradish peroxidase has a heme prosthetic group and is a hemoprotein. NADH peroxidase (EC 1.11.1.1) is a flavoprotein. Glutathione peroxidase (EC 1.11.1.9) is a selenium-containing protein. Chloride peroxidase (EC 1.11.1.10) is probably a heme-thiolate protein.

Thermal inactivation of horseradish peroxidase in buffer was studied by differential scanning calorimetry and residual activity determination. For both experimental systems, the apparent reaction order for thermal inactivation was 1.5. Since the four isoenzymes varied in heat resistance, the 1.5 order probably was caused by the heterogeneity of the enzyme. Sucrose at concentrations of 40% or greater stabilised the enzyme against thermal inactivation. At 10–20% sucrose, the thermal stability was reduced. At equimolar concentrations of glucose, fructose, or lactose, horseradish peroxidase was inactivated more rapidly than in the presence of sucrose. The thermal stability of immobilised horseradish peroxidase was tested in water-organic solvent systems.

8.4.1 Other analytical peroxidases

Peroxidase from *Coriolus versicolour* is now being used in analytical procedures as a substitute for horseradish peroxidase. The peroxidases from *C. versicolour* are excreted into the medium only when the carbon source is limited and veratryl alcohol is added to the medium as an enzyme inducer. In a medium with limited nitrogen, the excretion of enzyme is not dependent on veratryl alcohol. The optimum pH of the peroxidase is 4.0 and it is immunologically similar to horseradish peroxidase.

8.4.2 Reduction of peroxidase activity by other methods

Reduction of peroxidase activity was determined in tomato, carrot, or eggplant extracts using the naturally occurring antioxidants quercetin, rutinic acid, chlorogenic acid and α-tocopherol, alone or with heat treatment. Regeneration of enzyme at room temperature and after frozen storage (–18°C for 4 weeks) was tested also. Combining antioxidant (125 mg %) and heat (2 min at 75°C) resulted in almost complete inhibition of vegetable peroxidases. Regeneration of peroxidase in the extracts that were treated with heat and antioxidant was less than in extracts treated with heat or antioxidant alone. These same antioxidants without heat were tested as inhibitors of peroxidase in tomato, carrot, or eggplant extracts. Four concentrations of each were tested: 125, 250, 500 and 1000 mg %. Peroxidase activity of carrot and eggplant was inhibited more effectively than that of tomato by all four antioxidants. α-tocopherol was the least effective and chlorogenic acid was the most effective, followed by quercetin and rutinic acid. Gibberellic acid has been tested as a post-harvest treatment for mango fruit to reduce peroxidase activity. Gibberellic acid at the rate of 100–200 mg/litre significantly delayed ripening, reduced peroxidase and amylase activities and retarded the loss of weight, chlorophyll and ascorbic acid content.

8.5 Lactoperoxidase

Lactoperoxidase is a peroxidase enzyme secreted from mammary, salivary and other mucosal glands that functions as a natural antibacterial agent. Lactoperoxidase is a member of the heme peroxidase family of enzymes. In humans, lactoperoxidase is encoded by the LPO gene.

Lactoperoxidse catalyses the oxidation of a number of inorganic and organic substrates by hydrogen peroxide. These substrates include bromide and iodide and therefore lactoperoxidase can be categorised as a haloperoxidase. Another important substrate is thiocyanate. The oxidised products produced through the action of this enzyme have potent bactericidal activities. Lactoperoxidase together with its inorganic ion substrates, hydrogen peroxide and oxidised products is known as the lactoperoxidase system.

The lactoperoxidase system plays an important role in the innate immune system by killing bacteria in milk and mucosal (linings of mostly endodermal origin, covered in epithelium, which are involved in absorption and secretion) secretions hence augmentation of the lactoperoxidase system may have therapeutic applications.

Furthermore, addition or augmentation of the lactoperoxidase system has potential applications in controlling bacteria in food and consumer health care products. The lactoperoxidase system does not attack DNA and is not mutagenic. However, under certain conditions, the lactoperoxidase system may contribute to oxidative stress. Furthermore, lactoperoxidase may contribute to the initiation of breast cancer, through its ability to oxidise estrogenic hormones producing free radical intermediates.

The structure of lactoperoxidase consists mainly of alpha-helices plus two short antiparallel beta-strands and belongs to the heme peroxidase family of enzymes that also includes myeloperoxidase (MPO), eosinophil peroxidase (EPO), thyroid peroxidase (TPO) and prostaglandin H synthase (PGHS). A heme cofactor is bound near the centre of the protein.

8.5.1 Function of lactoperoxidase

Lactoperoxidase catalyses the hydrogen peroxide (H_2O_2) oxidation of several acceptor molecules:

$$\text{Reduced acceptor} + H_2O_2 \rightarrow \text{oxidised acceptor} + H_2O$$

Specific examples include:

1. Thiocyanate (SCN^-) → Hypothiocyanite ($OSCN^-$).
2. Bromide (Br^-) → Hypobromite (BrO^-).
3. Iodide (I^-) → Hypoiodite (IO^-).

Source of the hydrogen peroxide (H_2O_2) usually is the reaction of glucose with oxygen in the presence of the enzyme glucose oxidase (EC 1.1.3.4) that also takes place in saliva. Glucose, in turn, can be formed from starch in the presence of the saliva enzyme amyloglucosidase (EC 3.2.1.3).

These relatively short lived oxidised intermediates have potent bactericidal effects, hence lactoperoxidase is part of the antimicrobial defense system in tissues that express lactoperoxidase.

The lactoperoxidase system is effective in killing a range of aerobic and certain anaerobic micro-organisms. 'The effect of lactoperoxidase-thiocyanate-hydrogen peroxide mixtures on bacteria is dependent on experimental conditions.

If the bacteria are cultured after the exposure to lactoperoxidase-thiocyanate-hydrogen peroxide on nutrient agar under aerobic conditions, they may not grow, whereas they grow readily on blood agar under anaerobic conditions. In its antimicrobial capacity, lactoperoxidase appears to acts synergistically with lactoferrin and lysozyme.

8.5.2 Applications of lactoperoxidase

Lactoperoxidase is an effective antimicrobial agent. Consequently applications of lactoperoxidase are being found in preserving food, cosmetics and ophthalmic solutions. Furthermore lactoperoxidase have found application in dental and wound treatment. Finally lactoperoxidase may find application as anti-tumor and anti-viral agents.

Dairy products

Lactoperoxidase is an effective antimicrobial agent and is used as an antibacterial agent in reducing bacterial microflora in milk and milk products. Activation of the lactoperoxidase system by addition of hydrogen peroxide and thiocyanate extends the shelf life of refrigerated raw milk. It is fairly heat resistant and is used as an indicator of over pasteurisation of milk.

Oral care

A lactoperoxidase system is claimed to appropriate for the treatment of gingivitis and paradentosis. Lactoperoxidase has been used in toothpaste or a mouthrinse to reduce oral bacteria and consequently the acid produced by that bacteria.

Cosmetics

A combination of lactoperoxidase, glucose, glucose oxidase (GOD), iodide and thiocyanate is claimed to be effective in the preservations of cosmetics.

Cancer and viral infections

Antibody conjugates of glucose oxidase and to lactoperoxidase have been found to effective in killing tumor cells *in vitro*. In addition, macrophages exposed to lactoperoxidase are stimulated to kill cancer cells. Peroxidase-generated hypothiocyanite inhibits herpes simplex virus and human immunodeficiency virus.

8.5.3 Clinical significance

Innate immune system

Lactoperoxidase is an oxidoreductase secreted from salivary, mammary and other mucosal glands that functions as a natural antibacterial agent. Hence it plays an important role in the immune defence system.

Hypothiocyanite is one of the reactive intermediates produced by the activity of lactoperoxidase on thiocyanate and hydrogen peroxyde produced by Dual oxidase 2 proteins, also known as Duox2. Thiocyanate secretion in cystic fibrosis patients is decreased resulting in a reduced production of the antimicrobial hypothiocyanite and consequently contributes to the increased risk of airway infection.

The lactoperoxidase system efficiently inhibited helicobacter pylori in buffer, however, in whole human saliva, it had a weaker antibacterial effect. The lactoperoxidase system did not attack DNA and was not mutagenic for any of the tested strains. However, under certain conditions, the lactoperoxidase system may contribute to oxidative stress. It has been shown that lactoperoxidase in the presence of thiocyanate can trigger the bactericidal and cytotoxic effects of hydrogen peroxide under specific conditions, such as when hydrogen peroxide is present in the reaction mixtures in excess of thiocyanate.

Breast cancer

The oxidation of estradiol by lactoperoxidase is a possible source of oxidative stress in breast cancer. The ability of lactoperoxidase to propagate a chain reaction leading to oxygen consumption and intracellular hydrogen peroxide accumulation could explain the hydroxyl radical-induced DNA base lesions recently reported in female breast cancer tissue. Lactoperoxidase may be involved in breast carcinogenesis, because of its ability to interact with estrogenic hormones and oxidise them through two one-electron reaction steps. Lactoperoxidase reacts with the phenolic A-ring of estrogens to produce reactive free radicals. In addition, lactoperoxidase may activate carcinogenic aromatic and heterocyclic amines and increase binding levels of activated products to DNA, which suggests a potential role of lactoperoxidase catalysed activation of carcinogens in the causation of breast cancer.

Oral care

During the last decades, several clinical studies describing the clinical efficacy of the lactoperoxidase system in a variety of oral care products (tooth pastes, mouthrinses) have been published. After showing indirectly, by means of measuring experimental gingivitis and caries parameters, that mouth rinses containing amyloglucosidase (γ-amylase) and glucose oxidase activate the lactoperoxidase system, the protective mechanism of enzyme in oral care products has been partially elucidated. Enzymes such as lysozyme, lacto-peroxidase and glucose oxidase are transferred from the tooth pastes to the pellicle. Being components of the pellicle, these enzymes are catalytically highly active. Also, as part of tooth pastes, the lactoperoxidase system has a beneficial influence to avoid early childhood caries by reducing the number of colonies formed by the cariogenic microflora while increasing the thiocyanate concentration. With xerostomia patients, tooth pastes with the lactoperoxidase system are seemingly superior to fluoride containing tooth pastes with respect to plaque formation and gingivitis. More of such studies are required, also to examine further the protective mechanisms as there are more questions to answer. The application of lactoperoxidase is not restricted to caries, gingivitis and periodontitis. Thus, a combination of lysozyme and lacto-peroxidase can be applied to support the treatment of the Burning Mouth Syndrom (glossodynia). In combination with lactoferrin lactoperoxidase combats halitosis, in combination with lactoferrin and lysozyme lactoperoxidase helps to improve symptoms of xerostomia. Furthermore, gels with lactoperoxidase help to improve symptoms oral cancer patients are suffering from as their saliva production is compromised due to irradiation. In this case, also the oral bacterial flora is influenced favourably.

Lactoperoxidase is a haemoprotein that catalyses the oxidation of thiocyanate and iodide ions in the presence of hydrogen peroxide to produce highly reactive oxidising agents. These products have a broad spectrum of antimicrobial effects. They can kill viruses, gram positive and gram negative bacteria, fungi and probably also mycoplasmas and parasites. The molecular components that are oxidised in the cells are sulphydryl groups, NADH and NADPH and under some conditions, aromatic amino acids. The major products responsible for these effects are the hypothiocyanate ion ($OSCN^-$), hypothiocyanous acid (HOSCN) and iodine (I_2). Lactoperoxidase is found in milk, saliva, tears, cervical mucous and nasal glands. Human milk does not appear to contain lactoperoxidase. However, a more recent study confirmed the presence of lactoperoxidase in human colostrum. The human enzyme resembles bovine lactoperoxidase B in immunoreactive properties and has an M_w of 80000. The enzyme is a glycoprotein containing about 10% carbohydrate.

8.6 Catalase

Catalase is an enzyme that brings about (catalyses) the reaction by which hydrogen peroxide is decomposed to water and oxygen. Found extensively in organisms that live in the presence of oxygen, catalase prevents the accumulation of and protects cellular organelles and tissues from damage by peroxide, which is continuously produced by numerous metabolic reactions. In mammals, catalase is found predominantly in the liver.

Catalase has various industrial applications. In the food industry, it is used in combination with other enzymes in the preservation of foodstuffs and in the manufacture of beverages and certain food items. Commercial catalases also are used to break down hydrogen peroxide in wastewater.

8.6.1 Temperature affect and catalase enzyme activity

In general, the rates of enzyme-catalysed reactions are faster as temperature increases and slower as temperatures decrease below an optimal temperature level. When temperature increases, more molecules have more kinetic energy to react, so reaction speed increases. As temperature decreases, so does the available energy and then the reaction is slowed.

A variation in temperature as little as 1 or 2°C can increase an enzyme-catalysed reaction rate by 10 to 20%. Raising the temperature 10°C increases the activity rate of most enzymes by 50 to 100%. There is an exception when temperatures reach a certain threshold above the optimal temperature level. When this happens, the intermolecular attractions that maintain the shapes of proteins are broken and the enzyme molecule's shape changes. This results in decreased binding of reactants and a significant decrease in enzyme activity. At this point, the enzyme is said to be denatured. Many enzymes are denatured when temperatures exceed 40 to 50°C (104 to 122°F). Extremely cold temperatures also significantly slows reaction rate.

The effect of temperature on the rates of enzyme-catalysed reactions is exactly why food is refrigerated. A refrigerator's cooler environment slows the enzyme-catalysed reactions that result in food spoilage. Of course, freezing food slows those reactions even further. Other factors that influence the rates of enzyme-catalysed reactions include surface area, pH and light.

Catalase is a common enzyme found in nearly all living organisms exposed to oxygen (such as vegetables, fruit or animals). It catalyses the decomposition of hydrogen peroxide to water and oxygen. It is a very important enzyme in protecting the cell from oxidative damage by reactive oxygen species (ROS). Likewise, catalase has one of the highest turnover numbers of all enzymes, one catalase molecule can convert millions of molecules of hydrogen peroxide to water and oxygen each second.

Catalase is a tetramer of four polypeptide chains, each over 500 amino acids long. It contains four porphyrin heme (iron) groups that allow the enzyme to react with the hydrogen peroxide. The optimum pH for human catalase is approximately 7 and has a fairly broad maximum (the rate of reaction does not change appreciably at pHs between 6.8 and 7.5). The pH optimum for other catalases varies between 4 and 11 depending on the species. The optimum temperature also varies by species.

8.6.2 Reaction of catalase

The reaction of catalase in the decomposition of hydrogen peroxide in living tissue:

$$2\,H_2O_2 \rightarrow 2\,H_2O + O_2$$

The presence of catalase in a microbial or tissue sample can be tested by adding a volume of hydrogen peroxide and observing the reaction. The formation of bubbles, oxygen, indicates a positive result. This easy assay, which can be seen with the naked eye, without the aid of instruments, is possible because catalase has a very high specific activity, which produces a detectable response.

Molecular mechanism of catalase

While the complete mechanism of catalase is not currently known, the reaction is believed to occur in two stages:

$$H_2O_2 + \text{Fe(III)–E} \rightarrow H_2O + \text{O=Fe(IV)–E(+)}$$

$$H_2O_2 + \text{O=Fe(IV)–E(+)} \rightarrow H_2O + \text{Fe(III)–E} + O_2$$

Here Fe(IV)–E represents the iron center of the heme group attached to the enzyme. Fe(IV)–E(+) is a mesomeric form of Fe(V)–E, meaning the iron is not completely oxidised to +V, but receives some 'supporting electrons' from the heme ligand. This heme has to be drawn then as a radical cation (+).

As hydrogen peroxide enters the active site, it interacts with the amino acids Asn147 (asparagine at position 147) and His74, causing a proton (hydrogen ion) to transfer between the oxygen atoms. The free oxygen atom coordinates, freeing the newly formed water molecule and Fe(IV)=O. Fe(IV)=O reacts with a second hydrogen peroxide molecule to reform Fe(III)–E and produce water and oxygen. The reactivity of the iron center may be improved by the presence of the phenolate ligand of Tyr357 in the fifth iron ligand, which can assist in the oxidation of the Fe(III) to Fe(IV). The efficiency of the reaction may also be improved by the interactions of His74 and Asn147 with reaction intermediates. In general, the rate of the reaction can be determined by the Michaelis-Menten equation.

Catalase can also catalyse the oxidation, by hydrogen peroxide, of various metabolites and toxins, including formaldehyde, formic acid, phenols, acetaldehyde and alcohols. It does so according to the following reaction:

$$H_2O_2 + H_2R \rightarrow 2H_2O + R$$

The exact mechanism of this reaction is not known.

Any heavy metal ion (such as copper cations in copper(II) sulphate) can act as a noncompetitive inhibitor of catalase. Furthermore, the poison cyanide is a competitive inhibitor of catalase at high concentrations of hydrogen peroxide. Three-dimensional protein structures of the peroxidated catalase intermediates are available at the protein data bank. This enzyme is commonly used in laboratories as a tool for learning the effect of enzymes upon reaction rates.

8.6.3 Applications of catalase

Catalase is used in the food industry for removing hydrogen peroxide from milk prior to cheese production. Another use is in food wrappers where it prevents food from oxidising. Catalase is also used in the textile industry, removing hydrogen peroxide from fabrics to make sure the material is peroxide-free.

A minor use is in contact lens hygiene-a few lens-cleaning products disinfect the lens using a hydrogen peroxide solution, a solution containing catalase is then used to decompose the hydrogen peroxide before the lens is used again. Recently, catalase has also begun to be used in the aesthetics industry. Several mask treatments combine the enzyme with hydrogen peroxide on the face with the intent of increasing cellular oxygenation in the upper layers of the epidermis.

8.7 Sulphydryl oxidase

Sulphydryl oxidase catalyses the oxidation of thiols to their corresponding disulphides using molecular oxygen as the electron acceptor. The reaction requires that two moles of thiol are consumed for each mole of oxygen. A FAD-activated sulphydryl oxidase was isolated and identified from Whey obtained from bovine skim milk. The enzyme also has been identified in the spores of the fungus *Myrothecium werrucaria.* More recently, the enzyme was isolated and characterised from *Aspergillus sojae* and *Aspergillus niger*.

Mammalian: Sulphydryl oxidase from bovine milk was isolated and characterised. Cream was separated from the skim milk by centrifugation. Chymosin was added to the skim milk to precipitate the casein and the precipitate was separated from the whey by centrifugation. The Whey was saturated to 50% with ammonium sulphate and the precipitate collected by centrifugation. The precipitate was dissolved in phosphate buffer at pH 7.0 and dialysed against the same buffer. The dialysed enzyme was concentrated to about 3% protein and allowed to stand overnight at 4°C. The sedimented enzyme was centrifuged and the precipitate was dissolved in buffer, dialysed against deionised water and lyophilised. The sulphydryl oxidase was purified further by gel filtration and differential centrifugation. The enzyme was purified about 3200-fold with a 41% recovery. The specific activity was 104 μmol oxygen consumed/min/mg N. The reaction catalysed by bovine milk sulphydryl oxidase involves oxidation of sulphydryl groups to the disulphide in the presence of molecular oxygen. The sulphydryl groups in cysteine, glutathione, or even protein can be oxidised. Glutathione was chosen as the substrate because it is more stable than cysteine and reacts more quickly than protein-bound sulphydryl groups. This enzyme can be assayed by three techniques: oxygen consumption, disappearance of sulphydryl groups and hydrogen peroxide production. The oxygen consumption method usually is used because of its greater sensitivity and ease of use. The enzyme exhibited a sharp peak at 35°C and had only 10% activity at the relatively low temperature of 50°C. It had an optimal pH of 7.0 and showed very little activity at pH 4.5 and 8.5. The K_m for glutathione was 9.0×10^{-5} at 35°C and pH 7.0.

Sulphydryl oxidase is a glycoprotein containing about 11% carbohydrate, consisting of 7.2% total hexose, 2.9% *N*-acetyl hexosamine, 0.4% sialic acid and 0.44% fucose. Atomic absorption spectroscopy revealed the presence of 0.030% iron and 0.011% copper. Apparently, 1 gram-atom of iron may be required for two enzyme sub-units, each having a molecular mass of about 45 kDa, as determined by gel electrophoresis. The molecular mass may be inaccurate since it was difficult to determine because of the tendency of the enzyme to form aggregates. As expected, 0.1 mM EDTA inhibited its activity almost 100%. When the EDTA-treated enzyme was assayed in the presence of 1 μM iron, the activity was restored almost completely. Cupric ion restored its activity about 50% and manganese ion about 40%. Cobalt, zinc and molybdenum ions had little effect.

8.8 Glucose oxidase

The glucose oxidase enzyme (GOx) (EC 1.1.3.4) is an oxido-reductase that catalyses the oxidation of glucose to hydrogen peroxide and D-glucono-δ-lactone. In cells, it aids in breaking the sugar down into its metabolites.

Glucose oxidase is widely used for the determination of free glucose in body fluids (diagnostics), in vegetal raw material and in the food industry. It also has many applications in biotechnologies, typically enzyme assays for biochemistry including biosensors in nanotechnologies. It is often extracted from *Aspergillus niger*.

GOx is a dimeric protein, the 3D structure of which has been elucidated. The active site where glucose binds is in a deep pocket. The enzyme, like many proteins that act outside of cells, is covered with carbohydrate chains.

8.8.1 Activity of glucose oxidase

At pH 7, glucose exists in solution in cyclic hemiacetal form as 63.6% β-D-gluco-pyranose and 36.4% α-D-glucopyranose, the proportion of linear and furanose form being negligible. The glucose oxidase binds specifically to β-D-glucopyranose and does not act on α-D-glucose. It is able to oxidise all of the glucose in solution because the equilibrium between the α and β anomers is driven towards the β side as it is consumed in the reaction.

Glucose oxidase catalyses the oxidation of β-D-glucose into D-glucono-1,5-lactone, which then hydrolyses to gluconic acid. In order to work as a catalyst, GOx requires a cofactor, flavin adenine dinucleotide (FAD). FAD is a common component in biological oxidation-reduction (redox reactions). Redox reactions involve a gain or loss of electrons from a molecule. In the GOx-catalysed redox reaction, FAD works as the initial electron acceptor and is reduced to $FADH_2$. Then $FADH_2$ is oxidised by the final electron acceptor, molecular oxygen (O_2), which can do so because it has a higher reduction potential. O_2 is then reduced to hydrogen peroxide (H_2O_2).

8.8.2 Applications of glucose oxidase

Glucose oxidase is widely used coupled to peroxidase reaction that visualises colorimetrically the formed H_2O_2, for the determination of free glucose in sera or blood plasma for diagnostics, using spectrometric assays manually or with automated procedures and even point of use rapid assays. Similar assays allows to monitor glucose levels in fermentation, bioreactors and to control glucose in vegetal raw material and food products. In the glucose oxidase assay, the glucose is first oxidised by glucose oxidase to produce gluconate and hydrogen peroxide. The hydrogen peroxide is then oxidatively coupled with a chromogen to produce a coloured compound which may be measured spectroscopically. For example, hydrogen peroxide together with 4 amino-antipyrene (4-AAP) and phenol in the presence of peroxidase yield a red quinoeimine dye that can be measured at 505 nm. The absorbance at 505 nm is proportional to concentration of glucose in the sample.

Enzymatic glucose biosensors use an electrode instead of O_2 to take up the electrons needed to oxidise glucose and produce an electronic current in proportion to glucose concentration. This is the technology behind the disposable glucose sensor strips used by diabetics to monitor serum glucose levels. In manufacturing, GOx is used as an additive thanks to its oxidising effects: it prompts for stronger dough in bakery, replacing oxidants such as bromate. It also helps remove oxygen from food packaging, or D-glucose from egg white to prevent browning.

Glucose oxidase is found in honey and acts as a natural preservative. GOx at the surface of the honey reduces atmospheric O_2 to hydrogen peroxide (H_2O_2), which acts as an antimicrobial barrier. GOx similarly acts as a bactericide in many cells (fungi, immune cells).

Glucose oxidase (EC 1.1.3.4) has not lived up to its promise as an antioxidant in the food industry. It was modified by treatment with succinic anhydride to be much more acid stable. In fact, it had a half-life over 24 hr at pH 2.4 at room temperature. One of its major applications was seen to be the stabilisation of orange soft drinks against oxidative deterioration. This application failed because of the reversible inhibition of the glucose oxidase by the dyes in the orange soft drink below pH 3.0. There is considerable precedent for the formation of complexes between various chemical agents and the bound flavin moieties of flavoproteins. Another important reason for the lack of success in the food industry has been its inactivation by hydrogen peroxide, one of its products, when the enzyme is complexed with glucose.

Of Aspergillus niger: The production of glucose oxidase in *A. niger* was enhanced by treatment with mutagens. A number of mutants of *A. niger* affected in glucose oxidase expression are described. The overproducing mutants could be classified into seven groups. Conidia of *A. niger* strain G 13 were subjected to four stage mutagenisation using different combinations of mutagens. Each stage was treated with UV radiation and nitrosomethylurea, ethyleneimine, acryflavin, or *N*-methyl-*N*′-nitrosoguanidine. In all, 400 of 5208 colonies isolated after mutagenisation had higher glucose oxidase activity. The 9 most active showed an activity increase of almost 30%.

The formation and location of the enzyme was studied also in *A. niger* that was pre-grown under citric acid-producing conditions. Glucose oxidase could be induced by shifting the pH of the medium from 1.7 to 5.5. The induction required either glucose or glucose-6-phosphate. Rapid secretion into the medium of oxidase produced in this fashion by a mechanism other than autolysis was surprising, because in *A. niger* glucose oxidase is usually an intracellular enzyme and must be recovered from the mycelium by grinding or lysis. Histochemical evidence for the peroxisomal location of glucose oxidase in *A. niger* would lead one to propose an intracellular origin of gluconic acid. When O_2-enriched air was used to increase the dissolved oxygen concentration of an *A. niger* fermentation, the glucose oxidase activity increased about 3-fold. At the same time, the production of gluconic acid increased about 2-fold. The enzyme was produced in both endocellular and exocellular forms.

Glucose oxidase from *A. niger* was cloned from cDNA and genomic libraries using oligonucleotide probes. Two yeast expression plasmids containing the entire glucose oxidase coding sequence, the untranslated region and about 2 kb flanking gt10 sequences were transformed into yeast. One plasmid contained a yeast alcohol dehydrogenase II–glyceraldehyde-3-phosphate dehydrogenase promoter, glucose oxidase presequence, glucose oxidase coding sequence and glyceraldehyde-3-phosphate dehydrogenase terminator inserted into the *Bam* HI site of the yeast—*E. coli* shuttle vector pAB24. The second plasmid had the yeast α-factor leader substituted for the glucose oxidase presequence. The transformed yeast was grown in yeast extract peptone medium containing 4% glucose for 144 hr in shaker flasks at 300 rpm. When grown under derepressing conditions, transformants of these plasmids in yeast secrete large amounts of active glucose oxidase into the medium. Comparison of the yeast-derived glucose oxidase with purified *A. niger* glucose oxidase by polyacrylamide gel electrophoresis showed that the yeast glucose oxidase migrated more slowly than the *A. niger* enzyme. When the yeast enzyme was treated with endoglycosidase H, it migrated with about the same mobility as the *A. niger* glucose oxidase. Thus, it appears that the yeast enzyme has more extensive *N*-linked glycosylation. The yeast-derived enzyme had a turnover number of 17,000–20,000 min^{-1} compared with 16.200 $min.^{-1}$ for the wild-type enzyme.

8.9 Pyranose oxidase

Pyranose oxidase (EC 1.1.3.10) is an enzyme that catalyses the chemical reaction:

$$\text{D-glucose} + O_2 \rightleftharpoons \text{2-dehydro-D-glucose} + H_2O_2$$

Thus, the two substrates of this enzyme are D-glucose and O_2, whereas its two products are 2-dehydro-D-glucose and H_2O_2.

This enzyme belongs to the family of oxidoreductases, specifically those acting on the CH–OH group of donor with oxygen as acceptor. The systematic name of this enzyme class is pyranose:oxygen 2-oxidoreductase. Other names in common use include glucose 2-oxidase and pyranose-2-oxidase. This enzyme participates in pentose phosphate pathway. It employs one cofactor, FAD.

Pyranose oxidase or glucose-2-oxidase (EC 1.1.3.10) oxidises the C-2 atoms of D-glucose, D-xylose, D-sorbose and D-glucono-1,5-lactone to form their respective pyranosones. Hydrogen peroxide is formed as a by-product. It was studied in the 1980s as a more economical way to produce D-fructose from D-glucose. The production of D-fructose from D-glucose by glucose isomerase suffers from a drawback because the isomerisation reaction provides an equilibrium mixture typically containing only 42–50% D-fructose. Pyranose oxidase typically oxidises 95% of D-glucose to glucosone. D-Fructose can be prepared from the glucosone using catalytic hydrogenation. Pyranose oxidase free from glucosone utilising impurities was obtained from strains of *Coriolus versicolour*,

Lenzites betulinus and *Polyporus obtusus*. The enzyme can be used to make pure glucosone for use in making high purity food sugars, for example, fructose, mannitol and sorbitol. When pyranose oxidase is prepared from *P. obtusus*, it is a tetramer with a molccular mass of about 200 kDa with a sub-unit mass of about 69 kDa.

The enzyme is a flavo-protein with a sedimentation coefficient of 11.4. Its relative activities toward glucose, sorbose, xylose and δ-gluconolactone substrates are 100:80:48:12, respectively, at pH 5.0. The enzyme can be used to make D-glucosone from D-glucose, D-xylosone from D-xylose, 5-keto-D-fructose from L-sorbose, or a mixture of 2-keto-D-gluconic acid and D-isoacscorbic acid from δ-D-gluconolactone. Pyranose oxidase can be stabilised against heat inactivation partially by amidinating it with either dimethyl adipimidate or dimethyl suberimidate. The stabilised enzyme has about three times the heat stability of the native enzyme at 65°C. When the organisms are exposed to a mutagenic agent such as UV radiation, higher amounts of pyranose oxidase can be produced.

A new basidiomycete that produces pyranose oxidase has been isolated. Pyranose oxidase also can be used for the quantitative determination of glucose without the need for mutarotase, which might be needed by glucose oxidase if the reaction is expected to go to completion in a reasonable period of time. The enzyme also is useful for assaying 1,5-anhydroglucitol, a marker for the diagnosis of diabetes.

8.10 Xanthine oxidase

Xanthine oxidase (XO, sometimes 'XAO') is a form of xanthine oxidoreductase, a type of enzyme that generates reactive oxygen species. These enzymes catalyse the oxidation of hypoxanthine to xanthine and can further catalyse the oxidation of xanthine to uric acid. These enzymes play an important role in the catabolism of purines in some species, including humans.

Xanthine oxidase is defined as an enzyme activity (EC 1.17.3.2). The same protein, which in humans has the HGNC approved gene symbol XDH, can also have xanthine dehydrogenase activity (EC 1.17.1.4). Most of the protein in the liver exists in a form with xanthine dehydrogenase activity, but it can be converted to xanthine oxidase by reversible sulphydryl oxidation or by irreversible proteolytic modification.

8.10.1 Reaction of xanthine oxidase

The following chemical reactions are catalysed by xanthine oxidase:

1. Hypoxanthine + H_2O + O_2 $\rightleftharpoons$ Xanthine + H_2O_2.
2. Xanthine + H_2O + O_2 $\rightleftharpoons$ uric acid + H_2O_2.

3. Xanthine oxidase can also act on certain other purines, pterins and aldehydes. For example, it efficiently converts 1-methylxanthine (a metabolite of caffeine) to 1-methyluric acid, but has little activity on 3-methylxanthine.
4. Under some circumstances it can produce superoxide ion:

$$RH + H_2O + 2\ O_2 \rightleftharpoons ROH + 2\ O_2^- + 2\ H^+$$

8.10.2 Other reactions

Because xanthine oxidase (XO) is a superoxide-producing enzyme, with general low specificity, it can be combined with other compounds and enzymes and create reactive oxidants, as well as oxidise other substrates.

Bovine xanthine oxidase (from milk) was originally thought to have a binding site to reduce cytochrome *c* with, but it has been found that the mechanism to reduce this protein is through XO's superoxide anion by-product, with competitive inhibition by carbonic anhydrase. Another reaction catalysed by xanthine oxidase is the decomposition of S-Nitrosothiols (RSNO), a reactive nitrogen species, to nitric oxide (NO), which reacts with a superoxide anion to form peroxynitrite under aerobic conditions.

XO has also been found to produce the strong one-electron oxidant carbonate radical anion from oxidation with acetaldehyde in the presence of catalase and bicarbonate. It was suggested that the carbonate radical was likely produced in one of the enzyme's redox centers with a peroxymonocarbonate intermediate.

8.10.3 Catalytic mechanism

The active site of XO is composed of a molybdopterin unit with the molybdenum atom also coordinated by terminal oxygen (oxo), sulphur atoms and a terminal hydroxide. In the reaction with xanthine to form uric acid, an oxygen atom is transferred from molybdenum to xanthine, whereby several intermediates are assumed to be involved. The reformation of the active molybdenum center occurs by the addition of water. Like other known molybdenum-containing oxidoreductases, the oxygen atom introduced to the substrate by XO originates from water rather than from dioxygen (O_2).

Clinical significance

Xanthine oxidase is a superoxide-producing enzyme found normally in serum and the lungs and its activity is increased during influenza A infection.

During severe liver damage, xanthine oxidase is released into the blood, so a blood assay for XO is a way to determine if liver damage has happened. As well, because xanthine oxidase is a metabolic pathway for uric acid formation, the xanthine oxidase inhibitor allopurinol is used in the treatment of gout. Since xanthine oxidase is involved in the metabolism of 6-mercaptopurine, caution should be taken before administering allopurinol and 6-mercaptopurine, or its prodrug azathioprine, in conjunction.

Xanthinuria is a rare genetic disorder where the lack of xanthine oxidase leads to high concentration of xanthine in blood and can cause health problems such as renal failure. There is no specific treatment, sufferers are advised by doctors to avoid foods high in purine and to maintain a high fluid intake. Type I xanthinuria has been traced directly to mutations of the XDH gene which mediates xanthine oxidase activity. Type II xanthinuria may result from a failure of the mechanism which inserts sulphur into the active sites of xanthine oxidase and aldehyde oxidase, a related enzyme with some overlapping activities (such as conversion of allopurinol to oxypurinol).

Inhibition of xanthine oxidase has been proposed as a mechanism for improving cardiovascular health. A study found that patients with chronic obstructive pulmonary disease (COPD) had a decrease in oxidative stress, including glutathione oxidation and lipid peroxidation, when xanthine oxidase was inhibited using allopurinol. Oxidative stress can be caused by hydroxyl free radicals and hydrogen peroxide, both of which are by-products of XO activity.

Increased concentration of serum uric acid has been under research as an indicator for cardiovascular health factors and has been used to strongly predict mortality, heart transplant and more in patients. But it is not clear whether this could be a direct or casual association or link between serum uric acid concentration (and by proxy, xanthine oxidase activity) and cardiovascular health. States of high cell turnover and alcohol ingestion are some of the most prominent cases of high serum uric acid concentrations.

Reactive nitrogen species, such as peroxynitrite that xanthine oxidase can form, have been found to react with DNA, proteins and cells, causing cellular damage or even toxicity. Reactive nitrogen signalling, coupled with reactive oxygen species, have been found to be a central part of myocardial and vascular function, explaining, why xanthine oxidase is under research to links to cardiovascular health. Both xanthine oxidase and xanthine oxidoreductase are also present in corneal epithelium and endothelium and may be involved in oxidative eye injury. Xanthine oxidase (EC 1.1.3.22) is an iron-molybdenum flavoprotein that can oxidise xanthine, hypoxanthine, some other purines and pterins and aldehydes. It is unique among oxidases because it can produce superoxide anion O_2^- as well as H_2O_2. Superoxide dismutates two molecules of O_2^- to H_2O_2 and water.

8.11 Lipoxygenase

Lipoxygenases (EC 1.13.11) are a family of iron-containing enzymes that catalyse the dioxygenation of polyunsaturated fatty acids in lipids containing a *cis, cis*-1,4-pentadiene structure.

It catalyses the following reaction:

Fatty acid + O_2 = fatty acid hydroperoxide

Lipoxygenases are found in plants, animals and fungi. Products of lipoxygenases are involved in diverse cell functions.

8.11.1 Biological functions and classification

These enzymes are most common in plants where they may be involved in a number of diverse aspects of plant physiology including growth and development, pest resistance and senescence or responses to wounding. In mammals a number of lipoxygenases isozymes are involved in the metabolism of eicosanoids (such as prostaglandins, leukotrienes and nonclassic eicosanoids).

The plant lipoxygenases can be enzymatically cleaved into two fragments which stay tightly associated while the enzyme remains active, separation of the two domains leads to loss of catalytic activity. The C-terminal (catalytic) domain consists of 18–22 helices and one (in rabbit enzyme) or two (in soyabean enzymes) antiparallel β-sheets at the opposite end from the *N*-terminal β-barrel.

The iron atom in lipoxygenases is bound by four ligands, three of which are histidine residues. Six histidines are conserved in all lipoxygenase sequences, five of them are found clustered in a stretch of 40 amino acids. This region contains two of the three zinc-ligands, the other histidines have been shown to be important for the activity of lipoxygenases.

The two long central helices cross at the active site, both helices include internal stretches of π-helix that provide three histidine (His) ligands to the active site iron. Two cavities in the major domain of soyabean lipoxygenase-1 (cavities I and II) extend from the surface to the active site. The funnel-shaped cavity I may function as a dioxygen channel, the long narrow cavity II is presumably a substrate pocket. The more compact mammalian enzyme contains only one boot-shaped cavity (cavity II). In soyabean lipoxygenase-3 there is a third cavity which runs from the iron site to the interface of the β-barrel and catalytic domains. Cavity III, the iron site and cavity II form a continuous passage throughout the protein molecule.

Plant lipoxygenases have been known for over 50 years and have been purified and mechanistic details have been elucidated. Lipoxygenases play an important role in the stability of various processed foods. All lipoxygenases use molecular oxygen (dioxygen) to catalyse the oxidation of polyunsaturated fatty acids and initially form fatty acid peroxy free radicals that remove a hydrogen from another unsaturated fatty acid molecule to produce hydroperoxides. The hydroperoxides are converted into acids, aldehydes or ketones and other compounds. Lipoxygenases also can undergo an anaerobic reaction. Soyabean lipoxygenase-1 can use the hydroperoxide product as its second substrate. Although the mechanism of the reaction is simple, a complex pattern of products is formed: fatty acid dimers, oxodienoic acids and *n*-pentane. Lipoxygenases also show the phenomenon of self-inactivation. The instabilities of the lipoxygenases of pea seeds and wheat were confirmed in recent studies. In the kinetic study of the reaction of soyabean lipoxygenase with arachidonic acid as substrate, a self-catalysed destruction of the enzyme was observed that was involved in the interaction of the enzyme with the hydroperoxy fatty acid in only a minor way. Several types of lipoxygenases are classified by the enzyme commission.

The lipoxygenase that oxidises linoleate to 13-hydroperoxyoctadeca-9,11-dienoate has been assigned the enzyme commission number 1.13.11.12. Other lipoxygenases that have been studied are arachidonate 12-lipoxygenase (EC 1.13.11.31), arachidonate 15-lipoxygenase (EC 1.13.11.33) and arachidonate 5-lipoxygenase (EC 1.13.11.34). Many investigators have suggested that the physiological roles of lipoxygenases in plants might be lipid oxidation with respect to seed ageing and death, regulation of metabolism in germinating seeds, production of plant hormones and inhibitory effects on pathogenic organisms. These effects are not clearly understood. Defense against insect attack was studied for the lipid oxidation products of soyabean lipoxygenases. Bean bugs were repelled by products of linoleic acid oxidation such as linoleic acid monohydroperoxides and hexanal. The insects also preferred lipoxygenase null seeds more than normal soyabean seeds. The oxidation products also repelled two species of leaf beetles that usually do not feed on soyabean seeds. Although the repellant powers of the lipoxygenase products were not great, the results suggested a defensive role for lipoxygenases. More recently, lipoxygenase was implicated in carotenoid metabolism and synthesis of abscisic acid in *Lycopersicon* and *Phaseolus* seedlings.

8.12 Dehydrogenase

A dehydrogenase is an enzyme belonging to the group of Oxidoreductases that oxidises a substrate by a reduction reaction that transfers one or more hydrides (H^-) to an electron acceptor, usually NAD^+/$NADP^+$ or a flavin coenzyme such as FAD or FMN.

8.12.1 Examples of dehydrogenase

1. Aldehyde dehydrogenase.
2. Acetaldehyde dehydrogenase.
3. Alcohol dehydrogenase.
4. Glutamate dehydrogenase (an enzyme that can convert glutamate to α-Keto-glutarate and vice versa).

5. Lactate dehydrogenase.
6. Pyruvate dehydrogenase (a common enzyme that feeds the TCA Cycle in converting Pyruvate to Acetyl CoA).
7. Glucose-6-phosphate dehydrogenase (involved in the pentose phosphate pathway).
8. Glyceraldehyde-3-phosphate dehydrogenase (involved in glycolysis).
9. Sorbitol dehydrogenase.

TCA cycle examples:

1. Isocitrate dehydrogenase.
2. Alpha-ketoglutarate dehydrogenase.
3. Succinate dehydrogenase.
4. Malate dehydrogenase.

These oxidoreductases are categorised mostly as dehydrogenases, although 37 are categorised as reductases. This class of enzyme is very important to the metabolic functions of plants and animals but, in foods or food processing, only a few might be considered important. These enzymes are alcohol dehydrogenases, shikimate dehydrogenases, malate dehydrogenases, isocitrate dehydrogenases and lactate dehydrogenases.

In various plants: In peach, 18 dehydrogenase isoenzyme systems were examined. Among these were 3 isocitrate dehydrogenases (EC 1.1.1.41), 3 malate dehydrogenases (EC 1.1.1.37) and 2 shikimate dehydrogenases (EC 1.1.1.25).

A shikimate dehydrogenase from cucumber was extracted and purified 7-fold by precipitation with ammonium sulphate and elution from columns of Sephadex G-25, DEAE-cellulose and hydroxyapatite. Two activity bands were detected on polyacrylamide gel electrophoresis at the last purification step. The enzyme was inhibited competitively by protocatechuic acid. Heat inactivation at 50–55°C was biphasic. NADPH or shikimic acid protected the enzyme against heat inactivation.

An NAD-dependent alcohol dehydrogenase was isolated and purified from parsley leaves. The M_r of the enzyme was 37 kDa by SDS-polyacrylamide gel electrophoresis and 68 kDa by Sephadex G-100. The enzyme was assumed to consist of two equal-sized sub-units. The pH optimum was 8.9, K_m and V_{max} values for acetaldehyde were 0.6 mM and 279 U/mg, respectively. The nucleotide sequence of an alcohol dehydrogenase gene in octoploid strawberry was deduced.

In barley, alcohol dehydrogenase and lactate dehydrogenase were induced by anaerobiosis in both aleurone layers and roots. Under aerobic conditions, developing seeds accumulate alcohol dehydrogenase activity that survives seed drying and dehydration. Activity of lactate dehydrogenase also increases during seed germination, but the activity in dry or re-hydrated seeds is very low. Developmental expression of alcohol and lactate dehydrogenase was monitored from 0 to 24 days post-germination. Neither activity was induced to any extent in germinating seeds, however, both enzymes were highly induced by anoxia in root tissue during development.

Meat lactate dehydrogenase: Lactate dehydrogenase may be important in meat research. There is a relationship between the lactate dehydrogenase isoenzyme pattern and different types of muscle fibres. Differences exist in muscle lactate dehydrogenase activity and in isoenzymes among different species of animal.

9

Production of industrial enzymes

9.1 Introduction

Enzymes can be used as chemicals to determine the concentration of substrates, measure the catalytic activity of enzymes present in biological samples and serve as labels in immunoassays to determine the concentrations of enzymatically inert substances. For instance, enzymes are routinely used in determination of glucose (glucose oxidase, horse-radish peroxidase), urea (urease, glutamate dehydrogenase), triglycerides (lipase, carboxylesterase, glycerol kinase, etc.), in clinical diagnosis. Carbohydrates, organics acids, alcohols and other food ingredients are routinely determined in food analysis using enzymes.

Fermentation is a method of generating enzymes for industrial purposes. Fermentation involves the use of micro-organisms, like bacteria and yeast to produce the enzymes. There are two methods of fermentation used to produce enzymes. These are submerged fermentation and solid-state fermentation. Submerged fermentation involves the production of enzymes by micro-organisms in a liquid nutrient media. Solid-state fermentation is the cultivation of micro-organisms and hence enzymes on a solid substrate. Carbon containing compounds in or on the substrate are broken down by the micro-organisms, which produce the enzymes either intracellular or extracellular. The enzymes are recovered by methods such as centrifugation, for extracellularly produced enzymes and lysing of cells for intracellular enzymes.

9.2 Large scale fermentation of enzymes

Use of an aerobic submerged culture in a stirred tank reactor is the typical industrial process for enzyme production involving a micro-organism that produces an industrial enzyme. Figure 9.1 shows a flow diagram of a typical production process.

9.2.1 Organism and enzyme production

The basic mechanisms of enzyme synthesis, including transcription, translation and post-translational processing, seem to be highly conservative. However, several differences exist between various classes of organisms, as well as some fundamental differences between prokaryotic and eukaryotic organisms. The enzymes themselves differ enormously in molecular mass, number of

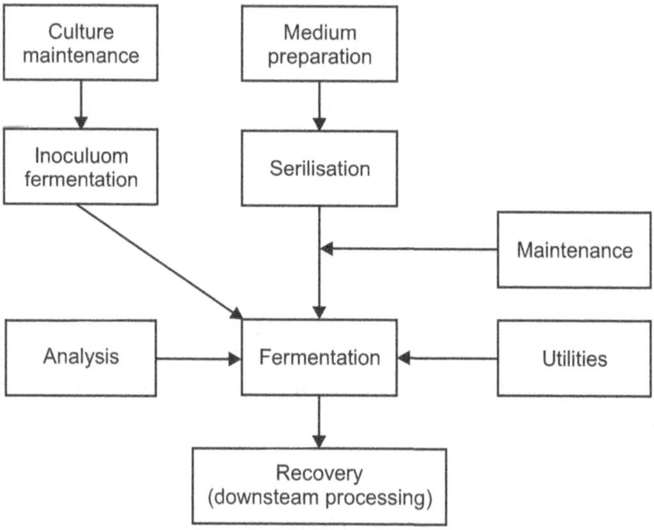

Figure 9.1: Flow diagram of a fermentation process.

polypeptide chains, isoelectric point and degree of glycosylation. In addition, a variety of enzyme-producing species exist. Although all the differences may influence the characteristics of synthetic patterns, the basic mechanisms underlying enzyme synthesis are similar enough to allow a general treatment of the microbiological production process. However, the differences in production kinetics among various species are large enough to make individual optimisation programmes necessary. Different organisms may also differ in their suitability for fermentation, such process characteristics as viscosity or recoverability, legal clearance of the organism and available knowledge about the selected organism, must be considered. As cellular enzyme expression is heavily influenced by regulatory mechanism, enzyme synthesis rates range from no synthesis to maximum synthesis allowed by the synthetic apparatus, as in a normal control loop. The complexity of mechanisms ranges from relatively simple and well-understood induction and repression systems, to very complex global regulatory networks. Process development must deal with the complexity of the enzyme synthesis system either by changing the structural characteristics, including structural elements of the regulatory systems, or by selecting optimal environmental conditions.

9.2.2 Media of enzyme production

Substrates for industrial applications are usually included in complex media, which must be supplemented with special compounds, such as a nitrogen source, various nutrient salts, or certain trace elements.

The main sources for microbial processes consist of sugars such as sucrose or hexoses. They represent the source for both carbon and energy. Typical feedstock materials are molasses, unrefined sugar, sulphite liquor from cellulose production plants, hydrolysates of wood and starch, or fruit juices, such as the grape juice used in wine making. In many cases starches from various cereals (barley, corn, rice, rye and wheat) or tubers (potatoes) are used as inexpensive raw materials. Starch-containing substrates from wheat are mostly used in raw form, i.e., in the form of grist and therefore contain a number of additional substances.

Nitrogen, phosphorus and potassium are important nutrients to maintain growth. They are added in the form of inorganic compounds such as phosphates, ammonium compounds, or potassium chloride. Soya meal, fish meal, cotton seed, low-quality protein materials such as casein or its hydrolysates, millet, stillage and especially corn steep liquor are also used as low-price nutrients. In addition, these chemically complicated mixtures contain trace elements and growth promoters. A number of trace elements are provided in sufficient quantity by the tap water that is used for the preparation of the medium.

In general, the raw materials are dissolved or suspended in water and the resulting medium is heated, filtered and sterilised. The complex composition of the media used in industry causes considerable problems. For downstream processing (harvest, concentration and purification of product) or for analytical assays during the process, additional pre-treatment of the raw material is needed to avoid unfavourable side effects.

9.2.3 Sterilisation

Most bioprocesses for the production of enzymes are based on pure cultures. Because contamination would prevent proper functioning of the process, bioprocessing must be carried out under aseptic conditions.

Solid substrates, such as grist used for amylase production or treated soil used in mushroom cultivation, are kept in rooms at elevated temperatures for an appropriate amount of time. Liquid media are sterilised *in situ*, e.g., in the reaction vessel or in separate containers, usually by external heating. In some cases, media are prepared in a concentrated form and steam is injected directly into the solution, with the resulting condensate making up the final volume.

The temperature for sterilisation is normally above 100°C for an appropriate period. Flow sterilisation through heat exchangers also applies temperatures far above 100°C, but residence times are shorter than in batch sterilisation. In general, the duration of the heat treatment does not only depend on the material to be processed but also on the pH of the medium and the initial number of viable germs or spores.

In aerobic processes the culture must be supplied with sterile air. Air is usually filtered through glass wool filters, sintered materials, or membranes of appropriate design.

9.2.4 Inoculation

After inoculation with the micro-organisms, the process should start immediately and the reaction should proceed fast. The amount of active cell culture added is therefore critical and depends on the size of the batch. Scaling-up from the original starter culture to the inoculation broth is done in several steps in large-scale industrial processes. The starter culture is kept deep frozen (–70 to –90°C) for preservation. In fungal inoculates proper wetting of the spores is achieved by adding small amounts of surfactants to the broth. If inoculation by spare suspensions is not optimal, mycelial pellets can be used for start-up. Bacterial spores must be activated by thermal treatment before they can be used for inoculation. During the exponential growth phase of the bioprocess cells can be harvested for following inoculations.

9.2.5 Fermentation

For enzyme production, economy of scale leads to the use of fermenters with a volume of 20–200 m^3. The higher energy yield from aerobic combustion results in the use of aerobic processes which require continuous transfer of poorly soluble oxygen into the culture broth.

Process design is the complicated task of choosing the optimal conditions for maximal process outcome. The number of interdependent factors is high and the available physiological knowledge is seldom complete. The relationship between synthesis rate and growth rate could be very complex and depends on the presence of inductors or the absence of repressors. The use of genetically developed strains may considerably ease process design. The designed process is usually first tested on a pilot-plant scale and optimised in a number of fermentation runs, it is then scaled up to production size. The total synthesis rate depends not only on growth rate but also on biomass concentration.

Two types of cultivation can be distinguished: In batch culture, the growth rate cannot be controlled by dosed feeding, because all substrates are added at the beginning. Batch processes are rarely used today.

In fed-batch processes, a low initial biomass concentration should be chosen to maintain the desired growth rate for a certain period, without exceeding the transport capacity of the equipment. In addition, fed-batch processes can be designed in a way that the enzyme concentration at harvesting is higher than in batch or continuous cultures and the productivity of fed-batch processes can be increased several-fold compared to batch processes. Fed-batch processes probably constitute the most frequently used process type.

Continuous cultures are ideally suited for high productivity because the excess of biomass is continuously withdrawn and both synthesis rate and biomass concentration will be optimal. Therefore, in principle, continuous culture is preferred for biotechnological production. However, enzyme concentrations are lower than those reached in fed-batch cultures. In addition, the use of continuous cultures is limited by technical reasons such as the higher contamination risk and the problem of strain degeneration.

9.3 Downstream processing

Downstream processing is a very important step in biotechnology because costs for collection, concentration and purification of the final product are substantial. High product concentrations in the supernatant or inside the cells and efficient purification are therefore important aspects in the overall economy of enzyme manufacture.

The degree of purity of commercial enzymes ranges from raw enzymes to highly purified forms and depends on the application. Raw materials for the isolation of enzymes are animal organs, plant material and micro-organisms.

Often enzymes may be purified several hundred-fold but the yield of the enzyme may be very poor, frequently below 10% of the activity of the original material. In contrast, industrial enzymes will be purified as little as possible, only other enzymes and material likely to interfere with the process which the enzyme is to catalyse, will be removed. Unnecessary purification will be avoided as each additional stage is costly in terms of equipment, manpower and loss of enzyme activity. As a result, some commercial enzyme preparations consist essentially of concentrated fermentation broth, plus additives to stabilise the enzyme's activity.

However, the content of the required enzyme should be as high as possible (e.g., 10% w/w of the protein) in order to ease the downstream processing task. This may be achieved by developing the fermentation conditions or, often more dramatically, by genetic engineering. It may well be economically viable, e.g., to spend some time cloning extra copies of the required gene together with a powerful promoter back into the producing organism in order to get over-producers. Downstream processing involves isolation and purification steps and ends up in the formulation of the enzyme preparation.

Enzymes are universally present in living organisms, each cell synthesises a large number of different enzymes to maintain its metabolic reactions. The choice of procedures for enzyme purification depends on their location. On the one hand, isolation of intracellular enzymes often involves the separation of complex biological mixtures. On the other hand, extracellular enzymes are generally released into the medium with only a few other components. Enzymes are very complex proteins and their high degree of specificity as catalysts is manifest only in their native state. The native conformation is attained under specific conditions of pH, temperature and ionic strength. Hence, only mild and specific methods can be used for enzyme isolation. Figure 9.2 shows the sequence of steps involved in the recovery of enzymes.

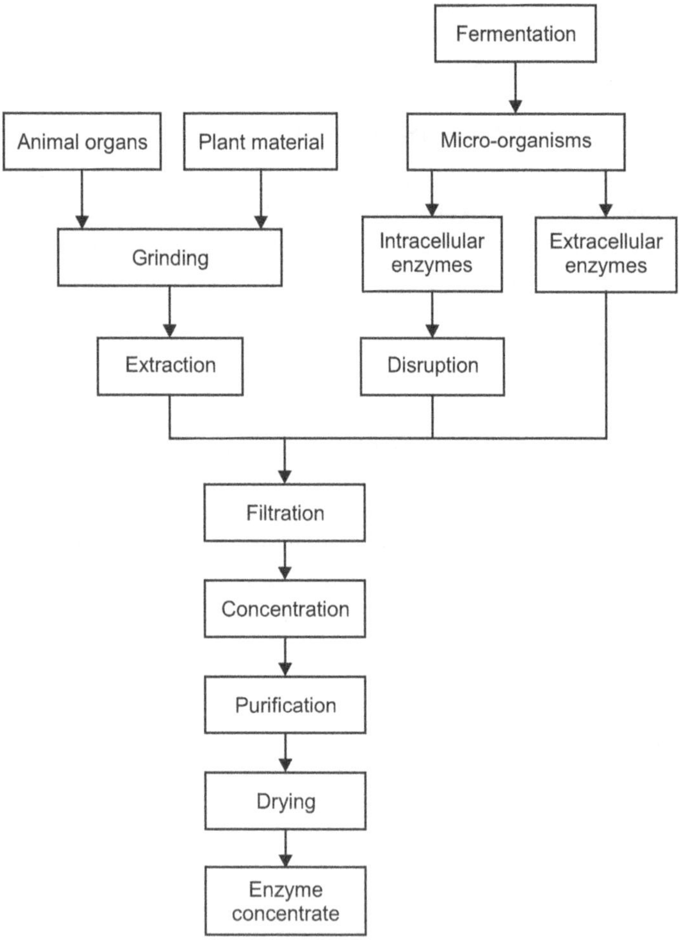

Figure 9.2: Flowchart of the downstream processing of enzymes.

9.3.1 Preparation of biological starting materials

Animal organs: Animal organs must be transported and stored at low temperature to retain enzymatic activity. Frozen organs can be minced with machines generally used in the meat industry and the enzymes can be extracted with a buffer solution. Besides mechanical grinding, enzymatic digestion can also be employed.

Plant material: Plant material can be ground with various crushers or grinders and the desired enzymes can be extracted with buffer solutions. The cells can also be disrupted by previous treatment with lytic enzymes.

Micro-organisms: Enzymes might accumulate inside the cells or be released into the medium. Most enzymes used commercially are extracellular enzymes and the first step in their isolation is the separation of the cells from the solution. For intracellular enzymes, which are being isolated today in increasing amounts, the first step involves grinding to rupture the cells. A number of methods for the disruption of cells are known, corresponding to the different types of cells and the problems involved in isolating intracellular enzymes. However, only a few of these methods are used on an industrial scale.

Cell disruption by mechanical methods

High-pressure homogenisation is the most common method of cell disruption. The cell suspension is pressed through a valve and hits an impact ring. The cells are ruptured by shearing forces and simultaneous decompression.

The wet grinding of cells in a high-speed bead mill is another effective method of cell disruption. Glass balls with a diameter of 0.2–1 mm are used to break the cells. The efficiency of this method depends on the geometry of the stirrer system.

Cell disruption by non-mechanical methods

Cells may frequently be disrupted by chemical, thermal, or enzymatic lysis. The drying of micro-organisms and the preparation of acetone powders are standard procedures in which the structure of the cell wall is altered to permit subsequent extraction of the cell contents.

9.3.2 Separation of solid matter

After cell disruption, the next step is separation of extracellular or intracellular enzymes from cells or cellular fragments, respectively. This operation is rather difficult because of the small size of bacterial cells and the slight difference between the density of the cells and that of the fermentation medium. Continuous filtration is used in industry. Large cells, e.g., yeast cells, can be removed by decantation. Today, efficient centrifuges have been developed to separate cells and cellular fragments in a continuous process. Residual plant and organ matter can be separated with simpler centrifuges or filters.

Filtration

Pressure filters: A filter press (plate filter, chamber filter) is used to filtrate small volumes or to remove precipitates formed during purification. The capacity to retain solid matter is limited and the method is rather work-intensive.

However, these filters are highly suitable for the fine filtration of enzyme solutions.

Vacuum filters: Vacuum filtration is generally the method of choice because biological materials are easily compressible. A rotary vacuum filter is used in the continuous filtration of large volumes. The suspension is usually mixed with a filter aid (e.g., kieselguhr) before being applied to the filter.

Cross-flow filtration: In recent years, a new method of filtration, cross-flow filtration, has been devised. In conventional methods, the suspension flows perpendicular to the filtering material. In cross-flow filtration, the input stream flows parallel to the filter area, thus preventing the accumulation of filter cake and an increased resistance to filtration. Cross-flow filtration can be conveniently used in recombinant DNA techniques to separate organisms in a closed system.

Centrifugation

The sedimentation rate of a bacterial cell with a diameter of 0.5 mm is less than 1 mm/h. An economical separation can be achieved only by sedimentation in a centrifugal field. The range of applications of centrifuges depends on the particle size and the content of the solids.

Decanters (scroll-type centrifuges) work with low centrifugal forces and are used in the separation of large cells or protein precipitates. Solid matter is discharged continuously by a screw conveyer moving at a differential rotational speed. Tubular bowl centrifuges are built for very high centrifugal forces and can be used to sediment very small particles. However, these centrifuges cannot be operated in a continuous process. Moreover, solid matter must be removed by hand after the centrifuge has come to a stop. A further disadvantage is the appearance of aerosols. Separators (disk stack centrifuges) can be used in the continuous removal of solid matter from suspensions. Solids are discharged by a hydraulically operated discharge port (intermittent discharge) or by an arrangement of nozzles (continuous discharge). Bacteria and cellular fragments can be separated by a combination of high centrifugal forces. Disk stack centrifuges that can be sterilised with steam are used for recombinant DNA techniques in a closed system.

Extraction

An elegant method used to isolate intracellular enzymes is liquid-liquid extraction in an aqueous two-phase system. This method is based on the incomplete mixing of different polymers, e.g., dextran and poly(ethylene glycol), or a polymer and a salt in an aqueous solution. The first extraction step separates cellular fragments. Subsequent purification can be accomplished by extraction or, if high purity is required, by other methods. The extractability can be improved by using affinity ligands or modified chromatography gels, e.g., phenyl-sepharose.

Flocculation and flotation

The flocculation of bacterial cells to form larger particles can be achieved by the addition of mineral colloids, salts, or organic polymers. The neutralisation of charges on the cell surface and the formation of bridges between individual cells lead to agglomeration. Agglomerates can then be removed by filtration or centrifugation.

9.3.3 Concentration

The enzyme concentration in starting material is often very low. The volume of material to be processed is generally very large and substantial amounts of waste material must be removed. Thus, if economic purification is to be achieved, the volume of starting material must be decreased by concentration. Only mild concentration procedures which do not inactivate enzymes can be employed. These procedures include thermal methods, precipitation and to an increasing extent, membrane filtration.

Thermal methods

Only brief heat treatment can be used for concentration because enzymes are thermolabile. Evaporators with rotating components which achieve a thin liquid film (thin-layer evaporator, centrifugal thin-layer evaporator) or circulation evaporators (long-tube evaporator) can be employed.

Precipitation

Enzymes are very complex protein molecules possessing both ionisable and hydrophobic groups which interact with the solvent. Indeed, proteins can be made to agglomerate and, finally, precipitate by changing their environment. Precipitation is actually a simple procedure for concentrating enzymes.

Precipitation with salts: High salt concentrations act on the water molecules surrounding the protein and change the electrostatic forces responsible for solubility. Ammonium sulphate is commonly used for precipitation, hence, it is an effective agent for concentrating enzymes. Enzymes can also be fractionated, to a limited extent, by using different concentrations of ammonium sulphate. Sodium sulphate is another precipitating agent used.

Precipitation with organic solvents: Organic solvents influence the solubility of enzymes by reducing the dielectric constant of the medium. The solvation effect of water molecules surrounding the enzyme is changed, the interaction of protein molecules is increased and therefore, agglomeration and precipitation occur. Commonly used solvents are ethanol and acetone.

Precipitation with polymers: The polymers generally used are polyethylenimines and polyethylene glycols of different molecular masses. The mechanism of this precipitation is similar to that of organic solvents and results from a change in the solvation effect of the water molecules surrounding the enzyme. Most enzymes precipitate at polymer concentrations ranging from 15 to 20%.

Precipitation at the isoelectric point: Proteins are ampholytes and carry both acidic and basic groups. The solubility of proteins is markedly influenced by pH and is minimal at the isoelectric point at which the net charge is zero. Because most proteins have isoelectric points in the acidic range, this process is also called acid precipitation.

Ultrafiltration

A semipermeable membrane permits the separation of solvent molecules from larger enzyme molecules, because only the smaller molecules can penetrate the membrane when the osmotic pressure is exceeded. This is the principle of all membrane separation processes including ultrafiltration. In reverse osmosis, used to separate materials with low molecular mass, solubility and diffusion phenomena influence the process, whereas ultrafiltration and cross-flow filtration are based solely on the sieve effect. In processing enzymes, cross-flow filtration is used to harvest cells, whereas ultrafiltration is employed for concentrating and desalting.

Desalting: The desalting of enzyme solutions can be carried out conveniently by diafiltration. The small salt molecules are driven through a membrane with the water molecules. The permeate is continuously replaced by fresh water.

9.3.4 Purification

For many industrial applications, partially purified enzyme preparations will suffice, however, enzymes for analytical purposes and for medical use must be highly purified. Special procedures employed for enzyme purification are crystallisation, electrophoresis and chromatography. However crystallisation and electrophoresis are not relevant for large scale purifications. Chromatography, in contrast, is of fundamental importance to enzyme purification. Molecules are separated according to their physical properties (size, shape, charge, hydrophobic interactions), chemical properties (covalent binding), or biological properties (biospecific affinity).

In gel chromatography (also called gel filtration), hydrophilic, cross-linked gels with pores of finite size are used in columns to separate biomolecules. In gel filtration, molecules are separated according to size and shape. Molecules larger than the largest pores in the gel beads, i.e., above the exclusion limit, cannot enter the gel and are eluted first. Smaller molecules, which enter the gel beads to varying extent depending on their size and shape, are retarded in their passage through the column and eluted in order of decreasing molecular mass. Gel filtration is used commercially for both separation and desalting of enzyme solutions.

Ion-exchange chromatography is a separation technique based on the charge of protein molecules. Enzyme molecules possess positive and negative charges. The net charge is influenced by pH and this property is used to separate proteins by chromatography on anion exchangers (positively charged) or cation exchangers (negatively charged). The ability to process large volumes and the elution of dilute sample components in concentrated form make ion exchange very useful.

For hydrophobic chromatography, media derived from the reaction of CNBr-activated Sepharose with aminoalkanes of varying chain length are suitable. This method is based on the interaction of hydrophobic areas of protein molecules with hydrophobic groups on the matrix. Adsorption occurs at high salt concentrations and fractionation of bound substances is achieved by eluting with a negative salt gradient. This method is ideally suited for further purification of enzymes after concentration by precipitation with such salts as ammonium sulphate. In affinity chromatography the enzyme to be purified is specifically and reversibly adsorbed on an effector attached to an insoluble support matrix. Suitable effectors are substrate analogues, enzyme inhibitors, dyes, metal chelates, or antibodies. The insoluble matrix is contained in a column. The biospecific effector, e.g., an enzyme inhibitor, is attached to the matrix. A mixture of different enzymes is applied to the column. The immobilised effector specifically binds the complementary enzyme. Unbound substances are washed out and the enzyme of interest is recovered by changing the experimental conditions, for example by altering pH or ionic strength.

Immunoaffinity chromatography occupies an unique place in purification technology. In this procedure, monoclonal antibodies are used as effectors. Hence, the isolation of a specific substance from a complex biological mixture in one step is possible. In this procedure, enzymes can be purified by immobilising antibodies specific for the desired enzyme. In this way, enzymes that usually do not bind to an antibody can be purified by immunoaffinity chromatography. Covalent chromatography differs from other types of chromatography by forming a covalent bond in between the required protein and the stationary phases.

9.4 Formulation of the final enzyme product

Once the enzyme has been purified to the desired extent and concentrated, the manufacturer's main objective is to retain the activity. Enzymes for industrial use are sold on the basis of overall activity. Often a freshly supplied enzyme sample will have a higher activity than that stated by the manufacturer. This is done to ensure that the enzyme preparation has the guaranteed storage life. The manufacturer will usually recommend storage conditions and quote the expected rate of loss of activity under those conditions. It is of primary importance to the enzyme producer and customer that the enzymes retain their activity during storage and use. Some enzymes retain their activity under operational conditions for weeks or even months. However, most enzymes do not. Most industrial enzyme preparations contain a relatively little amount of active enzyme, the rest being due to inactive protein, stabilisers, preservatives, salts and the diluent which allows standardisation between production batches of different specific activities.

The key to maintaining enzyme activity is maintenance of conformation, so preventing unfolding aggregation of the enzyme molecules and changes in the covalent structure. Three approaches are possible:

1. Use of additives.
2. The controlled use of covalent modification.
3. Enzyme immobilisation.

In general, proteins are stabilised by increasing their concentration and the ionic strength of their environment. Neutral salts compete with proteins for water and bind to charged groups or dipoles. This may result in the interactions between an enzyme's hydrophobic areas being strengthened causing the enzyme molecules to compress and making them more resistant to thermal unfolding reactions. Not all salts are equally effective in stabilising hydrophobic interactions, some are much more effective at their destabilisation by binding to them and disrupting the localised structure of water (the chaotropic effect). From this it can be seen why ammonium sulphate and potassium hydrogen phosphate are powerful enzyme stabilisers whereas sodium thiosulphate and calcium chloride destabilise enzymes. Many enzymes are specifically stabilised by low concentrations of cations which may or may not form part of the active site, for example Ca^{2+} stabilises α-amylases and Co^{2+} stabilises glucose isomerases. At high concentrations (e.g., 20% NaCl) salt discourages microbial growth due to its osmotic effect. In addition, ions can offer some protection against oxidation to groups such as thiols by salting-out the dissolved oxygen from solution.

Low molecular weight polyols (e.g., glycerol, sorbitol and mannitol) are also useful for stabilising enzymes, by repressing microbial growth due to the reduction in the water activity and by the formation of protective shells which prevent unfolding processes. Glycerol may be used to protect enzymes against denaturation due to ice-crystal formation at sub-zero temperatures. Some hydrophilic polymers (e.g., polyvinyl alcohol, polyvinylpyrrolidone and hydroxypropylcelluloses) stabilise enzymes by a process of compartmentalisation, whereby the enzyme-enzyme and the enzyme-water interactions are somewhat replaced by less potentially denaturing enzyme-polymer interactions. They may also act by stabilising the hydrophobic effect within the enzymes.

Many specific chemical modifications of amino acid side chains are possible which may (or, more commonly, may not) result in stabilisation. An useful example of this is the derivatisation of lysine side chains in proteases with *N*-carboxyamino acid anhydrides.

These derivatissation result in polyaminoacylated enzymes with various degrees of substitution and length of amide-linked side chains. This derivatisation is sufficient to disguise the proteinaceous nature of the protease and prevent autolysis. Enzymes are much more stable in the dry state than in solution. Solid enzyme preparations sometimes consist of freeze-dried protein. More usually they are bulked out with inert materials such as starch, lactose, carboxymethylcellulose and other poly-electrolytes which protect the enzyme during a cheaper spray-drying stage.

Other materials which are added to enzymes before sale may consist of substrates, thiols to create a reducing environment, antibiotics, benzoic acid esters as preservatives for liquid enzyme preparations, inhibitors of contaminating enzyme activities and chelating agents. Additives of these types must, of course, be compatible with the final use of the enzymes product.

In order to ensure safe handling, stability, suitable mixing, functionality, etc., in the various applications, most enzyme preparations are formulated in a variety of liquid and granular forms. Some enzyme preparations are immobilised. Often the precise details of the methods used to stabilise enzyme preparations are kept secret or revealed to customers as a confidential information only.

9.4.1 Nature of enzyme products

The nature of enzyme products is influenced by: (i) the enzyme itself and its properties as active compound, (ii) the enzyme source, the fermentation media and conditions and the purification steps (resulting in the enzyme concentrate), (iii) the additives used to formulate the final enzyme preparation. Thus, one has to differentiate between enzyme, enzyme concentrate and enzyme preparation. Table 9.1 describes the terminology used in enzyme manufacturing. Enzyme preparations vary considerably in terms of purity and formulation (additives) among companies and depending on the particular application. Most enzyme applications do not require high enzyme concentration of the isolate or the preparation. A look on the range of main components of industrial enzyme preparations (Table 9.2) reveals, that the contents of active enzyme protein in the final enzyme preparation is usually very low, typically in the order of 1–5%.

Table 9.1: Terminology used in enzyme manufacturing.

Nomenclature	*Description*
Enzyme	A pure protein with a specified activity, on basis of which the enzyme is identified. The activity has an IUB and a CAS number. Enzyme may also be referred to as active enzyme or active enzyme protein.
Enzyme concentrate	An enzyme-containing mixture in its most concentrated form as it occurs during commercial production, thus before formulation. Enzyme concentrates are not commercially available as such, but some may be used directly by the producer as a processing aid for food or to synthesise certain chemicals (i.e., captive use). Enzyme concentrate may also be referred to as enzyme 'isolate'. The enzyme concentrate is measured as total organic substance (TOS, TOS = 100% - (Water + Ash + Diluents).
Enzyme formulation	An enzyme concentrate, diluted to a standardised activity, stabilised and commercially available on the market.
Enzyme blend	A formulation of an intended mixture of enzyme concentrates originating from different sources.
Enzyme preparation	Enzyme formulation or enzyme blend.
Declared enzyme	The enzyme component(s) in the preparation of significance in terms of application and regulation.

Table 9.2: Typical range of main components of industrial enzymes in enzyme preparations.

Component	*Range of content (% dry solids)*
Proteins and amino acids	10–15
Active enzyme protein	1–5
Complex carbohydrates	5–12
Sugars (and sugar alcohols)	2–20
Inorganic salts	3–40
Preservatives	0–0.3

Sugars and inorganic salts are used for stabilising the finished product for storage and distribution. Salts and sometimes carbohydrates such as starch, maltodextrins and sugar alcohols are used to dilute extracted enzymes (the enzyme concentrate) to a standard activity. Preservatives are generally restricted to use in liquid preparations. Thus, by-products from the fermentation process (other proteins, carbohydrates, sugars, salts), additives and preservatives comprise up to 99% of the enzyme preparation. Manufacturing costs are not the only issue, in some cases limited purity and concentration are desirable in terms of formulation of the preparation, improving stability/performance, facilitating homogeneous blending, etc. The increasing use of genetically modified micro-organisms (GMM) and new recovery techniques have generally increased purity and amount of enzyme present in the enzyme concentrate.

9.4.2 Enzyme concentrate

Impurities in the enzyme isolate/concentrate result from the fermentation broth and subsequent purification and consist of proteins, peptides and amino acids, carbohydrates, minerals and other minor components. The relative amounts of these components vary considerably within and between categories of enzyme concentrates. Enzyme content and purity is similar for technical, food and feed enzymes. Enzymes used in personal care products, for therapeutic and analytical or diagnostic application may generally be of higher purity and concentration. Ash constituents comprise small amounts of minerals and diluents are additives for granulation, liquid formulation, stabilisation, preservation, etc. Table 9.3 and Table 9.4 shows the approximate ranges of impurities as stated by Association of Manufacturers and Formulators of Food Enzyme Preparations (AMFEP).

Table 9.3: Approximate ranges of impurities for technical, food and feed enzymes.

Fraction	*Range*
Proteins, peptides and amino acids	15–50%
Carbohydrates (excluding enzyme glycosylation)	10–20%
Minerals and other minor components	0–15%

Table 9.4: Approximate ranges of impurities of enzymes for personal care, therapeutic and analytical/diagnostic application

Fraction	*Range*
Proteins, peptides and amino acids	0–40%
Carbohydrates, minerals, other	0–10%

Enzymes used for food and feed have to comply with purity specifications comprising limits for heavy metals and contaminating micro-organisms and absence of mycotoxins, antibiotics and the production strain.

The type and range of impurities, which are present in the enzyme concentrate, is highly varying and depend on the production strain, the media used, the fermentation conditions and the subsequent purification steps as outlined in Section 9.3. Some of the impurities may also fulfil a technical function during enzyme application. This is particularly true for side activities.

An enzyme concentrate will typically contain other enzymes usually referred to as side activities (not to be confused with side activities of a particular enzyme protein). The type and range of these side activities are largely depending on the enzyme manufacturing conditions and the production strain.

These side activities may be more or less important in any given application. An example of this is the use of amylases in baking. Fungal alpha-amylase is added to dough as part of a flour improver. The objective is to hydrolyse starch to provide more fermentable sugars for the yeast. Different enzymes, however, result in different effects on the rheology of the dough and the final product quality. The reason for this is that the enzyme preparation contains a side activity of endo-beta-xylanase which cleaves the hemicellulose in the flour. This enzyme is now manufactured solely for this application.

Side activities can partly be inactivated during the enzyme manufacturing process. Nevertheless, it is quite normal to find a variety of different activities in an industrial enzyme preparation. If a preparation is marketed as a protease, e.g., glycosidases and lipases might be present as well.

9.4.3 Factors affecting the purity of enzyme concentrates

The purity of the enzyme concentrate is largely influenced by: (i) the fermentation and purification processing applied, (ii) the production organism used and (iii) the media used for fermentation.

Production organisms

The particular production strain is obviously affecting the nature as well as the range of by-products present in the enzyme concentrate. As micro-organisms are known to produce toxins, the production organisms may themselves be sources of hazardous materials and have therefore been a chief focus of attention by the regulatory authorities. Some of these toxins are quite well known. However, there is always the possibility of introducing new toxins. Therefore, production strains which are investigated not to produce toxins and which do have a long history of safe use are preferred.

Media for enzyme production

Media used have a bearing on the cost of the enzyme and media components often find their way into commercial enzyme preparations. Details of components used in industrial scale fermentation broths for enzyme production are not readily obtained. Not surprisingly, as manufacturers do not wish to reveal information that may be of technical or commercial value to their competitors. Also some components of media may be changed from batch to batch as availability and cost of, e.g., carbohydrate feedstock change. Such changes reveal themselves in often quite profound differences in appearance from batch to batch of a single enzyme from a single producer. The effects of changing feedstock must be considered in relation to downstream processing. If such variability is likely to reduce the efficiency of the standard methodology significantly, it might be economical to use a more expensive defined medium of easily reproducible composition.

Clearly defined media are usually out of question for large scale use on cost grounds but may be perfectly acceptable when enzymes are to be produced for high value uses, such as analysis or medical therapy where very pure preparations are essential. Less-defined complex media are composed of ingredients selected on the basis of cost and availability as well as composition. Waste materials and by-products from the food and agricultural industries are often major ingredients. Thus, molasses, corn steep liquor, distillers solubles and wheat bran are important components of fermentation media providing carbohydrate, minerals, nitrogen and some vitamins. Extra carbohydrate is usually supplied as starch, sometimes refined but often simply as ground cereal grains. Soyabean meal and ammonium salts are frequently used sources of additional nitrogen. Most of these materials will vary in quality and composition from batch to batch causing changes in enzyme productivity.

9.4.4 Enzyme preparations

Enzyme preparations contain a varying amount of additives for the purpose of stabilising the enzyme activity, preservation, granulation, coating or as (de)colouring aids. The choice of formulation ingredients (additives) is not specific for a given application category, but certain substances used for, e.g., technical enzymes may not comply with food, feed and cosmetics regulations, specifications, etc. Some applications may require special substances and technologies, e.g., in the case of immobilised enzymes or enteric coating of digestive aid preparations.

Formulation additives used in enzyme preparations are shown in Table 9.5.

Table 9.5: Formulation additives used in enzyme preparations.

Type of formulation	*Type of additive (typical examples)*	*Purpose*	*Range (%)*
Liquid/slurry	Sucrose, sorbitol, glycerol, propylene glycol	Stabiliser	20–50%
Liquid/slurry	Ethoxylated fatty alcohols (technical products only)	Stabiliser	20–40%
Liquid/slurry	Sodium chloride	Stabiliser	10–20%
Liquid/slurry	Sodium benzoate, potassium benzoate, BIT (isothiazoline derivative, technical products only)	Preservative	0.1–0.5%

(Cont'd...)

Type of formulation	*Type of additive (typical examples)*	*Purpose*	*Range (%)*
Granular	Inorganic salts, e.g., sodium chloride, sulphate, calcium carbonate	Stabilising, granulation	50–75%
Granular	Cellulose, dextrin, sucrose	Stabilising, granulation, coating	10–75%
Granular	Kaolin, titanium dioxide (Technical products only)	(De)colouring	0–20%
Granular	Polyethylene glycol (all PEG variants, Technical products only)	Stabilising, granulation, coating	0–10%
Granular	Vegetable oil (e.g., hydrogenated palm oil)	Stabilising, coating	5–15%

9.5 Recent developments in enzyme manufacturing

The exploitation of new types of enzymes, improvements of enzyme properties and of the production process are overall goals of innovation in the enzyme manufacturing industry. Although considerable improvements had been made in process engineering, large scale fermentation and downstream purification of enzymes remained to be a significant bottleneck in innovation. Until the mid 1980s, screening for interesting enzymes was mainly confined to groups of organisms which promised a realistic chance of developing a cost efficient production process. Micro-organisms which are adapted to extreme environments and therefore, are difficult to be accessed and cultured, but which are producing enzymes with promising properties, could not be exploited. Moreover, strain improvement in general as well as of enzymes properties in particular were largely restricted to statistical methods such as induced mutagenesis. These methods are both time-consuming and undirected. Only when genetic engineering techniques was introduced into routine research and development of enzyme manufacturing companies a promising tool became available to circumvent these problems. The introduction of genetic engineering is most probably the most important breakthrough in enzyme manufacturing during the last 30 years. Much progress have been made since the first enzyme from genetically modified micro-organisms (GMM) was introduced onto the market in 1987.

This section will therefore focus on the state of the art in genetic engineering techniques applied in the improvement of both production strains and enzyme properties. Furthermore, some related technologies such as chemical modification of enzymes will be briefly described. Finally, potential impacts for enzyme regulation of the widespread application of this technology in enzyme manufacturing process will be discussed.

9.5.1 Genetic engineering

From the perspective of an enzyme-producing company, genetic engineering serves as a core technology which is offering the following advantages:

1. The exploitation of new types of enzymes and new source organisms, even enzymes from organisms which are difficult to handle or non-culturable (e.g., extremophiles).
2. Drastic shortening of development times from screening to marketing.
3. Significant cost reductions in the development and production process of enzymes.
4. Genetic engineering is a prerequisite for the optimisation of enzyme molecular properties by protein engineering.
5. Improved product safety and fewer production risks, due to the production of enzymes from a wide variety of different source organisms in a small number of well-characterised enzyme production organisms.

Table 9.6 gives an overview how these advantages are achieved by genetic engineering techniques. Important techniques and most recent trends are briefly described in the following section.

Table 9.6: Contribution of genetic engineering techniques to the goals of innovation in enzyme manufacturing

Goals of innovation	*Technical approaches due to genetic engineering*
Reduction of manufacturing costs	Increase of enzyme yield by *increasing enzyme expression* in the production organism. No need for a *de novo* design of a production process for a new production organisms. Instead, the enzyme gene of interest is cloned into a well known production strain.
New enzymes	Increase of accessibility of new enzymes especially from extremophiles. Isolation of the respective genes and expression in known production strains.
Improved enzyme properties	Rational protein engineering/directed molecular evolution.
Improved product safety	Use of well-characterised production strains instead of new less characterised strains and sometimes strains that might be less safe.

Increasing enzyme expression

Before the era of genetic engineering, an increase in enzyme yield could only be attempted by traditional strain selection from natural sources or by random mutagenesis induced by mutagenic agents. Using genetic engineering techniques enzyme yield could be further increased dramatically. This increase is made technically feasible either by multiplying the enzyme gene or by constructing an *artificial expression system*. Both approaches aim at increasing the transcription/translation of the enzyme gene(s) into proteins on the cellular level.

Multiplying enzymes genes could be done just by amplifying the number of copies of the enzyme gene in the source organisms thereby not necessarily ending up in a GMM as defined in by the regulator. Alternatively, the gene could be isolated from the source organism, cloned onto a plasmid which is then introduced in a production strain. Whereas, normally only one set of genomic genes (two sets in the case of diploid organisms) is present in a micro-organism, up to 1000 and more copies of plasmids and consequently of plasmid related enzyme genes could be present in one cell.

The continuous improvement of artificial expression systems is sparked by the fast growing knowledge on regulatory DNA sequences. Strong promoters, enhancing sequences and efficient terminators often derived from different organisms are used in order to enhance the transcription rate of the enzyme gene. Frequently, ribosome binding sites are also optimised in order to increase the translation rate. However, this approach could lead to changes in the amino acid sequence of the enzyme protein. Applying these techniques enzyme yield could be enormously increased, e.g., heterologous expression of a xylanase in *E. coli* lead to an accumulation of the enzymes up to 70% of total cell protein. An increase in enzyme yield of 3 to 50 was reported in several studies. In case of a yeast-glucosidase a factor of 300 was achieved. Table 9.7 shows an overview on enzymes from extremophiles and their industrial importance.

Table 9.7: Industrial applications of enzymes isolated from extremophiles.

Extremophile	*Habitat*	*Enzymes*	*Representative applications*
Thermophile	High temperature	Amylases	Glucose, fructose for sweeteners
	Moderate thermophiles (45–65°C)	Xylanases	Paper bleaching
	Thermophiles (65–85°C)	Proteases	Baking, brewing, detergents
	Hyperthermophiles (<85°C)	DNA polymerases	Genetic engineering
Psychrophile	Low temperature	Proteases	Cheese maturation, dairy production
		Dehydrogenases	Biosensors
		Amylases	Polymer degradation in detergents
Acidophile	Low pH	Sulphur oxidation	Desulphurisation of coal
		Chalcopyrite concentrate	Valuable metals recovery

(Cont'd...)

Extremophile	*Habitat*	*Enzymes*	*Representative applications*
Alkalophile	High pH	Cellulases	Polymer degradation in detergents
Halophile	High salt concentration		Ion exchange resin regenerant disposal, producing poly (γ-glutamic acid) (PGA) and poly(β-hydroxy butyric acid) (PHB)
Piezophile	High pressure	Whole micro-organism	Formation of gels and starch granules
Metalophile	High metal concentration	Whole micro-organism	Ore-bioleaching, bioremediation, biomineralisation
Radiophile	High radiation levels	Whole micro-organism	Bioremediation of radionuclide contaminated sites
Microaerophile	Growth in $<21\ \%\ O_2$		

Thermophilic enzymes

These enzymes are well adjusted to temperature above 50°C. As perhaps an outstanding example amylase from *Pyrococcus furiosus* is one of the most stable enzymes presently known with significant half-life at 130°C. Whether this already represents the upper limits of thermal stability, remains however a question to be solved. Physical limits seem to be reached around 250°C. At this temperature, peptide bond hydrolysis reactions rapidly occur in (denatured) proteins. For comparison, the temperature optima of normal mesophilic enzymes are in the range of 30° to 40°C with denaturation starting around 50 to 60°C. Interestingly, it could be shown that resistance to high temperature is often accompanied with resistance to proteases, chaotropic agents, low pH, oxidation and high salt concentration. The high stability of enzymes from extreme thermophiles generally resides in the amino acid sequence and three-dimensional structure. This is shown by stability which is retained upon purification and when genes for stable enzymes are expressed in mesophiles. A few proteins might also require post-translational modification to become fully thermostable. As a consequence, these enzymes are highly attractive for biotechnological applications, such as starch processing, leaching of low grade ores, sugar conversion, detergents, or as a tool for molecular biology (PCR)—as these processes are carried out at elevated temperatures.

Psychrophilic bacteria

These group of organisms optimally grow at or below 15°C, having an upper limit of growth of about 20°C and a lower limit of growth of 0°C or below.

Psychrophilic enzymes have a high specific activity at low and moderate temperatures and are inactivated easily by a moderate increase in temperature. Typically, the specific activity of these cold adopted enzymes is higher than that of their mesophilic counterparts at temperatures of approximately 0–30°C. This increased activity is accompanied by lower thermal stability.

These properties could be of interest for various applications. According to Gerday cold-adapted cellulase could be used in biopolishing and stone-washing of cotton garments. In fabric production, tissues often have cotton fibre ends protruding from the main fibres which reduce smoothness and alter the appearance of the garment. Treatment with cellulases could excise protruding ends. The current treatment, however, is accompanied by a loss of mechanical resistance. A cold-adapted enzyme would enable the decrease of processing temperature and rapid inactivation as a result of thermal liability would be possible. The mechanical resistance of the final product would also be improved as a result of rapid inactivation of the enzyme.

In the beverage industry, pectinases are added in the juice extraction process in order to reduce viscosity and to clarify the final product. In the meat industry, proteases are applied to tenderise meat and in baking processes amylases, proteases and xylanases are used to reduce the dough fermentation time. The use of psychrophilic enzymes in all these processes could be advantageous not only because of their high specific activity, thereby reducing the amount of enzyme needed, but also for their easy inactivation. Psychrophilic enzymes could therefore become interesting alternatives to mesophilic enzymes in brewing and wine industries, cheese manufacturing, animal feed and other applications.

Improving enzyme properties – protein engineering

Basically three approaches exist for generating enzymes with improved properties:

1. First, the enzyme properties can be modulated by an exchange of a single or a few amino acid residues. This is mainly achieved by *oligonucleotide-directed mutagenesis*.
2. Second, by exchanging functional domains between related enzymes, which lead to hybrid enzymes. This could technically be achieved by DNA shuffling.
3. Third, the active site of an enzyme can be introduced into a small protein fragment scaffold of the enzyme that is devoid of its original active site.

These modifications could be made on a rational basis directed to particular changes which impacts on the enzyme properties (rational protein design). Alternatively, modifications are made randomly, however, restricted to the gene of interest and not affecting the whole genome as described above (directed evolution, molecular evolution of enzymes).

As a prerequisite of rational protein design some hypothesis on the functional effects of well defined alterations is needed—usually affecting one or a few amino acids. The latter approach takes into account the limited knowledge of structure—function relations in enzymes. Virtually, no information on how enzyme structure relates to function is needed. This strategy is even more useful as the problems and demand from enzyme application are usually complex and multifactorial. According to industry, the directed evolution approach might therefore be the more promising one.

Major goals for improving enzyme properties are:

1. Increase in catalytic activity (under conditions applied in practice).
2. Enzyme stability.

Improved substrate range: Stability to pH and temperature are important factors, e.g., for detergent enzymes. Starch conversion into sugars needs low pH and high temperature and so the demand for thermostable α-amylase and amyloglucosidase from thermophile organisms is a logical consequence. Enzymes which are able to do their job at lower temperatures are also of interest. For example, in diary industry, α-galactosidase is used at low temperature to reduce the amount of lactose.

Subtilisin, a bacterial serine protease often used in detergents, might illustrate the manifold goals of protein engineering. Up to present, mutations in well over 50% of the 275 amino acids have been reported in the scientific literature. Although enhanced stability has been the predominant target, these alterations also resulted in changes in catalytic mechanism, substrate specificity, surface activity, folding mechanisms and also in new activities.

In the following section the different approaches and methods in use to modify protein structure will be briefly described.

Oligonucleotide-directed mutagenesis

This method is based on computer modelling of individual amino acid changes, followed by site-directed mutagenesis of the corresponding DNA with pre-designed oligonucleotide primer and expression of the recombinant protein in mutants for testing and evaluation. This method makes it possible to modify one or a few amino acids in the protein.

DNA (gene) shuffling

The technique of DNA shuffling has been derived from imitating natural recombination by allowing *in vitro* homologous recombination of DNA. With this method, a population of related genes is randomly fragmented and subjected to denaturation and hybridisation, followed by extension with PCR reaction. As a result of repeated PCR cycles, the length of fragments is expanded. DNA recombination occurs when a fragment derived from one template primes a template with a different sequence. Combined with well focused selection procedures, this technique makes it possible to rapidly develop a huge variety of enzyme variants.

Improved product safety

In the course of screening for new enzyme producing micro-organisms, it sometimes turns out that growing of technically interesting micro-organisms under industrial conditions will not be possible. Either the micro-organism is pathogenic or toxic or not safe to handle. Enzyme purification could be prohibitively expensive, e.g., because the enzyme was cell associated or contaminated with undesirable compounds. Using production strains without long-term experience will anyhow be accompanied by a higher risk of harmful by-products in the final enzyme isolate. The cloning of enzymes following heterologous expression makes it possible to confine oneself to a small number of production strains which already have been used for enzyme production for decades and even might be considered as GRAS.

Chemical modification of enzymes

Chemical modification of enzymes was in fact the first method available to alter enzyme properties in the sixties. Since the mid eighties, interests on chemical modification have been rekindled. One major goal for chemical modification is to increase enzyme stability and activity, particularly in non-aqueous applications for biocatalysis in organic synthesis.

Examples: Various approaches were reported on covalent modification of enzymes: incorporation of vanadium into phytase for example, which has the *in vivo* role of catalysing phosphate ester hydrolysis is sufficient to convert phytase to a vanadium-dependent peroxidase that could catalyse enantio-selective sulphoxidation.

Section III

Marine, microbial, fungal and soil enzymes

10

Marine microbial enzymes

10.1 Introduction

Several industrial enzymes are derived from terrestrial sources. Whereas, marine environment, which encompass about 71% of the earth's surface and a vast resource for useful enzymes, remain unexplored. Marine micro-organisms take active part in the mineralisation of complex organic matter through degradative pathways of their metabolism in the marine environments and contribute to the secondary production in sea. Micro-organisms as the fastest means to search for food may often use cell-free extracellular enzymes.

Bacteria and fungi from marine environments secrete different enzymes based on their habitat and their ecological functions. Marine microbial enzymes have become the focal point of interest and several enzymes have drawn the attention of the microbial prospectors and a few enzymes were isolated from seawater and marine sediments, purified and characterised for their properties and applications. Enzymes known so far from marine micro-organisms include α-amylase, α-glucosidase, pullulanase, cyclomaltodextrin-glucanotransferase, agarase, alginate-lyase, κ-carrageenanase, α-galactosidase, β-galactoside-α-2, 6-sialyltransferase, cellulases, endo-1, 4-β-xylanase, β-galactosidase, β-glucosidase, β-fucosidase, endo-1,3-β-D-glucanase (I), endo-1,6-β-D-glucanase, chitinase, chitobiose-deacetylase, *N*-acetylglucosamine-deacetylase, *N*-acetylglucosamine-6-phosphate-deacetylase, lignin modifying enzymes (LMEs): manganese-dependent peroxidase (MnP), lignin peroxidase (LiP) and laccase, catechol-oxidase, cresolase (monophenolmono-oxygenase), β-1,4-mannanases, poly(3-hydroxybutyrate)-depolymerase, polybeta-hydroxyalkanoate (PHA)-depolymerase, alkaline protease, subtilisin (EC 3.4.21.62)-like serine protease, neutral protease, metal neutral proteinases, thiol protease, collagenase, lipase, L-asparaginase, L-glutaminase, tyrosinase, hydrogenases, superoxide-dismutase, glucose-dehydrogenase, etc.

Extremophiles are the primary source of enzymes that are active at extreme conditions of life. Harsh marine environments, such as deep-ocean hydrothermal vents, polar oceans and extremely saline bodies of water, have yielded valuable extremophilic micro-organisms. Of particular interest are the enzymes that help extremophiles to function in markedly hot, cold, acidic, basic, pressurised, saline or mineral rich environments. Extremozymes from extremophiles isolated so far include thermostable DNA polymerases-*Taq* polymerase and

Pfu polymerase, glutamate-dehydrogenase, beta-glucosidase, thermostable esterase, thiol protease, adenylate-kinase, hydrogenases, α-glucosidase, lipase, α-amylase, β-lactamase, protease, endo-1,4-β-D-xylanase, cellulase and xylanase, extremely thermostable protease, thermostable serine protease, etc., psychrostable metallo protease (almelysin), protease, chitinase, glucanase, esterase, lipase, phospholipase and DNA-degrading enzymes have been obtained from marine bacteria isolated from different sites in permanently cold arctic and antarctic habitats.

Some of the cold adapted enzymes from bacteria include α-amylase, isocitrate dehydrogenase, lipase, β-lactamase, triose phosphate isomerase, subtilisin. Alkaline protease, alkaline cyclomaltodextrin-glucanotransferase, lipases, alkaline cellulases, alkaline phosphatase, endo-1,3-β-D-glucanase and alkaline metallo endopeptidase and halotolerant amylases, proteases lipases, Nuclease H, cyclophilin-type peptidyl-prolylcis-*trans*-isomerase (PPIase) are also known from extremophiles.

Enzymes that are capable of carrying out very specific molecular tasks, usually related to the modification of DNA or RNA, for the creation of genetically modified organisms or for diagnostic procedures include restriction endonucleases, RNA and DNA polymerases, DNA ligases, alkaline phosphatases, kinases, reverse transcriptases, several such enzymes have been isolated.

In general traditional microbial enzyme technology ventures are undergoing rapid transformation through the process of evolution of innovative technologies facilitated through techniques like molecular gene cloning, protein and enzyme engineering, metabolic engineering and immobilisation of enzymes.

Despite the fact that more and more novel enzymes are yet to be discovered from marine micro-organisms from diverse marine environments around the world, the prospects of isolating the genes coding for the novel enzymes with potential applications have already received the attention of the genetic engineers and biotechnologists. The art of gene cloning have paved the way for easy isolation of the concerned genes, characterisation and cloning and expression in hosts such as *E. coli* successfully.

Extremozymes can be produced through recombinant DNA technology without massive culturing of the source extremophiles. Some of the success stories of cloning and expression of few genes of novel enzymes of industrial importance include alkaline serine protease (*apr*II, subtilase), endo-1,4-β-D-mannanase, salt-tolerant glutaminase-I, agarase (*pja*A), *Age*I methylase, DNA-polymerase, *lux* genes encoding luciferase, tyrosinase, α-amyalse, pullulanase type II, β-galactosidase, β-glucosidase, acid protease (thermopsin), S-adenosyl-homocysteine hydrolase, glyceraldehyde-3-Pdehydrogenase and glutamine synthetase. Nitrite reductase genes (*nir*K and *nir*S), structurally different but functionally equivalent single-copy genes coding for nitrite reductases, a key enzyme of the denitrification process, were used as a molecular marker for denitrifying bacteria in pacific northwest marine sediment communities.

While considering the enormous microbial diversity native to the vast marine environments of this planet earth the efforts channelled into discovery of novel enzymes from marine microbes are inadequate and warrant launching of intensive screening programmes by the scientists at global level. Such a mammoth attempt alone can return large number of novel enzymes for varied purposes and services of humanity, for the simple reason that marine environments are rich treasure of novel enzymes which probably may even avoid the need for enzyme engineering or molecular cloning for designing novel enzymes for specific requirements. In the long run, probably, marine microbial enzyme based processes would substitute several of the current chemical processes under practice.

10.2 Advantages of microbial enzymes over enzymes derived from plant and animals

Microbial enzymes have several advantage over the enzymes derived from plant or animal sources by virtue of their great variety of catalytic activities, cheaper in cost, regular abundant supplies at even quantity and relatively more stability. Major targets of modern enzyme technology continue to be preservation of foods and food components, efficient use of raw materials, improvement of food quality such as texture and taste, manufacture of dietetic foods, eliminating antinutritive substances from certain nutritional raw materials, utilisation of raw materials for preparation of animal feed and optimisation of process to reduce process costs. Enzymes are used as cost-effective and environmentally sensitive substitutes for chemical processing in several industries including pharmaceuticals, food, starch, laundry, detergents, for processing textiles, leather, wood pulp and paper and for the production of fine and specialty chemicals and industrial catalysis, organic synthesis and transformation of compounds and bioremediation.

Enzymes are finding applications as research tools in biotechnology and molecular biology. Further, several industrial applications demand that industrial enzymes must be stable at extremes of temperature, pH and salt concentration. For this reason, the isolation and characterisation of enzymes from micro-organisms known as 'extremophiles' may yield useful new enzymes for this purposes. For instance, the deep seabed provides many of these extremophiles and consequently may be of interest to companies involved in developing enzymes for this sector.

Despite the fact that several industrial enzymes are derived from terrestrial sources, marine micro-organisms are yet to be exploited to its full potential and consequently warrant immediate attention for industrial exploitation. Nevertheless, marine micro-organisms have drawn the attention of investigators in the last decade and there is lot of interest on microbial enzymes now.

Microbes inhabit various habitats of marine environment that include neuston, plankton, nekton, seston and epibiotic, endobiotic, pelagic and benthic environments. These habitats harbour a diverse range of microbes including archaebacteria, cyanobacteria, eubacteria, actinomycetes, yeasts, filamentous fungi, microalgae, algae and protozoa. Almost all of these groups are potential source of useful enzymes that remain unexplored.

Enzymes catalyse not only biochemical reactions in living cells but also in the mineralisation processes and cycling of elements in various environments. Hence, every marine micro-organism should act as a dependable source of useful enzymes such as protease, amylase, lipases, chitinase, cellulase, ligninase, pectinase, xylanase, nucleases (DNAses, RNAses, restriction enzymes), etc. Production of copious quantities of amylase, alginate lyases, chitinases, glutaminase, asparaginase, arylsulphatase, phosphatase and beta lactamase by marine bacteria are known. With the advent of biotechnology, enzyme engineering and introduction of other innovative technologies there is plenty of scope for efficient management of our rich marine microbial biodiversity towards deriving novel enzymes could also be recovered from marine micro-organisms and efficiently exploited not only as a cost effective biocatalyst but also as an ecofriendly reagent in the coming years.

10.3 Role of microbial enzymes in marine environment

Marine micro-organisms take active part in the mineralisation of complex organic matter through degradative pathways of their metabolism in the marine environments and contribute to the secondary production in sea. Complex polysaccharides such as cellulose, lignin, pectin, xylan, starch, proteins, fats, sugar, urea, aromatic and aliphatic hydrocarbons and several other organic compounds which reach marine environments, besides the dead plants and animal residues, are degraded by marine micro-organisms. Their participation in the degradation of organic compounds and retting of ropes and fibres testify their potential as rich source of hydrolytic enzymes of industrial importance. Micro-organisms as the fastest means to search for food may often use cell-free enzymes. Heterotrophic bacteria living in particle aggregates and sediments therefore must depend largely on extracellular enzymolysis and (or) desorption of particulate and particle-sorbed organic material (OM) to generate the dissolved, low-molecular-weight compounds required for uptake and metabolism. Transport of OM into bacterial cells is limited by cell permeability, probably to molecules smaller than about 600 Daltons. Both substrate uptake and bacterial growth may be coupled with extracellular hydrolysis. Bacterial extracellular enzymes (EE) can effect partitioning of OM between particulate and dissolved pools in the pelagic ocean and probably contribute to significant dissolved OM fluxes out of the sediment. Studies with size-fractionated samples and with radio labelled substrates suggest that cell-attached EE dominate extracellular hydrolysis by free-living marine bacteria, perhaps due to dilution of cell-free EE below detection limits. However, significant dissolved EE activity is sometimes observed in environmental samples and for some aquatic and sedimentary environments, cell-free EE are produced by most of the cultivable bacteria. Cell-free EE foraeging should be a powerful bacterial feeding mechanism in high surface-area, organic-rich, liquid-bathed environments, with the potential to support maximal growth rates. Where net energetic gain is the purpose of producing cell-free EE, bacteria can usually be expected to release enzyme at high rates. Many organisms may use cell-free EE to search for food, even in environments in which actual feeding is optimised by some other mechanism.

Polysaccharases release micro-organisms from their natural seat, marine sediments for example. The enzymatic activity works both on the microbial adherence polysaccharides and on the support surfaces (cellulose, pectin, etc.). Dosages of glucose confirm polysaccharase activity. An association of bacitracine, thiophenicol and a few enzymes: cellulase, pectinase, amylo-glucosidase, α-amylase, hyaluronidase, release a considerable number of bacteria. High activities of exoenzyme aminopeptidase and alkaline phosphatase were detected even in a 124000-year-old sapropel layer.

10.4 Enzymes from marine micro-organisms

Bacteria and fungi from marine environments secrete different enzymes based on their habitat and their ecological functions. Marine microbial enzymes have become the focal point of interest and several enzymes have drawn the attention of the microbial prospectors and a few enzymes were isolated from seawater and marine sediments, purified and characterised for their properties and applications. Interestingly the enzymes reported from marine environments belonged to one or more of the major classes of enzymes viz., oxidoreductase, transferases, hydrolases, lyases, isomerases and ligases.

10.5 Polysaccharases

10.5.1 Starch hydrolysing enzymes

The enzymes involved in the conversion of starch to low molecular-weight compounds such as glucose, maltose and oligosaccharides are α-amylase, β-amylase, glucoamylase, debranching enzymes (pullulanase) and α-glucosidase. α-amylase, pullulanase and α-glucosidase from archaea are all active in the same pH and high temperature range and hence they could be used in a one-step process for the industrial bioconversion of starch. Improvement of the starch-conversion process using new efficient and thermoactive enzymes would significantly lower the cost of sugar syrup production.

α-amylase (EC 3.2.1.1)

α-amylase has a wide range of industrial applications particularly in the manufacture of alcoholic beverages like beer, alcohol, preparation of animal feed, laundry, detergent, in starch industry, confectioneries, as desizing agent in textile industry and in the production of sugar syrup. This is one among the few major industrial enzymes, which is in great demand and mainly produced using microbial sources.

α-amylase has been obtained from thermophilic archaea *Pyrococcus woesei, Pyrococcus furiosus, Thermococcus celer, Fervidobacterium pennavorans, Desulphurococcus mucosus* and *Thermotoga maritime, Psychrotrophic Vibrio* isolated from a deep-sea mud, *Vibrio gazogenes, Alteromonas rubra* and *Mucor* sp., *Pseudomonas*-like strain MS300, from deep-sea, produce two major and two miner maltotetraohydrolases (G4-amylase). Under high hydrostatic pressure, the strain MS300 produced more amylase than under atmospheric pressure.

α-glucosidase (EC 3.2.1.2)

α-glucosidase (EC 3.2.1.2) is mainly used in the starch industry along with α-amylase. An extremely thermostable α-glucosidase (EC 3.2.1.2) was isolated from the hyperthermophilic marine archaebacterium, *Poyrococcus furiosus* and *Pyrococcus wesei.* The enzymes from these micro-organisms appear to be the most thermostable α-glucosidases.

Pullulanases (EC 3.2.1.41)-debranching enzymes

Similar to α-amylases and α-glucosidase, pullulanase are mainly used in starch industry. Pullulanase type I is a typical bacterial enzyme that is specific for α-1, 6-linkages in branched oligosaccharides. It is unable to attack α-1,4-linkages in α-glucan.

The production of pullulanases type I seem to be very rare amongst thermophilic micro-organisms that produce a thermoactive pullulanase with a temperature optimum of 90°C. This enzyme has been reported only in *Fervidobacterium pullulanolyticum.* Pullulanase type II (amylo pullulanase), which is capable of hydrolysing the α-1,4-linkages and branching points (α-1,6-linkages) in polysaccharides and limit dextrins is mainly found in anaerobic bacteria and is widely distributed among thermophilic bacteria and archaea. Marine thermophilic *Thermococcus litoralis* (Tl) DSM 5473 (optimal growth temperature 90°C) and *Pyrococcus furiosus* (Pf) DSM 3638 (98°C) produce pullulanase type II whose thermoactivity and thermostability were enhanced in the presence of 5 mM Ca^{2+} and under these conditions, enzyme activity could be measured at temperature of up to 130–140°C. Each of these enzymes was able to hydrolyse, in addition to the alpha-1,6-linkages in pullulan, alpha-1,4-linkages in amylose and soluble starch. The enzymes appear to represent highly thermostable amylopullulanase versions of those which have been isolated from less thermophilic organisms. *Pyrococcus woesei*, *Thermococcus celer*, *Fervidobacterium pennavorans* and *Desulfurococcus mucosus* have also been reported to produce pullulanase type II.

Cyclomaltodextrin-glucanotransferase (CGase EC 2.4.1.19)

This enzyme has potential application in cyclodextrin production, was isolated from alkalophilic *Bacillus subtilis* sp., from a deep-sea mud.

10.5.2 Agarase (EC 3.2.1.81)

Agarase has been the subject of investigations for quite some time owing to its immediate applications in gene technology for the elution and isolation of DNA fragments from agarose gels after electrophoresis, in the preparation of algal protoplasts such as from red alga *Gelidium robustum*, seaweed polysaccharide characterisation, in the production of simple sugars, including neoagarobiose, neoagarotetraose and neoagarohexaose, degradation of agarose to oligosaccharides, facilitating the liquefaction of agar and agarose gels and in the defouling of fermentors and bioreactors. Moreover, the purified enzyme could effectively control red algae bloom contaminations, prevent the biofouling of submerged marine surfaces or pipes by contaminating complex polysaccharide layers, or treat such biofouled surfaces after contamination. Marine Bacterial agarase has a high level of activity for the depolymerisation of complex polysaccharides, including agar and agarose. Agarase is obtained from few bacteria. Agarase, named agarase-0107, an endo-type β-agarase that hydrolysed the β-1,4-linkage of agarose to yield neoagarotetraose and neoagarobiose at a pH of around 8, have been isolated from *Vibrio* sp., JT0107, that requires seawater salts for growth. *Pseudomonas stutzeri*, *Aeromonas* sp., and *Vibrio* sp., isolated from sea produce agarase, which have been characterised.

10.5.3 Alginate-lyase (EC 4.2.2.3)

The alginate-lyase may be useful for the conversion of brown algal biomass to methane. This observation has led to the isolation and characterisation of an intracellular inducible alginate-lyase from marine *Bacillus* sp., *Alteromonas and Photobacterium.*

10.5.4 κ-carrageenanase (EC 3.2.1.83)

A Cytophaga-like, carrageenan-degrading bacterium, referred to as strain Dsij, isolated from the marine red alga *Delesseria sanguinea*, simultaneously produce extracellular κ-carrageenanase (EC 3.2.1.83) and iota-carrageenanase in the presence of crude λ-carrageenan.

10.5.5 α-Galactosidase (EC 3.2.1.22)

α-galactosidase is used in the sugar industry for enhancing the yield of sucrose by the hydrolysis of raffinose, an α-galactoside, α-galactosidase was obtained from marine *Alteromonas spp.*, isolated from sponges and alga *Polysiphonia* sp., and from bacteria associated with mussel (*Crenomytilus grayanus*) and scallop (*Patinopecten jessoensis*). A novel β-galactoside-α-2, 6-sialyltransferase (EC 2.4.99.1) is produced by *Photobacterium damsela* JT0160 (FERM BP-4900), ATCC 33539 or ATCC 35083. The enzyme should be useful in modification of glycoproteins and industrial-scale production of sialosides.

10.5.6 Cellulases and related enzymes

Cellulases are used in the manufacture of alcohol, flavours, maize gluten, silage, in laundry and detergents and in wastewater treatment. A symbiotic bacterium found in the gland of *Deshayes* of the marine shipworm and *Aspergillus terreus* isolated from the seawater produce cellulase. Thermophilic *Thermotoga* sp., produces thermostable exo-1,4-β-cellobiohydrolase.

11

Microbial enzymes in bioremediation of pollution

11.1 Introduction

A large number of enzymes from bacteria, fungi and plants have been reported to be involved in the biodegradation of toxic organic pollutants. Bioremediation is a cost effective and nature friendly biotechnology that is powered by microbial enzymes. This chapter provides descriptive information on the enzymes from various micro-organisms involved in the biodegradation of wide range of pollutants, applications and suggestions required to overcome the limitations of their efficient use.

The quality of life on the Earth is linked inextricably to the overall quality of the environment. Unfortunately the progress in science, technology and industry a large amount ranging from raw sewage to nuclear waste is let out or dumped into the ecosystem there by posing a serious problem for survival of mankind itself on earth. In the past, wastes were traditionally disposed by digging a hole and filling it with waste material. This mode of waste disposal was difficult to sustain owing to lack of new place every time to dump. New technologies for waste disposal that use high-temperature incineration and chemical decomposition (e.g., base-catalysed dechlorination, UV oxidation) have evolved. Although they can be very effective at reducing wide a range of contaminants but at the same time have several drawbacks. These methods are complex, uneconomical and lack public acceptance. The associated deficiencies in these methods have focused efforts towards harnessing modern day bioremediation process as a suitable alternative.

Bioremediation is a micro-organism mediated transformation or degradation of contaminants into nonhazardous or less-hazardous substances. The employability of various organisms like bacteria, fungi, algae and plants for efficient bioremediation of pollutants has been reported. The involvement of plants in the bioremediation of pollutants is called as phytoremediation. The process of phytoremediation is an emerging green technology that facilitates the removal or degradation of the toxic chemicals in soils, sediments, groundwater, surface water and air (RTDF). Genetically, engineered plants are also in use. For instance arsenic is phytoremediated by genetically modified plants such as *Arabidopsis thaliana* which expresses two bacterial genes. One of these genes allows the plant to modify arsenate into arsenite and the second one binds the modified arsenite and stores it in the vacuoles.

The process of bioremediation mainly depends on micro-organisms which enzymatically attack the pollutants and convert them to innocuous products. As bioremediation can be effective only where environmental conditions permit microbial growth and activity, its application often involves the manipulation of environmental parameters to allow microbial growth and degradation to proceed at a faster rate (Fig. 11.1).

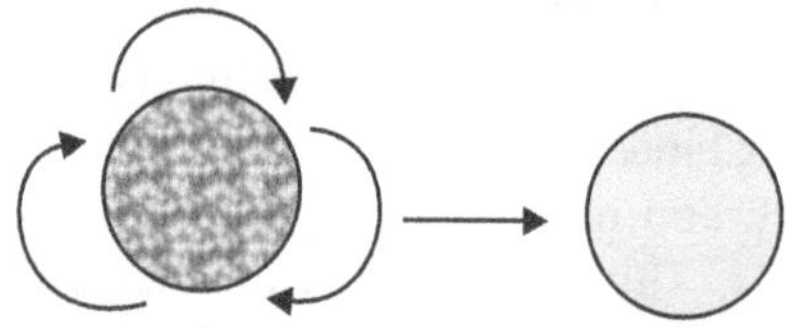

Figure 11.1: The process of waste bioremediation.

The process of bioremediation is a very slow process. Only certain species of bacteria and fungi have proven their ability as potent pollutant degraders. Many strains are known to be effective as bioremediation agents but only under laboratory conditions. The limitation of bacterial growth is under the influence of pH, temperature, oxygen, soil structure, moisture and appropriate level of nutrients, poor bioavailability of contaminants and presence of other toxic compounds. Although micro-organisms can exist in extreme environment, most of them prefer optimal condition a situation that is difficult to achieve outside the laboratory. Most bioremediation systems operate under aerobic conditions, but anaerobic environments may also permit microbial degradation of recalcitrant molecules. Both bacteria and fungi rely on the participation of different intracellular and extracellular enzymes respectively for the remediation of recalcitrant and lignin and organo pollutants.

11.2 Microbial enzymes in bioremediation

11.2.1 Microbial oxidoreductases

The detoxification of toxic organic compounds by various bacteria and fungi and higher plants through oxidative coupling is mediated with oxidoreductases. Microbes extract energy via energy yielding biochemical reactions mediated by these enzymes to cleave chemical bonds and to assist the transfer of electrons from a reduced organic substrate (donor) to another chemical compound (acceptor). During such oxidation-reduction reactions, the contaminants are finally oxidised to harmless compounds. The oxidoreductases participate in the humification of various phenolic substances that are produced from the decomposition of lignin in a soil environment. In the same way, oxidoreductases can also detoxify toxic xenobiotics, such as phenolic or anilinic compounds, through polymerisation, copolymerisation with other substrates, or binding to humic substances. Microbial enzymes have been exploited in the decolourisation and degradation of azo dyes.

Many bacteria reduce the radioactive metals from an oxidised soluble form to a reduced insoluble form. During the process of energy production, bacterium takes up electrons from organic compounds and use radioactive metal as the final electron acceptor. Some of bacterial species reduce the radioactive metals indirectly with the help of an intermediate electron donor. Finally precipitant can be seen as the result of redox reactions within the metal-reducing bacteria.

Chlorinated phenolic compounds are among the most abundant recalcitrant wastes found in the effluents generated by the paper and pulp industry. These compounds are produced upon the partial degradation of lignin during pulp bleaching process. Many fungal species are considered to be suitable for the removal of chlorinated phenolic compounds from the contaminated environments. The activity of fungi is mainly due to the action of extracellular oxidoreductase enzymes, like laccase, manganese peroxidase and lignin peroxidase, which are released from fungal mycelium into their nearby environment. Being filamentous, fungi can reach the soil pollutants more effectively than bacteria.

Water polluted with phenolic compounds can be decontaminated by plants with the help of enzymes exuded by their roots. The plant families of *Fabaceae, Gramineae* and *Solanaceae* are found to release oxidoreductases which take part in the oxidative degradation of certain soil constituents. Phytoremediation of organic contaminants has been generally focused on three classes of compounds: chlorinated solvents, explosives and petroleum hydrocarbons.

Microbial oxygenases

Oxygenases belong to the oxidoreductase group of enzymes. They participate in oxidation of reduced substrates by transferring oxygen from molecular oxygen (O_2) utilising FAD/NADH/NADPH as a cosubstrate. Oxygenases are grouped into two categories, the mono-oxygenases and dioxygenases on the basis of number of oxygen atoms used for oxygenation. They play a key role in the metabolism of organic compounds by increasing their reactivity or water solubility or bringing about cleavage of the aromatic ring. Oxygenases have a broad substrate range and are active against a wide range of compounds, including the chlorinated aliphatics. Generally the introduction of O_2 atoms into the organic molecule by oxygenase results in cleavage of the aromatic rings. Historically, the most studied enzymes in bioremediation are bacterial mono- or dioxygenases. Halogenated organic compounds comprise the largest groups of environmental pollutants as a result of their widespread use as herbicides, insecticides, fungicides, hydraulic and heat transfer fluids, plasticisers and intermediates for chemical synthesis. The degradation of these pollutants is achieved by specific oxygenases. Oxygenases also mediate dehalogenation reactions of halogenated methanes, ethanes and ethylenes in association with multifunctional enzymes.

Mono-oxygenases

Mono-oxygenases incorporate one atom of the oxygen molecule into the substrate. Mono-oxygenases are classified into two subclasses based on the presence cofactor: flavin-dependent mono-oxygenases and P450 mono-oxygenases. Flavin-dependent mono-oxygenases contain flavin as prosthetic group and require NADP or NADPH as coenzyme. P450 mono-oxygenases are hemecontaining oxygenases that exist in both eukaryotic and prokaryrotic organisms. The mono-oxygenases comprise a versatile superfamily of enzymes that catalyses oxidative reactions of substrates ranging from alkanes to complex endogenous molecules such as steroids and fatty acids. Mono-oxygenases act as biocatalysts in bioremediation process and synthetic chemistry due to their highly region selectivity and stereo selectivity on wide range of substrates. Majority of mono-oxygenase studied previously are having cofactor, but there are certain mono-oxygenases which function independent of a cofactor. These enzymes require only molecular oxygen for their activities and utilise the substrate as reducing agent.

The desulphurisation, dehalogenation, denitrification, ammonification, hydroxylation, biotransformation and biodegradation of various aromatic and aliphatic compounds are catalysed by mono-oxygenases. These properties have been explored in recent years for important application in biodegradation and biotransformation of aromatic compounds. Methane mono-oxygenase enzyme is the best characterised one, among mono-oxygenases. This enzyme is involved in the degradation of hydrocarbon such as substituted methanes, alkanes, cycloalkanes, alkenes, haloalkenes, ethers and aromatic and heterocyclic hydrocarbons. Under oxygen-rich conditions, mono-oxygenase catalyses oxidative dehalogenation reactions, whereas under low oxygen levels, reductive dechlorination takes place. Oxidation of substrate can lead to de-halogenation as a result of the formation of labile products that undergo subsequent chemical decomposition.

Microbial dioxygenases

Dioxygenases are multicomponent enzyme systems that introduce molecular oxygen into their substrate. Aromatic hydrocarbon dioxygenases, belong to a large family of Rieske nonheme iron oxygenases. These dioxygenases catalyse enantiospecifically the oxygenation of wide range of substrates. Dioxygenases primarily oxidise aromatic compounds and, therefore, have applications in environmental remediation. All members of this family have one or two electron transport proteins preceding their oxygenase components. The crystal structure of naphthalene dioxygenase has confirmed the presence of a Rieske (2Fe–2S) cluster and mononuclear iron in each alpha subunit.

The catechol dioxygenases serve as part of nature's strategy for degrading aromatic molecules in the environment. They are found in the soil bacteria and involved in the transformation of aromatic precursors into aliphatic products. The intradiol cleaving enzymes utilise Fe(III), while the extradiol cleaving enzymes utilise Fe(II) and Mn(II) in a few cases.

11.2.2 Microbial laccases

Laccases (*p*-diphenol:dioxygen oxidoreductase) constitute a family of multicopper oxidases produced by certain plants, fungi, insects and bacteria, that catalyse the oxidation of a wide range of reduced phenolic and aromatic substrates with concomitant reduction of molecular oxygen to water. Laccases are known to occur in multiple isoenzyme forms each of which is encoded by a separate gene and in, some cases, the genes have been expressed differently depending upon the nature of the inducer.

Many micro-organisms produce intra and extracellular laccases capable of catalysing the oxidation of ortho and paradiphenols, aminophenols, polyphenols, polyamines, lignins and aryl diamines as well as some inorganic ions. Laccases not only oxidise phenolic and methoxyphenolic acids, but also decarboxylate them and attack their methoxy groups (demethylation). These enzymes are involved in the depolymerisation of lignin, which results in a variety of phenols. In addition, these compounds are utilised as nutrients for micro-organisms or repolymerised to humic materials by laccase. Among the biological agents, laccases represent an interesting group of ubiquitous, oxidoreductase enzymes that show promise of offering great potential for biotechnological and bioremediation applications. The substrate specificity and affinity of laccase can vary with changes in pH. Laccase can be inhibited by various reagents such as halides (excluding iodide), azide, cyanide and hydroxide. Different laccases appear to have differing tolerance toward inhibition by halides, indicating differential halide accessibility. Laccase production is sensitive to the nitrogen concentration in fungi. High nitrogen levels are usually required to obtain greater amounts of laccase. Recombinant laccase can be produced by either homologous or heterologous means.

11.2.3 Microbial peroxidases

Peroxidases (donor: hydrogen peroxide oxidoreductases) are ubiquitous enzymes that catalyse the oxidation of lignin and other phenolic compounds at the expense of hydrogen peroxide (H_2O_2) in the presence of a mediator. These peroxidases can be haem and nonhaem proteins. In mammals, they are involved in biological processes such as immune system or hormone regulation. In plants, they are involved in auxin metabolism, lignin and suberin formation, cross-linking of cell wall components, defense against pathogens, or cell elongation.

The hemeperoxidases have been classified into two distinct groups as found only in animals and found in plants, fungi and prokaryotes. The second group peroxidases have been subdivided into three classes on the basis of sequence comparison. Class I is intracellular enzymes including yeast cytochrome *c* peroxidase, ascorbate peroxidase (APX) from plants and bacterial gene-duplicated catalase peroxidases. Class II consists of the secretory fungal peroxidases such as lignin peroxidase (LiP) and manganese peroxidase (Mnp) from *Phanerochaete chrysosporium* and *Coprinus cinereus* peroxidase or *Arthromyces ramosus* peroxidase (ARP). The main role of class II peroxidases appears to be the degradation of lignin in wood. Class III contains the secretory plant peroxidases such as those from horseradish (HRP), barley, or soyabean. These peroxidases seem to be biosynthetic enzymes involved in processes such as plant cell wall formation and lignifications.

Nonhaem peroxidases are not evolutionarily linked and form five independent families. They are thiol peroxidase, alkylhydroperoxidase, nonhaem haloperoxidase, manganese catalase and NADH peroxidase. Among all these thiol peroxidase is the largest and having two subfamilies such as glutathione peroxidases and peroxy redoxins.

Classification of peroxidase enzymes

Peroxidases have been classified into many types based on its source and activity. Among peroxidases, lignin peroxidase (LiP), manganese-dependant peroxidase.

Microbial lignin peroxidases: Lignin peroxidases are heme proteins secreted mainly by the white rot fungus during secondary metabolism. In the presence of cosubstrate H_2O_2 and mediator like veratryl alcohol LiP degrade lignin and other phenolic compounds. During the reaction, H_2O_2 gets reduced to H_2O with the gaining of electron from LiP, (which itself gets oxidised). The LiP (oxidised) with gaining an electron from veratryl alcohol returns to its native reduced state and veratryl aldehyde is formed. Veratryl aldehyde then again gets reduced back to veratryl alcohol by gaining an electron from substrate. This result in the oxidation of halogenated phenolic compounds, polycyclic aromatic compounds and other aromatic compounds followed by a series of nonenzymatic reactions. Lignin peroxidase (LiP) plays a central role in the biodegradation of the plant cell wall constituent lignin. LiP is able to oxidise aromatic compounds with redox potentials higher than 1.4V (NHE) by single-electron abstraction, but the exact redox mechanism is still poorly understood.

Microbial manganese peroxidases: MnP is an extracellular heme enzyme from the lignin-degrading basidiomycetes fungus, that oxidises Mn^{2+} to the oxidant Mn^{3+} in a multistep reaction. Mn^{2+} stimulates the MnP production and functions as a substrate for MnP. The Mn^{3+}, generated by MnP, acts as a mediator for the oxidation of various phenolic compounds. The resulting Mn^{3+} chelate oxalate is small enough to diffuse into areas inaccessible even to the enzyme, as in the case of lignin or analogous structures such as xenobiotic pollutants buried deep within the soil, which are not necessarily available to the enzymes.

Microbial versatile peroxidases (MVP): MVP enzymes are able to directly oxidise Mn^{2+}, methoxybenzenes, phenolic aromatic substrates like that of MnP, LiP and horseradish peroxidase. MVP has extraordinary broad substrate specificity and tendency to oxidise the substrates in the absence of manganese when compared to other peroxidases. It has also been demonstrated that MVP is able to oxidise both phenolic and nonphenolic lignin model dimers. Therefore, a highly efficient MVP overproduction system is desired for biotechnological applications in industrial processes and bioremediation of recalcitrant pollutants.

11.3 Microbial hydrolytic enzymes

The pollution of soil and water by industrial chemicals and petroleum hydrocarbons is a serious problem of the modern world. Due to their extensive use, they are found as environmental contaminants in numerous aquatic and terrestrial ecosystems. The use of bioremediation technologies for removing these contaminants provides a safe and economic alternative to commonly used physical-chemical treatment.

Bacterial activity is the major process involved in the hydrolysis of organic pollutants. Extracellular enzyme activity is a key step in degradation and utilisation of organic polymers, since only compounds with molecular mass lower than 600 daltons can pass through cell pores.

Hydrolytic enzymes disrupt major chemical bonds in the toxic molecules and results in the reduction of their toxicity. This mechanism is effective for the biodegradation of oil spill and organo phosphate and carbamate insecticides. Organochlorine insecticides such as DDT and heptachlor are stable in well-aerated soil but readily degrade in anaerobic environments. Hydrolases also catalyse several related reactions including condensations and alcoholysis. The main advantages of this enzyme class are ready availability, lack of cofactor stereoselectivity and tolerate the addition of water-miscible solvents. Hydrolases belong to group 3 of enzyme class and may further be classified according to the type of bond hydrolysed.

Extracellular hydrolytic enzymes such as amylases, proteases, lipases, DNases, pullulanases and xylanases have quite diverse potential usages in different areas such as food industry, feed additive, biomedical sciences and chemical industries. The hemicellulase, cellulase and glycosidase are of much importance due to its application in biomass degradation.

11.3.1 Microbial lipases

Lipase degrades lipids derived from a large variety of micro-organisms, animals and plants. Recent works have shown that lipase is closely related with the organic pollutants present in the soil. Lipase activity was responsible for the drastic reduction total hydrocarbon from contaminated soil. Research undertaken in this area is likely to progress the knowledge in the bioremediation of oils spill. Lipases have been extracted from bacteria, plant, actinomycetes and animal cell. Among these microbial lipases are more versatile because of their potent application in industries. These enzymes can catalyse various reactions such as hydrolysis, interesterification, esterification, alcoholysis and aminolysis.

Lipases are ubiquitous enzymes which catalyse the hydrolysis of triacylglycerols to glycerol and free-fatty acids. Lipolytic reactions occur at the lipid-water interface, where lipolytic substrates usually form equilibrium between monomeric, micellar and emulsified states. Lipases have been classified into two types on the basis of criteria such as: (a) enhancement in enzyme activity as soon as the triglycerides form an emulsion and (b) lipases with a loop of protein (lid) covering on the active site.

Triglyceride is the main component of natural oil or fat. This can hydrolyse consecutively to diacylglycerol, monoacylglycerol, glycerol and fatty acids. Glycerol and fatty acids are widely used as raw materials, for instance, monoacylglycerol is used as an emulsifying agent in the food, cosmetic and pharmaceutical industries. The study made on triolein hydrolysis from *Candida rugosa* lipase in the biphasic oil-water system as proven to be effective. The lipase adsorbs on to the oil-water interface in the bulk of the water phase.

The lipase then breaks the ester bonds of triolein to produce consecutively diolein, mono-olein and glycerol. During the catalysis oleic acid is formed at each consecutive reaction stage. The glycerol formed is hydrophilic and thus dissolves into the water phase. Lipase activity was found to be the most useful indicator parameter for testing hydrocarbon degradation in soil. Lipase is of much interest in the production of regiospecific compounds which are employed in pharmaceutical industry. Along with its diagnostic usage in bioremediation, lipase has many potential applications in food, chemical, detergent manufacturing, cosmetic and paper making industries, but its production cost has restricted its industrial use.

11.3.2 Microbial cellulases

Cellulases now promise the potential of converting waste cellulosic material into foods and have been the subject of intense research. Some organisms produce cell bound, cell envelope associated and some extra cellular cellulases. Extracellular cellulases, hemicellulases and pectinases have been shown to be constitutively expressed at very low levels by some bacteria and fungi.

Cellulases are usually a mixture of several enzymes. At least: three major groups of cellulases are involved in the hydrolysis process: (i) endoglucanase (EG, endo- 1,4-D-glucanohydrolase) which attacks regions of low crystallinity in the cellulose fibre, creating free chain ends, (ii) exoglucanase or cellobiohydrolase (CBH, 1,4-β-D-glucan cellobiohydrolase) which degrade the cellulose molecule further by removing cellobiose units from the free chain ends and (iii) β-glucosidase which hydrolyses cellobiose to glucose units. Along with major enzymes, some ancillary enzymes are also present. During the enzymatic hydrolysis, cellulose is degraded by the cellulases to reducing sugars (Fig. 11.2) that can be fermented by yeasts or bacteria to ethanol.

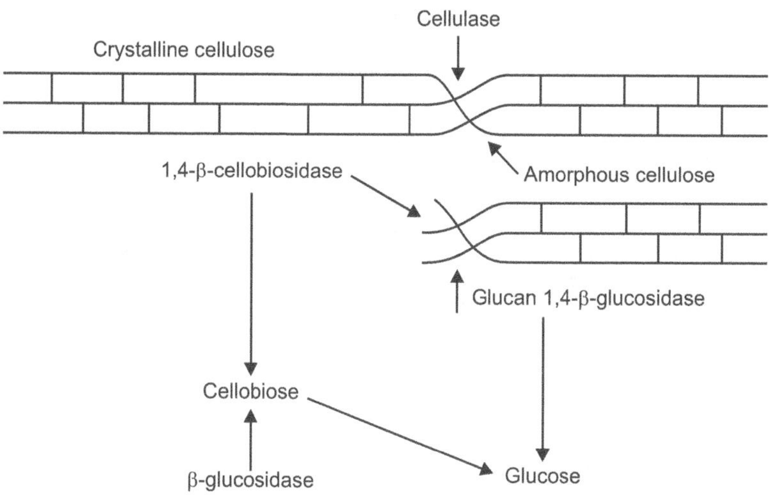

Figure 11.2: Proposed mechanism for the hydrolysis of cellulose by the fungal cellulase enzyme system.

Cellulase enzymes are capable of degrading crystalline cellulose to glucose. Cellulases have been used in the manufacture of detergents since early 1990s. Cellulases remove cellulose microfibrils, which are formed during washing and the use of cotton-based cloths. This is also observed as the colour brightening and material softening in the textile industry. Alkaline cellulases are produced by *Bacillus* strains and neutral and acidic cellulases by *Trichoderma* and *Humicola* fungi. In paper and pulp industry, cellulases have been employed for the removal of ink during recycling of paper. The cellulases are added during brewing to increase the juice liberation from fruit pulp and for the production of ethanol from cellulosic biomass.

11.3.3 Microbial proteases

Proteases hydrolyse the breakdown of proteinaceous substance which enter atmosphere due to shedding and moulting of appendages, death of animals and also as by-product of some industries like poultry, fishery and leather. Proteases belong to group of enzymes that hydrolyse peptide bonds in aqueous environment and synthesise them in nonaqueous environment. The enzymes whose reaction mechanism is completely elucidated are grouped under various categories as shown in Fig. 11.3.

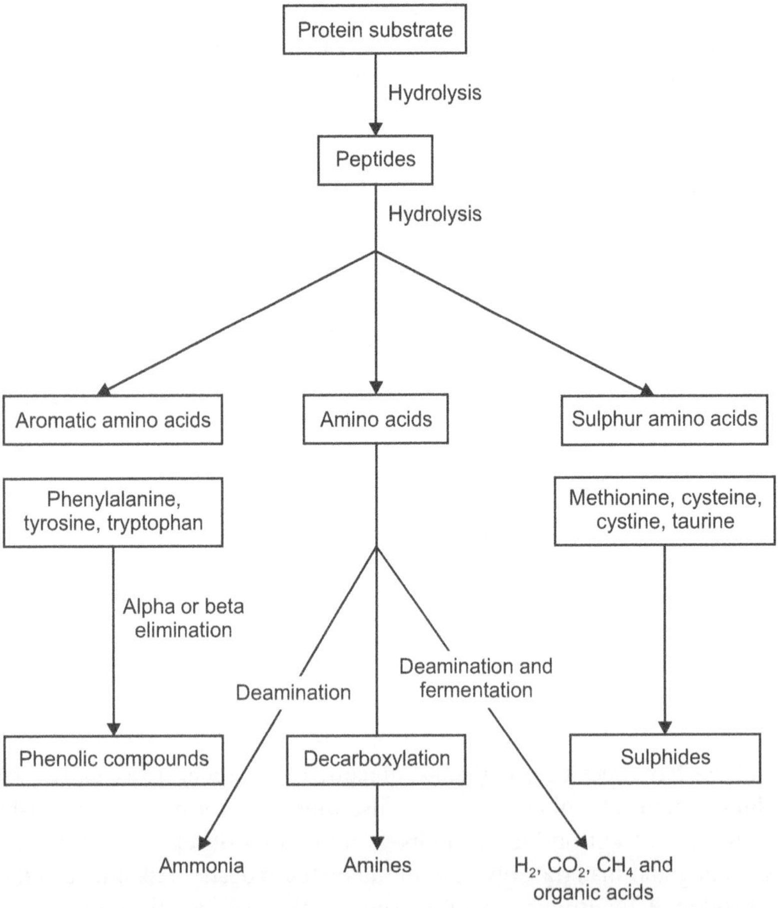

Figure 11.3: Proposed pathway for protease hydrolysis.

Proteases have wide range of applications in food, leather, detergent and pharmaceutical industry. Proteases are divided as endopeptidases and exopeptidases based on the catalysis of peptide chain. Endo peptidases further grouped based on the position of active site such as serine endopeptidase, cysteine peptidase, aspartic endopeptidases and metallopeptidases.

The exopeptidases act only near the terminal amino or carboxylic position of chain. The protease that acts on free amino and carboxyl terminals are called as aminopeptidase and carboxypeptidase, respectively.

The endopeptidase acts on the inner regions of peptide chain. The presence of free amino and carboxyl terminal will have negative impact on enzyme activity.

Proteases have been used in the manufacture of cheese and detergent manufacturing industry since many years.

The alkaline proteases are used in leather industry for the removal of hairs and parts which are present on the animal skin. Proteases have been employed for the production of dipeptide aspartame, which is a noncalorific artificial sweetener. In the pharmaceutical industry, a varying and specific proteases are used in developing effective therapeutic agents. Clostridial collagenase or subtilisin is used in combination with broad-spectrum antibiotics in the treatment of burns and wounds.

12

Fungal laccase enzyme for biotechnological application

12.1 Introduction

Laccase belongs to the small group of enzymes called the blue multi copper oxidases. Laccase is widely distributed in higher plants and fungi. In fungi, laccase is present in Ascomycetes, Deuteromycetes, *Basidiomycetes* and is particularly abundant in many white-rot fungi that degrade lignin. Laccases have been subject of intensive research in the last decades due to their broad substrate specificity. In the recent years, their uses span from the textile to the pulp and paper industries and food applications to bioremediation processes. Laccases also have uses in organic synthesis, where typical substrates are phenols and amines and the reaction products are dimers and oligomers derived from the coupling of reactive radical intermediates. More recently, they have found applications in other field such as in the design of biosensors and biofuel cells. In this chapter, the occurrence, mode of action, general properties, production and immobilisation of laccases are discussed.

As discussed above laccase (benzenediol:oxygen oxidoreductase, EC 1.10.3.2) belongs to the small group of enzymes called the blue copper proteins or the blue copper oxidases along with the plant ascorbate oxidase and the mammalian plasma protein ceruloplasmin among others. These proteins are characterised by containing 4 catalytic copper atoms. One copper is placed at the T1 site, where reducing substrate binds and it is responsible in the characteristic blue-greenish colour in the oxidising resting state Cu^{2+}. The other three coppers are clustered in the called T2/T3 site in which molecular oxygen binds.

Laccase is widely distributed in higher plants and fungi and has been found also in insects and bacteria. Recently a novel polyphenol oxidase with laccase like activity was mined from a metagenome expression library from bovine rumen microflora.

Fungal laccases have higher redox potential than bacterial or plant laccases (up to +800 mV) and their action seems to be relevant in nature finding also important applications in biotechonology. Thus, fungal laccases are involved in the degradation of lignin or in the removal of potentially toxic phenols arising during lignin degradation. In addition, fungal laccases are hypothesised to take part in the synthesis of dihydroxynaphthalene melanins, darkly pigmented polymers that organisms produce against environmental stress or in fungal morphogenesis by catalysing the formation of extracellular pigments.

Concerning their use in the biotechnology area, fungal laccases have widespread applications, ranging from effluent decolouration and detoxification to pulp bleaching, removal of phenolics from wines, organic synthesis, biosensors, synthesis of complex medical compounds and dye transfer blocking functions in detergents and washing powders, many of which have been patented. The biotechnological use of laccase has been expanded by the introduction of laccase-mediator systems, which are able to oxidise non-phenolic compounds that are otherwise hardly or not oxidised by the enzyme alone.

12.2 Occurrence of laccases

Laccase is the most widely distributed of all the large blue copper-containing proteins, as it is found in a wide range of higher plants and fungi and recently some bacterial laccases have also been characterised from *Azospirillum lipoferum*, *Bacillus subtilis*, *Streptomyces lavendulae*, *S. cyaneus* and *Marinomonas mediterranea*. The occurrence of laccases in higher plants appears to be far more limited than in fungi. Laccases in plants have been identified in trees, cabbages, turnips, beets, apples, asparagus, potatoes, pears and various other vegetables. The classical demonstration of laccase in *R. vernicifera* is well documented. In addition, the lacquer tree is a member of the Anacardiaceae family, appear to contain laccase in the resin ducts and in the secreted resin. Cell cultures of *Acer pseudoplatanus* have been shown to contain eight laccases, all expressed predominantly in xylem tissue. Other reports are those of Wosilait and others on the presence of a laccase in leaves of *Aesculus parviflora* and in green shoots of tea. Other higher plant species also appear to contain laccases, although their characterisation is less convincing. Laccases have been isolated from Ascomyceteous, Deuteromyceteous and *Basidiomyceteous* fungi. In the fungi, Ascomycetes and Deuteromycetes have not been a clear focus for lignin degradation studies as much as the white-rot *Basidiomycetes*. Laccase from *Monocillium indicum* was the first laccase to be characterised from an Ascomycete showing peroxidative activity.

The white-rot basidiomycetes are the most efficient degraders of lignin and also the most widely studied. The enzymes implicated in lignin degradation are: (i) lignin peroxidase, which catalyses the oxidation of both phenolic and non-phenolic units, (ii) manganese-dependent peroxidase, (iii) laccase, which oxidises phenolic compounds to give phenoxy radicals and quinones, (iv) glucose oxidase and glyoxal oxidase for H_2O_2 production and (v) cellobiose-quinone oxidoreductase for quinone reduction. The veratryl alcohol oxidase and some esterases may also play roles in the complex process of natural wood decay. The different degrees of lignin degradation with respect to other wood components depend on the environmental conditions and the fungal species involved. It has been demonstrated that there is no unique mechanism to achieve the process of lignin degradation and that the enzymatic machinery of the various micro-organisms differ. *Pleurotus ostreatus,* for instance, belongs to a subclass of lignin–degrading micro-organisms that produce laccase, manganese peroxidase and veratryl alcohol oxidase but no lignin peroxidase. *Pycnoporus cinnabarinus* has been shown to produce laccase as the only ligninolytic enzyme and *Pycnoporus sanguineus* produces laccase as the sole phenol oxidase. In plants, laccase plays a role in lignification, whereas in fungi laccases have been implicated in many cellular processes, including delignification, sporulation, pigment production, fruiting body formation and plant pathogenesis. Only a few of these functions have been experimentally demonstrated.

Ligninolytic enzymes have mostly been reported to be extracellular but there is evidence in literature of the occurrence of intracellular laccases in white–rot fungi. Intracellular as well as extracellular laccases were identified for *Neurospora crassa* by Froehner and Eriksson, who suggested that the intracellular laccase functioned as a precursor for extracellular laccase as there were no differences between the two laccases other than their occurrence.

12.3 Mode of action of the laccase enzyme

Laccases contain 4 copper atoms termed Cu T1 (where the reducing substrate place) and trinuclear copper cluster T2/T3 (where oxygen binds and is reduced to water). As a one-electron substrate oxidation is coupled to the four-electron reduction of oxygen the reaction mechanism cannot be entirely straight forward. Laccase can be thought to operate as a battery, storing electrons from individual oxidation reactions in order to reduce molecular oxygen. Hence the oxidation of four reducing substrate molecules is necessary for the complete reduction of molecular oxygen to water.

In general terms, substrate oxidation by laccase is a one-electron reaction generating a free radical. The initial product is typically unstable and may undergo a second enzyme-catalysed oxidation or otherwise a non-enzymatic reaction such as hydration, disproportionation or polymerisation. The bonds of the natural substrate, lignin, that are cleaved by laccase include, Cα- oxidation, Cα-Cβ cleavage and aryl-alkyl cleavage (Fig. 12.1a).

Laccases are similar to other phenol-oxidising enzymes, which preferably polymerise lignin by coupling of the phenoxy radicals produced from oxidation of lignin phenolic groups. Due to this specificity for phenolic subunits in lignin and its restricted access to lignin in the fibre wall, laccase has a limited effect on pulp bleaching unless redox mediators [e.g., 2,2′-azino-*bis* (3-ethy-benzthizoline-6-sulphonic acid (ABTS)] will be introduced in the reaction (Fig. 12.1b).

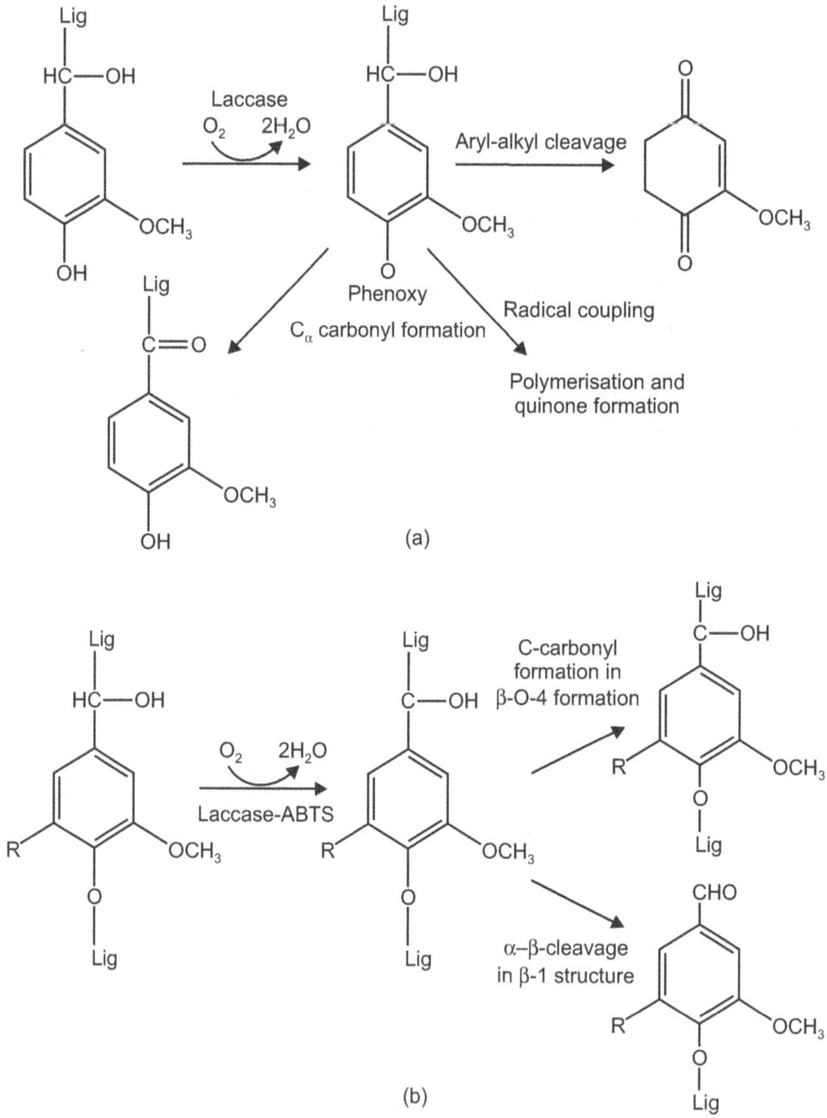

Figure 12.1: Oxidation of (a) phenolic subunits of lignin by laccase and (b) non-phenolic lignin model compounds by a laccase mediator system.

12.4 Laccase mediator system

With respect to other ligninolytic enzymes, laccase can oxidise only phenolic fragments of lignin due to the random polymer nature of lignin and to the laccase lower redox potential. Small natural low molecular weight compounds with high redox potential than laccase itself (> 900 mV) called mediators may be used to oxidise the non-phenolic part of lignin (Fig. 12.1b). Recently the discovery of new and efficient synthetic mediators extended the laccase catalysis towards xenobiotic substrates.

A mediator is a small molecule that acts as a sort of 'electron shuttle': once it is oxidised by the enzyme generating a strongly oxidising intermediate, the co-mediator (oxidised mediator), it diffuses away from the enzymatic pocket and in turn oxidises any substrate that, due to its size could not directly enter into the active site. Futhermore, the use mediators allows the oxidation of polymers by side-stepping the inherent steric hindrance problems (enzyme and polymer do not have to interact in a direct manner) (Fig. 12.2).

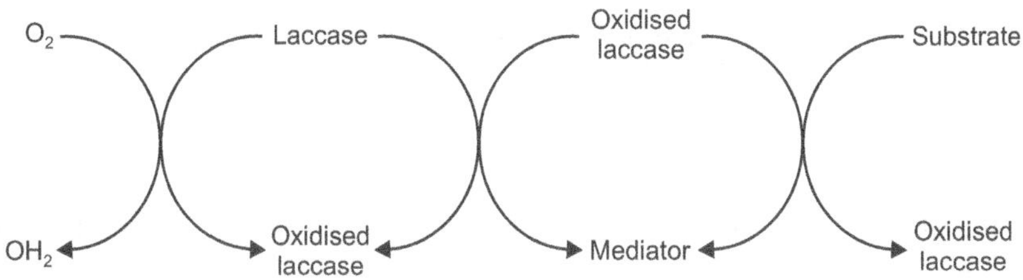

Figure 12.2: Catalytic cycle of a laccase-mediator oxidation system.

Alternatively, the oxidised mediator could rely on an oxidation mechanism not available to the enzyme, thereby extending the range of substrates accessible to it. It is therefore of primary importance to understand the nature of the reaction mechanism operating in the oxidation of a substrate by the oxidised mediator species derived from the corresponding mediator investigated. In the laccase-dependent oxidation of non-phenolic substrates, previous evidence suggests an electron-transfer (ET) mechanism with mediator ABTS, towards substrates having a low oxidation potential. Alternatively, a radical hydrogen atom transfer (HAT) route may operate with *N*-OH type mediators, if weak C-H bonds are present in the substrate.

More than 100 mediator compounds have been described but the most commonly used are the ABTS and the triazole 1-hydroxybenzotriazole (HBT). It is well known that cation radicals represent an intermediate oxidation step in the redox cycle of azines and upon extended oxidation and abstraction of the second electron, the corresponding dications can be obtained. The redox potentials of $ABTS^{+}$ and $ABTS^{2+}$ were evaluated as 0.680 V and 1.09 V respectively.

HBT belongs to the *N*-heterocyclic coumpounds bearing *N*–OH groups mediators. Consuming oxygen HBT is converted by the enzyme into the active intermediate, which is oxidised to a reactive radical (R–NO·) and HBT redox potential has been estimated as 1.1–1.2 V. Mediated laccase catalysis has been used in a wide range of applications, such as pulp delignification, textile dye bleaching, polycyclic aromatic hydrocarbon degradation, pesticide or insecticide degradation and organic synthesis. In pulp and paper industry, novel enzymatic bleaching technologies are attracting increasing attention because of concerns regarding the environmental impact of the chlorine-based oxidants currently being used in delignification or bleaching.

However, synthetic mediators are toxic, expensive and generally at concentrations above 1 mM inactivate the laccase. Novel approaches to over come this hurdles are coming up (from searching for natural mediators such as *p*-coumaric acid, 4-hidroxybenzoic acid, syringaldehyde, etc.), to the directed evolution of laccases.

12.5 General properties of laccase enzymes

Most fungal laccases are monomeric, dimeric or tetrameric glycoproteins. Glycosylation of fungal laccase is believed to play a role in secretion, susceptibility to proteolytic degradation, copper retention and thermal stability. Upon purification, laccase enzymes demonstrate considerable heterogeneity. Glycosylation content and composition of fungal glycoproteins can vary with growth medium composition. For this reason data can be heterogeneous. The molecular mass of the monomer ranges from about 50 to 100 kDa. An important feature is a covalently-linked carbohydrate moiety (10–45% of total molecular mass), which may contribute to the high stability of the enzyme. The sugar composition has been analysed in several examples, such as *Podospora ansenna* and *Botrytis cinerea, Trametes hirsuta, Trametes ochracea, Cerrena maxima* and *Coriolopsis fulvocinerea* and *Melanocarpus albomyces*.

12.5.1 Isozymes

Many laccase producing fungi secrete isoforms of the same enzyme. These isozymes have been found to originate from the same or different genes encoding for the laccase enzyme. The number of isozymes present differs between species and also within species depending on whether they are induced or non-induced. They can differ markedly in their stability, optimal pH and temperature and affinity for different substrates.

Furthermore, these different isozymes can modulate different roles in the physiology of different species or in the same species under different conditions. *Cerrena unicolour* secreted two laccase isoforms with different characteristics during the growth in a synthetic low-nutrient nitrogen/glucose medium. Various laccase encoding gene sequences have been reported from a range of ligninolytic fungi, these sequences encode for proteins between 515 and 619 amino acid residues and close phylogenetic proximity between them is indicated by sequence comparisons.

12.5.2 Substrate specificity of laccase

Laccases are remarkably non-specific as to their reducing substrates and the range of substrates oxidised varies from one laccase to another. These enzymes catalyse the one-electron oxidation of a wide variety of organic and inorganic substrates, including polyphenols, methoxy-substituted phenols, aromatic amines and ascorbate with the concomitant four-electron reduction of oxygen to water. Laccase has broad substrate specificity towards aromatic compounds containing hydroxyl and amine groups and as such, the ability to react with the phenolic hydroxyl groups found in lignin.

Kinetic data of laccases from different sources were reported. K_m values are similar for the cosubstrate dissolved oxygen (about 5–10 M), but V_{max} varies with the source of laccase (50–300 M/s). The turnover is heterogeneous over a broad range depending on the source of enzyme and substrate/type of reaction. The kinetic constants differ in their dependence on pH. K_m is pH-independent for both substrate and co-substrate, while K_{cat} is pH-dependent.

12.5.3 Influence of pH on laccase activity and stability

The pH optima of laccases are highly dependable on the substrate. When using ABTS as substrate the pH optima are more acidic and are found in the range 3.0–5.0. In general, laccase activity has a bell shaped profile with an optimal pH that varies considerably. This variation may be due to changes in the reaction caused by the substrate, oxygen or the enzyme itself. The difference in redox potential between the phenolic substrate and the T1 copper could increase oxidation of the substrate at high pH values, but the hydroxide anion (OH-) binding to the T2/T3 coppers results in an inhibition of the laccase activity due to a disruption of the internal electron transfer between the T1 and T2/T3 centres. These two opposing effects can play an important role in determining the optimal pH of the bi-phasic laccase enzymes. Laccase produced by *Trametes modesta* was fully active at pH 4.0 and very stable at pH 4.5 but its half-life decreased to 125 min at pH 3.0.

12.5.4 Influence of temperature on laccase activity and stability

The optimal temperature of laccase can differ greatly from one strain to another. The laccases isolated from a strain of *Marasmius quercophilus* were found to be stable for 1 h at 60°C. Farnet and others further found that pre-incubation of enzymes at 40°C and 50°C greatly increased laccase activity. Another technique that can be used to increase the stability of laccase is to immobilise the enzyme on glass powder by means of air-drying. This technique also has potential for the enzyme to be used on the glass powder matrix in specific biotechnology applications where stability is required. The laccase from *P. ostreatus* is almost fully active in the temperature range of 40–60°C, with maximum activity at 50°C. The activity remains unaltered after prolonged incubation at 40°C for more than 4 hr. Nyanhongo and others showed that laccase produced by *T. modesta* was fully active at 50°C and was very stable at 40°C but half-life decreased to 120 min at higher temperature (60°C).

12.5.5 Influence of inhibitors on enzyme activity

In general, laccases responds similarly to several inhibitors of enzyme activity. Many ions such as azide, halides, cyanide, thiocyanide, fluoride and hydroxide bind to the type 2 and type 3 Cu, resulting in the interruption of internal electron transfer and accordingly therefore inhibition of activity. Other inhibitors include metal ions (e.g., Hg^{+2}), fatty acids, sulphydryl reagents, hydroxyglycine, kojic acid, desferal and cationic quaternary ammonium detergents, the reactions with which may involve amino acid residue modifications, confirmational changes or Cu chelation. Confirmational changes are highly dependent on the state of oxidation of the copper atoms. This is one of the reasons for the sensitivity towards chelating agents. The selective removal of Cu by chelating agents (EDTA, dimethyl glyoxime, *N,N′*-dimethyldithiocarbamate, NTA) leads to a loss of catalytic activity.

12.6 Production of fungal laccases

Laccase activity was detected in the cultures of a wide range of fungi, from Ascomycetes to *Basidiomycetes* and from wood- and litter-decomposing fungi to ectomycorrhizal fungi. White-rot fungi have been studied extensively for application in biological pulping and bleaching because they are of the only organisms that are able to degrade lignin efficiently. White-rot fungi, such as *Coriolus versicolour* and *P. sanguineus*, are known producers of lignolytic enzymes that are involved in the natural delignification of wood. This group of fungi is the only known micro-organisms that have evolved complex enzymatic systems that enable them to degrade lignin.

In general, laccases occur as extracellular glyco-proteins, which allows for rapid removal from fungal biomass. One of the major limitations for the large scale applications of fungal laccases is the low production rates by both wild type and recombinant fungal strains according to Galhaup and others.

White-rot fungi constitutively produce low concentrations of various laccases when they are cultivated in submerged culture or on wood. Higher concentrations can be induced by the addition of various aromatic compounds such as 2,5-xylidine and ferulic acid. High concentrations of laccase have also been observed in old non-induced cultures. The mechanisms of metabolism in micro-organisms are used and controlled by its environmental conditions and medium composition. There are various response element sites in the promoter regions of laccase genes that can be induced by certain xenobiotic compounds, heavy metals or heatshock treatment.

12.6.1 Induction of laccase production

Laccase production has been found to be highly dependent on the conditions for the fungus cultivation and media supporting high biomass did not necessarily support high laccase yields. Ligninolytic systems of white-rot fungi were mainly activated during the secondary metabolic phase and were often triggered by nitrogen concentration or when carbon or sulphur became limiting. Laccases were generally produced in low concentrations by laccase producing fungi, but higher concentrations were obtainable with the addition of various supplements to media. The addition of aromatic compounds such as 2,5-xylidine, lignin and veratryl alcohol is known to increase and induce laccase activity. Many of these compounds resemble lignin molecules or other phenolic chemicals. Veratryl alcohol is an aromatic compound known to play an important role in the synthesis and degradation of lignin. The addition of veratryl alcohol to cultivation media of many white-rot fungi has resulted in an increase in laccase production. Some of these compounds affect the metabolism or growth rate while others, such as ethanol, indirectly trigger laccase production.

Eggert and others found that the addition of 2,5-xylidine as inducer had the most pronounced effect on laccase production. The addition of 10 μM 2,5-xylidine after 24 h of cultivation gave the highest induction of laccase activity and increased laccase activity nine-fold. At higher concentrations the 2,5-xylidine had a reduced effect, probably due to toxicity.

The promoter regions of the genes encoding for laccase contain various recognition sites that are specific for xenobiotics and heavy metals. These can bind to the recognition sites when present in the substrate and induce laccase production. White-rot fungi were very diverse in their responses to tested inducers for laccase. The addition of certain inducers can increase the concentration of a specific laccase or induce the production of new isoforms of the enzyme. Some inducers interact variably with different fungal strains.

Lee and others investigated the inducing effect of alcohols on the laccase production by *Trametes versicolour.* The enhanced laccase activity was comparable to those obtained using 2,5-xylidine or veratryl alcohol. It was postulated that the addition of ethanol to the cultivation medium caused a reduction in melanin formation. The monomers, when not polymerised to melanin, then acted as inducers for laccase production. The addition of ethanol as an indirect inducer of laccase activity offers a very economical way to enhance laccase production.

Lu and others found that there is a strong correlation between hyphal branching and the expression and secretion of laccase. The addition of cellobiose can induce profuse branching in certain *Pycnoporus* species and consequently increase laccase activity. The addition of cellobiose and lignin can increase the activity of extracellular laccases without an increase in total protein concentration. Osma and others showed that soya oil was the best inducer of laccase activities, attaining 4-fold higher than those obtained in the reference cultures.

The addition of low concentrations of Cu^{+2} to the cultivation media of laccase producing fungi stimulates laccase production. Palmieri and others found that the addition of 150 μM copper sulphate to the cultivation media can result in a fifty-fold increase in laccase activity compared to a basal medium. Employing copper sulphate as laccase inducer or supplementing the culture medium with veratryl alcohol, led to maximum values of laccase activity. A new basidiomycete, *Trametes* sp. 420, produced laccase in glucose medium and in cellobiose medium with induction by 0.5 mM Cu^{+2} and 6 mM *o*-toluidine.

12.6.2 Influence of carbon sources on laccase production

The carbon sources in the medium play an important role in ligninolytic enzyme production. Mansur and others showed that fructose induced 100-fold increase in laccase production of *Basidiomycete* sp. I-62. *T. versicolour* is an excellent producer of laccase in fermentation of mandarin peels. Glucose and cellobiose were efficiently and rapidly utilised by *Trametes pubescens* with high laccase activity. Similarly, the replacement of crystalline cellulose or xylan by cellobiose increased laccase activity of *C. unicolour* by 21- and 70-fold, respectively. Furthermore, in *T. versicolour* lignocellulosic material (barly bran) increased almost 50-fold laccase activity compared to the control culture with glucose. In the medium with the best carbon sources (mandarine peels and grapevine sawdust), both *Pleurotus eryngii* and *P. ostreatus* strain No. 493, showed the highest laccase activity. Glucose showed the highest potential for the production of laccase.

12.6.3 Influence of nitrogen sources on laccase production

White-rot fungi ligninolytic systems are mainly activated during the secondary metabolic phase of the fungus and are often triggered by nitrogen depletion. Monteiro and De Carvalho reported high laccase activity with semi-continuous production in shake-flasks using a low carbon to nitrogen ratio (7.8 g/g). Buswell and others found that laccases were produced at high nitrogen concentrations, although it is generally accepted that a high carbon to nitrogen ratio is required for laccase production. Laccase was also produced earlier when the fungus was cultivated in a substrate with a high nitrogen concentration and these changes did not reflect differences in biomass. Elisashvili and others observed highest laccase activity in *C. unicolour* IBB 62 in a medium with ammonium sulphate as the nitrogen source. D'Souza-Ticlo and others showed that well defined organic nitrogen sources such as glutamic acid and glycine were better than beef extract and corn steep liquor for laccase production. Heinzkill and others also reported a higher yield of laccase using nitrogen rich media rather than the nitrogen-limited media usually employed for induction of oxidoreductases.

12.6.4 Influence of pH on laccase production

There is not much information available on the influence of pH on laccase production, but when fungi are grown in a medium of which the pH is optimal for growth (pH 5.0), the laccase will be produced in excess. Most reports indicated initial pH levels set between pH 4.5 and pH 6.0 prior to inoculation, but the levels are not controlled during most cultivations. Nyanhongo and others reported that an initial pH of 7.0 was the best for optimal growth and laccase production by a newly isolated strain of *T. modesta*.

12.6.5 Influence of temperature on laccase production

It has been found that the optimal temperature for fruiting body formation and laccase production is 25°C in the presence of light, but 30°C for laccase production when the cultures are incubated in the dark. In general the fungi were cultivated at temperatures between 25°C and 30°C for optimal laccase production. When cultivated at temperatures higher than 30°C the activity of ligninolytic enzymes was reduced.

12.6.6 Inhibition of laccase production

It seems as if the use of excessive concentrations of glucose as carbon source in cultivation of laccase producing fungal strains has an inhibitory effect on laccase production. An increase in the amount of glucose in the media resulted in a delay of the laccase production. An excess of sucrose or glucose in the cultivation media can reduce the production of laccase, as these components allow constitutive production of the enzyme, but repress its induction when applicable. A simple but effective way to overcome this problem is the use of cellulose as carbon source during cultivation.

12.6.7 Heterologous expression

Most commercial laccases are produced in *Aspergillus* hosts. The functional expression of the *Myceliophthora thermophila* laccase in *Saccharomyces cerevisiae* by directed molecular evolution was recently reported, becoming the mutant T2 an idoneous scaffold for further improvements towards biotechnological applications. Another efficient expression system was developed for the basidiomycete *P. cinnabarinus* and this was used to transform a laccase-deficient monokaryotic strain with the homologous laccase gene. The yield was above 1.2 g of laccase per litre and represents the best laccase production reported for recombinant fungal strains.

12.7 Laccase immobilisation

Enzymes exhibit a number of features that make their use advantageous as compared to conventional chemical catalysts. However, a number of practical problems exist that reduce their operational life-time, such as their high cost of isolation and purification, their non-reusability, the instability of their structures and their sensitivity to harsh process conditions. Many of these undesirable limitations may be overcome by the use of immobilised enzymes.

Immobilisation is achieved by fixing enzymes to or within solid supports, as a result of which heterogeneous immobilised enzyme systems are obtained. By mimicking the natural mode of occurrence in living cells, where enzymes for the most cases are attached to cellular membranes, the systems stabilise the structure of enzymes, hence their activities. In the immobilised form enzymes are more robust and more resistant to environmental changes allowing easy recovery and multiple reuse.

Compared with the free enzyme, the immobilised enzyme has usually its activity lowered and the Michaelis constant increased. These alterations result from structural changes introduced to the enzyme by the applied immobilisation procedure and from the creation of a microenvironment in which the enzyme works, different from the bulk solution. Enzymes may be immobilised by a variety of methods (adsorption, entrapment, cross-linking and covalent bonding) mainly based on chemical and/or physical mechanisms.

Since the methods for the immobilisation procedures greatly influence the properties of the resulting biocatalyst, immobilisation strategy determines the process specifications for the catalyst.

Laccase immobilisation was extensively studied with a wide range of different methods and substrates. The adsorption of chromophoric-oxidised products on the surface of the immobilisation support often leads to enzyme inactivation phenomena.

12.8 Applications of fungal laccase in biotechnology

A number of industrial applications for fungal laccases have been proposed and they include paper processing, prevention of wine decolouration, detoxification of environmental pollutants, oxidation of dye and their precursors, enzymatic conversion of chemical intermediates and production of chemicals from lignin. Before laccases can be commercially implemented for potential applications, however, an inexpensive enzyme source needs to be made available. Two of the most intensively studied areas in the potential industrial application of laccase are the delignification and pulp bleaching and the bioremediation of contaminating environmental pollutants.

12.8.1 Delignification and pulp bleaching

In the industrial preparation of paper the separation and degradation of lignin in wood pulp are conventionally obtained using ClO_2 and O_3. Oxygen delignification process has been industrially introduced in the last years to replace conventional and polluting chlorine-based methods. In spite of this new method, the pre-treatments of wood pulp with laccase can provide milder and cleaner strategies of delignification that also respect the integrity of cellulose. Lignocellulose is a common substrate for laccase and the laccase ability to break down nonphenolic ligno-cellulose is provided by certain phenolic compounds acting as mediators. More recently, the potential of this enzyme for cross-linking and functionalising lignocellulose compounds was discovered.

Laccases can be used for binding fibre-, particle- and paper-boards. However, different wood decaying basidiomycetes have shown a highly variable pattern of laccase formation and this subject requires more detailed experiments.

12.8.2 Bioremediation

Laccases have also shown to be useful for the removal of toxic compounds through oxidative enzymatic coupling of the contaminants, leading to insoluble complex structures. The degradation of a variety of persistent environmental pollutants, in particular phenols, was also observed.

Phenolic compounds are present in wastes from several industrial processes, as coal conversion, petroleum refining, production of organic chemicals and olive oil production among others. Immobilised laccase was found to be useful to remove phenolic and chlorinated phenolic pollutants. Laccase was found to be responsible for the transformation of 2,4,6-trichlorophenol to 2,6-dichloro-1,4-hydroquinol and 2,6-dichloro-1,4-benzoquinone. Laccases from white rot fungi have been also used to oxidise alkenes, carbazole, *N*-ethyl-carbazole, fluorene and dibenzothiophene in the presence of HBT and ABTS as mediators. Isoxaflutole is an herbicide activated in soils and plants to its diketonitrile derivative, the active form of the herbicide. Laccases are able to convert the diketonitrile into the acid. The study of the laccase-mediator system in the bioremediation of polycyclic aromatic hydrocarbons (PAHs) has been extensively reported. In particular, the combination of several mediators looking for synergetic effects along with the use of natural mediators open new alternatives in this field.

12.8.3 Alternative applications

Laccases do not only show potential for biological delignification of pulp but also for other applications. Laccases can be applied for the treatment of and detoxification of soils containing phenolic pollutants as well as other polluted systems due to the broad substrate range of the enzyme. The application of laccase for dyeing of materials with sulphur and reduced vat dyes has been patented. The use of laccase for the treatment of textile and bleach-plant effluents has also been investigated with success. Recently, increasing interest has arises on the application of laccase as a new biocatalyst in organic synthesis. The use of laccase for the production and treatment of beverages and as biosensor for the estimation of phenol or other enzymes in fruit juice has also been proposed. Recently, they have applications in other field such as in the design of biofuel cells.

12.9 Fungal enzymes in occupational disease

The industrial utility of fungi has been well known since antiquity. In addition to the role of fungi as saprophytes in the environment, many species have commercial use, for example, mushrooms as food sources, ingredients in food preparation (cheese flavouring *Penicillium roqueforti*), alcoholic fermentation and the conversion of sugars in bread dough to carbon dioxide (*Saccharomyces cerevisiae*). In Asia, *Aspergillus oryzae* is an essential ingredient for the production of soya sauce and the fermented drink, sake. *Rhizopus* spp. secrete a wide variety of enzymes including cellulolytic, proteolytic, lipolytic and pectinolytic enzymes that are used in the production of various foods such as Tempe from Indonesia. *Rhizopus oryzae* has also been identified as a biocatalyst for biodiesel fuel production.

Other fungi such as *Yarrowia lipolytica* have more recent applications in the biodegradation of industrial products. Advances in industrial enzymology following World War II have enabled researchers to identify and utilise various enzymes and proteases that fungi produce to break down carbohydrate and lignin containing plant material in the environment. To date, close to 200 fungal enzymes have been purified from fungal cultures and the biochemical and catalytic properties characterised. These enzymes have great utility in pharmaceutical, agricultural, food, paper, detergent, textile, waste treatment and the petroleum industries. Industrial fungal enzymes are high-molecular-weight proteins that are catalysts. A description of the common enzymes used in various industries is presented in Table 12.1. The most widely used enzymes of occupational importance are derived from the genus *Aspergillus* and include α-amylase, xylanase and cellulase. Other enzymes are also utilised from rhizosphere fungal species belonging to the genera *Rhizopus* and *Humicola*.

Table 12.1: Fungal enzymes utilised in different industries and associations with occupational sensitisation.

Industry	*Fungal enzyme*	
	Characterised occupational allergen	*Uncharacterised occupational allergen*
Agriculture	Protease, lipase	
Animal feed	α-amylase, cellulase, lipase, phytase, protease and xylanase	Beta-glucanase, endo-xylanase
Pulp and paper production	Cellulase, hemicellulase, lipase and xylanase	Esterase, laccase, lignin peroxidase, manganese peroxidase, pectinase and mannose
Waste management	Lipase	Esterase, cytochrome P450, laccase, lignin peroxidase, manganese peroxidase and monooxygenase
Biotechnology	α-amylase, cellulase, glucoamylase, hemicellulase and protease	Cytochrome P-450 monooxygenase, glucose oxidase, glutathione-transferase, lignin peroxidase and manganese peroxidase
Detergent	α-amylase, cellulase, lipase and protease	
Food processing	α-amylase, cellulase, glucoamylase, lactase, lipase, protease and xylanase	Glucose isomerase, invertase and pectinase
Biofuels	α-amylase, cellulase, glucoamylase, protease and xylanase	
Bakery	α-amylase, cellulase, glucoamylase, hemicellulase, lipase, protease and xylanase	Glucose oxidase, lipoxygenase
Brewing and wine production	Cellulase, glucosidase, protease and xylanase	Alpha-acetolactate decarboxylase, beta-glucanase and pectinase
Pharmaceutical	Lactase, lipase and protease	Alpha-galactosidase, catalase, cytochrome P450 oxygenase and glutathione transferase
Textile	α-amylase, cellulase, lipase, protease and xylanase	Catalase
Leather processing	Lipase, protease	
Hygiene products	Glucoamylase, protease	Catalase, glucose oxidase

These enzymes usually have intracellular or other functional roles associated with apical hyphal growth. It is uncommon for individuals in the general population to be exposed and sensitised to these antigens. In fact, in the general population, the prevalence of sensitisation to fungal enzymes has been reported to be as low as 1% and as high as 15%. However, in the occupational environment, workers that handle purified fungal enzymes are at an increased risk of becoming sensitised to enzymes. This is especially the case for workers whose occupation requires debagging, sieving, weighing, dispensing and mixing enzymes. Eight-hour time-weighted average exposures demonstrate that occupations weighing the enzyme preparations have the lowest average exposure compared to those workers that sieve. These workers are often exposed to levels of dust that exceed 4 mg m^{-3}, the threshold limit value (TLV) for inhalable dust. For other industrial environments that use lipase and cellulose in production, occupational exposure is highest in production areas and laboratories.

Adverse health effects associated with enzyme exposure are well characterised in the baking industry. In some countries, bakery exposures to enzymes are one of the leading causes of occupational allergy. Fungal enzymes are commonly used as baking additives to improve the dough, increase shelf life and decrease production time. Airborne concentrations ranging from 5.3 ng m^{-3} to 200 ng m^{-3} have been reported in occupational environments. Occupational sensitisation to fungal enzymes was first reported by Flindt. Later, Baur and others demonstrated IgE sensitisation in workers handling these products. Since the original study, fungal enzymes have been identified as potent allergens in the occupational environment. Prevalence of sensitisation to *Aspergillus* enzymes ranges from 8% for glucoamylase, 11% for xylanase, 13% for cellulase and up to 34% for α-amylase. Sensitisation to α-amylase in bakery workers results in decreased peak expiratory flow and OA.

In one report, workers exposed to fungal enzymes induced an immediate bronchospastic reaction. In the United States, the prevalence of work-related wheeze, runny nose, frequent sneesing and specific IgE to fungal enzymes was significantly higher among highly exposed workers. However, other irritantinduced mechanisms associated with high total dust levels have also been reported in a cohort of British bakers. To date, atopy has been hypothesised to be an important risk factor for OA to fungal enzymes.

Occupational exposure to enzymes has been demonstrated in other industries including manufacturing, pharmaceutical, food processing, animal feed and biotechnology. Like in baking environments, workers handling or in direct contact with fungal enzymes and with a history of atopy are at increased risk of becoming sensitised. Sensitisation to proteolytic enzymes has also been demonstrated in the manufacture of detergents. In the future, additional uses for fungal enzymes in industrial environments will be identified. Recent examples include the use of α-amylase and glucoamylase for the production of ethanol in the biofuel industry. If proper methods of exposure prevention are not followed and exposure is not monitored in these industries, it is possible that new groups of workers will suffer adverse health outcomes and become sensitised to enzymes. In the following sections we describe the major fungal enzymes, prevalence of sensitisation and occupational environment that they are most likely to be encountered.

12.10 Emerging occupational fungal enzyme exposures

The utility of fungal enzymes to degrade xenobiotics and organic compounds in the industrial sector continues to be recognised. Fungal enzymes are now being used for a variety of purposes across many different industries. Improved biochemical and molecular technologies have enabled the production of other potentially allergenic proteins.

Interestingly, alpha-galactosidase has been associated with delayed anaphylaxis, angioedema, or urticaria in sensitised patients following the ingestion of beef, pork, or lamb. Although the role of alpha-galactosidase and these other enzymes following occupational exposure remains unclear, these studies provide preliminary insight into the possible potency of these allergens in industrial environments.

12.11 Immunodiagnostic detection methodologies

Occupational allergic sensitisation to fungal enzymes is diagnosed clinically using available *in vivo* SPT reagents, or *in vitro* assays such as Phadia ImmunoCap. However, SPT reagents for most of the fungal enzymes used in industrial settings are not commercially available and have to be either custom ordered or prepared individually by the investigator. Methods for SPT extract preparation that are used by investigators in the field have been previously described by Quirce and others. *In vitro* diagnostic tools that can quantify the amount of specific IgE to an occupational allergen are not readily available except in research laboratories where investigators prepare their own inhibition or radioallergosorbent enzyme-linked immunosorbent assay (ELISA) to quantify specific IgE. To date, α-amylase (k87) is the only fungal enzyme available on the Phadia ImmunoCap testing panel. To confirm OA caused by fungal enzymes, bronchial provocation tests can be undertaken to document immediate or late-phase responses to fungal enzymes. Positive immediate response criteria used in workers exposed to enzymes include a greater than 20% fall in FEV1, whereas a late-phase response has been considered positive when there is a 30% or greater fall in peak expiratory flow rate.

In order to better understand the relationships between occupational fungal enzyme exposure and clinical symptomology, accurate information on the distribution and quantity of the fungal enzyme in the occupational environment will be required. Immunodiagnostic methods that utilise antibodies could provide standardised methods for quantifying fungal enzyme biomarkers in a variety of occupational environments. Following validation and interlaboratory comparison, the assays could be used for exposure assessment to determine the existence of exposure-response relationships. This information is critical for the development of future threshold limit values (TLVs) and other occupational standards.

Several antibodies and immunodiagnostic methods have been produced to detect industrial fungal enzymes, in particular α-amylase. These methods have been employed in field investigations and used to quantify the concentration of the enzyme from collected air samples. Bogdanovic and others used an enzyme immunoassay with a sensitivity of 25 pg/mL to quantify α-amylase in airborne and surface dust samples collected from five bakeries. In the same study, a lateral flow immunoassay for α-amylase was compared to the reference enzyme immunoassay. The sensitivity of the lateral flow assay was 1–10 ng/mL and extracts with >5 ng/mL allergen were positive in the lateral flow assay. In a study of 507 personal air samples, Houba and colleagues used a rabbit IgG capture immunoassay to quantify α-amylase in specific baking job category. Concentrations of α-amylase up to 40 ng m^{-3} were quantified and workers directly involved with dough preparation had the highest exposures.

Using the same rabbit IgG sandwich assay, Nieuwenhuijsen and others identified dispensing and mixing areas to have the highest α-amylase exposure in British bakeries and flour mills.

Two monoclonal antibody- (mAb-) based ELISAs have been developed for the detection of α-amylase in the occupational environment. Assay sensitivities ranged from 0.2 ng/mL to 0.6 ng/mL. A quantitative mAb mediated dot blot assay has also been previously described for cellulase and xylase, the detection limits reported were 20 ng m^{-3} and 2 ng m^{-3}, respectively. mAbs to other fungal enzymes, such as xylanase have been produced and reported in the literature. Similarly, the detergent industry has produced antibodies and immunoassays for several common fungal enzymes and these have been utilised in industrial hygiene safety programmes to mitigate worker exposures. The development of fungal enzyme specific mAbs in combination with immunodiagnostic techniques will further our knowledge of the exposure-response relationships in occupational environments. Using these methods will also help enable the development of standards and focus on the prevention of sensitisation in heavily contaminated work environments.

12.12 Allergen avoidance and directions for the future

Exposure to fungal enzymes, in particular α-amylase, is a considerable health risk in a number of industries. Cross-sectional studies have shown that processing workers in high-exposure categories who handled fungal enzymes are up to ten times more likely to be sensitised to fungal enzymes than workers in the low-exposure category. Highest concentrations of enzymes in the inhalable fraction were encountered among workers located in dispensing, mixing, weighing and sieving occupations.

Airborne concentrations as high as 40 ng m^{-3} and in some cases even higher (200 ng m^{-3}) have been reported for sensitised workers located in these handling areas. Concentrations in the low ng m^{-3} range have been associated with an increased frequency of sensitisation. For other fungal enzymes, such as phytase, similar findings have been reported.

The continued utilisation of other previously overlooked enzymes as well as new genetically engineered enzymes in various industries will continue to provide diagnostic challenges, even for the most seasoned occupational medicine professional. It is likely that new cases of occupational allergic disease will emerge following exposure to fungal industrial enzymes during the next decade. In response, identification of exposure-response relationships will be critical for the development of TLVs and occupational exposure levels. However, this will depend on the development of suitable diagnostic antibodies and immunoassays. Currently, subtilisin, a sereine endopeptidase derived from *Bacillus subtilis*, is the only enzyme for which the American Conference of Governmental Industrial Hygienists (ACGIH) has established a TLV value (60 ng m^{-3}). The European Union Directive also classifies the fungal enzymes cellulase and α-amylase with the risk phrase R42 (may cause sensitisation by inhalation). There are currently no consensus standards for other industrially utilised fungal enzymes.

As a precautionary measure, it has been concluded that all enzymes should be regarded as an allergen that can exacerbate respiratory sensitisation in susceptible populations. Baur has further proposed that all enzymes should be classified as R42 according to the European Union Directive criteria. Although intervention in the bakery industry has had little to no effect, installation of engineering controls and implementation of personal protective equipment programmes in animal feed workers exposed to phytase was shown to result in the immediate cessation of hypersensitivity symptoms. Improvements in biotechnology have also included the encapsulation of some enzymes and proteins. These engineering controls have been proposed to reduce occupational exposure to enzymes, however, encapsulation alone may not completely prevent enzyme-induced allergy and OA. To date, the detergent industry has implemented a derived minimal effect level (DMEL) of 60 ng m^{-3} for pure enzyme proteins. Although this DMEL was provided as guidance by the ACGIH, other manufacturers have implemented their own occupational exposure guidelines (OEGs) for fungal enzymes such as α-amylase (5–15 ng m^{-3}), lipase

(5–20 ng m^{-3}) and cellulase (8–20 ng m^{-3}). In addition, the detergent industry has developed a medical surveillance programme to identify and correct elevated exposures before occupational illnesses occur. As a result, the incidence of occupational allergy has dropped substantially. Implementation of DMELs and OEGs will further assist in the reduction of occupational exposure. Reducing worker exposure to fungal enzymes in industry by the implementation of engineering controls and other allergen avoidance strategies will continue to mitigate personal exposure and further reduce the occupational health risk.

To sum up, laccases are widespread in nature, being produced by a wide variety of plants, fungi and also bacteria. The functions of the enzyme differ from organism to organism and typify the diversity of laccase in nature. Laccases catalyse the oxidation of phenolic compounds whilst simultaneously reducing molecular oxygen to water. The catalytic ability of laccases has, not surprisingly, led to diverse biotechnological applications of this enzyme.

The introduction of the laccase-mediator system provides a biological alternative to traditional chlorine bleaching processes. The laccase enzyme has a wide field of application including the pulp and paper industries, the treatment of various industrial effluents, enzymatic decolouring of material and bioremediation of soils. One of the limitations to the large-scale application of the enzyme is the lack of capacity to produce large volumes of highly active enzyme. These problems can be solved with the use of recombinant organisms or screening for natural hypersecretory strains. Environmental factors influence the ability of fungi to produce high titres of laccase and different strains react differently to these conditions. One should thus select a strain capable of producing high concentrations of a suitable enzyme and then optimise conditions for laccase production by the selected organism. It is therefore not surprising that this enzyme has been studied intensively since the nineteenth century and yet remains a topic of intense research today.

13

Soil enzymes

13.1 Introduction

Soil enzyme activities are of paramount importance to evaluate functions of micro-organisms, cycling of nutrients and carbon-sources decomposition. Combination of emerging technologies with information gained by current techniques will allow the development of new, more microbially explicit biogeochemical models that will be better able to predict the impacts of enzyme mediated soil processes. Moreover, to achieve an enhanced comprehension of carbon cycling and its availability dynamics among ecosystems, major attention has to be devoted to extracellular enzyme (EE) activities operating not only in terrestrial environment but also in freshwater and marine environments, where they are alike the main actors of nutrient cycling. Several features are common to these different environmental enzyme groups. As underlined by Arnosti and others 'While the relative importance of specific structuring factors (e.g., pH, presence of surfaces, etc.), varies dramatically among terrestrial, freshwater and marine environments, the underlying process is the same: micro-organisms produce EEs in order to gain a selective advantage, those enzymes subsequently catalyse biogeochemical cycles'. Therefore, integrated enzymatic studies across these different environments may greatly assist researchers to achieve a thorough comprehension 'of the ultimate controls and biogeochemical consequences of EEs across environments'.

Soil enzymes assist agricultural sciences: Soil enzymes are involved and assist all activities fundamental to agricultural sciences. Indeed, all biochemical transformations taking place in soil are dependent on the presence of enzymes. They are a composite of different intracellular and extracellular enzymatic components, produced by microbial organisms (bacteria, fungi), or derived from animal and plant sources (plant roots, lysed plant residues, digestive tracts of small animals, etc.). Soil enzymes are located and/or distributed in the soil matrix and their spatial variability in a soil profile and localisation in soil structural fractions of different nature and size will influence their activity and performance.

Soil enzymes: Soil enzymes increase the reaction rate at which plant residues decompose and release plant available nutrients. The substance acted upon by a soil enzyme is called the substrate. For example, glucosidase (soil enzyme) cleaves glucose from glucoside (substrate), a compound common in plants.

Enzymes are specific to a substrate and have active sites that bind with the substrate to form a temporary complex. The enzymatic reaction releases a product, which can be a nutrient contained in the substrate.

13.2 Sources of soil enzymes

Sources of soil enzymes include living and dead microbes, plant roots and residues and soil animals. Enzymes stabilised in the soil matrix accumulate or form complexes with organic matter (humus), clay and humus-clay complexes, but are no longer associated with viable cells. It is thought that 40 to 60% of enzyme activity can come from stabilised enzymes, so activity does not necessarily correlate highly with microbial biomass or respiration. Therefore, enzyme activity is the cumulative effect of long term microbial activity and activity of the viable population at sampling. However, an example of an enzyme that only reflects activity of viable cells is dehydrogenase, which in theory can only occur in viable cells and not in stabilised soil complexes.

13.2.1 Categories of enzymes recognised in soil

Soil is the place where all natural and anthropogenic activities necessary to the life of humans, animal and plants occur. Forests, plants, grasses, crops grow on soil, superficial water basins and rivers are and flow on soil, animals spend their life and move on soil, between cycles of incorporation of food and release of droppings, these latter contributing to the normal recycling of soil and its nutrients. Soil is where humans have established their residence and built all structures necessary for their life.

Soil is also the place where hazardous residues deriving from anthropic activities are very often released with dangerous and frequently irreversible effects on its safety and consequently human health. Often these human activities have produced desertification, deficit or even loss of biodiversity, alteration of the soil matrix, deficiency of organic matter and nutrients. It appears therefore evident as what happens on and within soil is of paramount importance for maintaining soil health and productivity and consequently for a balanced and integrated life on earth.

An equilibrated, safe and productive soil exists when its biological, biochemical, physical and chemical properties, all correlated to each other, contribute to sustain all activities occurring in it. Among these properties the activities of soil enzymes play a fundamental role because they mediate numerous chemical reactions involved in soil nutrient cycling, transformation of plant and microbes debris, mineralisation and transformation of organic matter within the carbon cycle, transformation and degradation of potentially hazardous pollutants, thus contributing to the restoration and remediation of polluted soils. Therefore, soil enzymes are involved and assist all activities fundamental to agricultural sciences and as such they may be used as useful and suitable indicators of microbial nutrient demand and soil health and quality.

Two main categories of enzymes can be recognised in a soil: Intracellular enzymes (IEs) and extracellular enzymes (EEs). All enzymes occurring and functioning in living cells, i.e., microbial, plant and animal cells residing in soil belong to the first category and as such they are considered as a whole of microbial, plant and animal metabolism. Enzymes produced by living cells but secreted outside in their neighbouring environment can be considered EEs. They may explicate their activities still bound or associated to their producing cell or at a distance from the parent cell, being free in the liquid phase. EEs can be in turn be divided in more categories that do not have a defined borderline being interrelated with each other and possibly transforming to each other.

An interesting group of soil EEs is those classified in the past by Gianfreda and Bollag as 'naturally immobilised enzymes' and recently renamed as stabilised enzymes (SEs). Once released into the soil from their originating cells, these enzymes are likely adsorbed, linked, anchored or embedded on/in/to solid supports such as clays, clay minerals, organic matter and organo-mineral complexes.

Although a huge amount of findings and related publications are available in literature on soil enzymes, there is still a consistent number of open, not yet resolved questions about several aspects regarding EEs such as: localisation of EEs in the soil matrix, their contribution to substrates turnover and global biogeochemical processes, their regulation by biotic and abiotic soil factors and consequent effects on soil organic carbon dynamics, their relationship with producing organisms and relative counter-effects, development of suitable models to describe both the function and biomics of ecosystems, the possible development of a suitable soil enzyme index of soil status and health and role that soil enzymes may play in the dynamic of plant nutrients, in the control and restoration of environments polluted by both pollutants and pathogenic diseases.

13.3 Problems with soil enzymatic methods

Problems with soil enzymatic methods are summarised as under:

1. Current and standard soil enzymatic methods usually provide limited information on the real activity of soil enzymes. Assays performed under laboratory conditions measure only potential and not real enzyme activities without giving any information on *in situ* activities or on the relationship between the two types of activities, moreover they take also into account the contribution of stabilised enzymes that could not be active under *in situ* conditions.

2. These assays do not provide information neither on the production of enzymes or their producing organisms, neither on their turnover rates or their meaning.
3. Substrates used in these assays are simple, soluble compounds whereas substrates of enzyme activities in soil are usually larger, complex insoluble polymers that are degraded in simpler monomeric products.
4. Assays under lab conditions are usually carried out at fixed pH, temperature, soil moisture, etc., and thus they do not provide information on sensitivity of enzyme activities to changes of these parameters.
5. Minor modifications occurring in experimental procedures can lead to errors and affect not only the final values of measured activities but also their comparison with other data collected by different researchers.
6. Several enzymes are involved in nutrient dynamics, consequently a single enzyme activity cannot be used as valuable indicator of such dynamics.
7. Since soil enzyme activities can be influenced by several factors such as changes in soil management, plant cover, changes in environmental conditions, presence or addition of fertilisers, pesticides or contaminants, current measurement assays are often not suitable to allow a right interpretation of the response of the enzyme activities to these factors.

13.4 Soil enzyme activities: Measurement, location and distribution in the soil matrix

13.4.1 Measurement of soil enzymes

As outlined by several authors and continuously stated in various literatures, the most, still present, important shortcoming, limiting a right interpretation and utilisation of soil enzyme activities (and their changes by both abiotic and biotic processes) is the lacking of universal, standardised, uniform and optimised methodologies to measure the activities of enzymes in soil. Therefore, it is difficult to use absolute soil enzyme activity values as a powerful tool for understanding soil biological processes.

It is well established that 'maximum potential' and non-actual or 'realised' enzymatic activities are measured with assays now available. Enzymes are usually measured indirectly by determining their activity in the laboratory using biochemical assays that, therefore, do not represent true *in situ* activity levels. Methodological aspects to be considered in the evaluation of soil enzymatic activity are shown in Table 13.1.

In situ measurement: The most realistic method to detect enzyme activities in soil would be their direct visualisation in undisturbed soils, i.e., *in situ* (intended as on site) measurements.

Table 13.1: Methodological aspects to be considered in the evaluation of soil enzymatic activity.

How to measure a soil enzyme activity?	*In situ* measurements: • Histochemical techniques • Electron microscopy • Zymography *In vitro* measurements: • On soils in toto (soil slurry) • On soil extracts		
Activity measurement problems	Sample preparation: • Collection • Sterilisation Sample storage: • Air-dried or moist • Refrigerated • Room temperature		
Assay conditions	Determination of product formation or substrate disappearance under defined reaction conditions: • Saturating substrate concentrations • Optimal pH • Optimal temperature • Stirring		
Differences between In vitro and field conditions		***In vitro assays***	***Field conditions***
	Substrate	Excess	Limiting
		Homogenous	Heterogeneous
		Soluble	Insoluble
		Artificial	Natural
	Buffer	Present	Absent
	Temperature	Optimal	Variable
	Flora and fauna	Absent	Present
	Shaking	Often	Stationary
	Reproducibility	High	Low
Difficulties in the determination under of soil enzymatic activity	Contribution by inorganic catalysts to the reaction investigation Possible growth of micro-organisms and *de novo* synthesis of enzymes during the assay Stability of the reaction product in the reaction mixture Influence of the soil treatment on the activity assay Contribution of the different soil enzymatic categories		

Attempts in this way were made early by Ladd and others by using electron microscopic observations of soil sections, after their treatment with suitable compounds, which allow enzymes in soil structure to be localised, or reaction with substrates giving rise to products easily visualisable as coloured compounds. Histochemical staining was also considered a sensitive and suitable method to detect enzyme activity in soil as in the case of alkaline phosphatase in extra radical mycelium of *Arbuscular mycorrhizal* (AM) fungi. These methods, however, had not a large application for the measurements of enzymes associated with soil particles because of possible interferences by soil minerals and humic substances that are electron-dense components. Moreover, in the field diffusion rates may reduce the rate at which enzymes and substrate can interact to each other.

Visualisation *in situ* of the activities of hydrolases (e.g., acid phosphatase, amino peptidase, chitinase and β-glucosidase) in forest soils was achieved by a modified root window-based, enzyme-imprinted, membrane system. Substrates giving rise to coloured or fluorescent products were utilised with a good response in the field.

Recently, a different approach based on an *in situ* zymography for analysis of the two-dimensional distribution of enzyme activities in soil was utilised. By means of thin gels with embedded substrates it was possible to quantify the activity of protease and amylase in the rhizosphere of lupine (*Lupinus polyphyllus*) grown in rhizoboxes. Smith and others concluded that since zymography 'does not require destruction of soil structure, it likely pictures enzyme activities more realistically than standard enzyme assays', thus offering 'a promising tool for mapping distributions of enzyme activities in soils in a work- and cost-efficient way'.

A very new method that allowed to rapidly measuring multiple soil biological properties simultaneously was proposed and described by Dick and others. It utilised a near infrared spectroscopy (NIRS) method requiring not expensive equipment and usable for a very large numbers of samples.

Laboratory conditions: The majority of assays described in literature are, however, usually carried out under laboratory conditions. Soil are sampled at 0–10/20 cm layer and stored refrigerated at 4°C for 10–15 days, at the most. Debate is still present whether it is much more convenient to use soil slurries or soil extracts, to store the soil sample in the refrigerator or in freezer or also after air-drying at room temperature. Since all manipulations occurring in the sampling, preparation, handling and storage of soil samples may disturb the soil, altering several factors connected to enzyme activity (e.g., activity of micro-organisms, status of the enzyme, whether immobilised or free, substrate availability, etc.), it is important to well define and describe how the soil has been sampled, handled, treated and stored prior to the determination of the enzyme activity. This allows to reproduce as much as possible the procedure and to compare with other results reported in literature.

Without entering in detail, soil sampling strategy (i.e., the sampling time, the number of samples, their horizontal and vertical distribution, the sampling procedure and device) should be representative of the natural situation of studied soils and should consider temporal and spatial dynamics of biochemical and microbiological soil properties to be measured. Indeed enzyme activities usually display evident temporal and spatial variations and generally decrease with increasing soil depth. An accurate sampling procedure can be easily assured if simple rules well described in literature are followed.

Soil storage: As regards storage, there are several findings demonstrating that a short storage (at the most 10–15 days or less) of field moist soils at 4°C is preferable as respect to the other storage procedures that may affect also the response of florescent substrates.

DeForest evaluated whether soil storage and processing methods significantly influenced measurements of potential *in situ* enzyme activity in acidic forest soils. Six soil EEs (NAGase, phosphatase, phenol oxidase, β-glucosidase, β-xylosidase and peroxidase) were measured using 4-4-methyl-umbelliferone (MUF)-linked substrates and L-dihydroxyphenylalanine (L-DOPA) 2, 7, 14 and 21 days after collection on soil stored at both 4 and 20°C. Results indicated that storage temperature did not influence enzyme activity values whereas extended time in buffer did. Analyses within 2 hr were considered the best measure of potential *in situ* enzyme activity and the benchmark for all statistical comparisons.

Moreover, the activities of β-glucosidase, β-xylosidase and peroxidase were insensitive to storage and processing methods.

Later, Wallenius and others showed that, 16 weeks-storage of frozen or air-dried soil, humus or compost differently affected the assay of ten hydrolysing enzyme activities determined by artificial fluorogenic substrates and the impact depended on the soil matrix and the assayed enzyme. In particular, freezing affected soil enzyme activities less than air-drying that decreased compost activity more than 50%. Similar results were obtained by Peoples and Koide.

Two soils with significantly different enzyme activities were tested for 1,4-β-cellobiohydrolase, acid phosphatase and β-*N*-acetylglucosaminidase as fresh, frozen or dried samples and the ratios of each soil activity to that the other soil for the three storage treatments were determined. Freezing and drying significantly affected activity ratios when compared to the fresh control for all three enzymes, being the effect of freezing much less than that measured for drying.

Soil storage affects: Each investigator should consider whether soil storage affects within-experiment treatment comparisons of soil enzyme activity.

Therefore, the type of storage has to be considered being often dependent on the physical-chemical characteristics of the investigated materials and very often site and enzyme-specific.

Soil slurry and soil extract: Soil slurry is usually preferred to soil extract being this latter possibly disruptive of natural soil conditions and reductive as respect to the whole enzymatic activities of the investigated soil. Not-with-standing, Vancon and Keen demonstrated that a new method for extracting soil β-1,4-glucanases (cellulases) and β-1,3-glucanases (laminarinases) was reproducible, could be completed in 1 day and measured twice as much enzyme activity than the standard passive soil enzyme extraction procedure. It consisted in the mechanical disruption of soil samples with acid washed ceramic and glass beads, successively processed six times with phosphate buffer (pH 7.0) by bead-beating for 30 s and 1 min incubation in an ice bath. Crude enzyme extracts were used for measuring enzyme activities. The method allowed multiplex enzymatic assays from several soil samples at one time.

Assay conditions: As previously discussed, assay conditions such as presence or absence of a buffer, pH, substrates and their concentration, temperature, shaking of soil, inhibitors of microbial proliferation, etc., markedly affect the measured activity. Usually a soil enzyme assay is based on the use of a buffered solution of a synthetic, artificial substrate at a concentration high enough (saturation concentration) to be assumed constant throughout the time course of the enzymatic reaction and assuring a zero-order kinetics. Moreover, substrate concentration should be very much larger than that of the enzyme to allow a reaction rate proportional to the enzyme concentration. A buffered vs. an unbuffered condition is usually chosen when the disappearance of substrate and/or the formation of reaction products may change the pH of the soil slurry, since an optimal pH value is required during the assay.

These *in vitro* optimised conditions, very different from those existing *in vivo*, are usually well target for studying enzymes characteristics. By contrast, conditions simulating as much as possible those present in nature, i.e., soil conditions, should be preferred when the goal is to study soil enzymes from an ecological point of view.

Substrates commonly used: Substrates commonly used in the past and very often still used in recent research, either as bench or microplate assays, are those including in their molecular structure a chromophore component which, after the enzymatic reaction, achieves spectroscopic properties, i.e., becomes detectable and measurable by spectroscopic determination, or releases as product molecules itself coloured or colourable under particular chemical conditions.

Classical examples of those substrates are *p*-nitrophenyl derivatives, used for assaying the activity of hydrolases such as phosphatases, glucosidases and sulphatases. The main product of the enzymatic hydrolysis is *p*-nitro phenol that under alkaline conditions acquires a yellow colour, whose intensity is detectable at 400–405 nm wavelengths.

An example of the use of these substrates was the high-throughput microplate assay implemented for simultaneous colorimetric quantification of multiple enzyme activities in soil. The activities of β-*N*-acetyl-glucosaminidase, β-glucosidase, β-glucosidase, β-galactosidase and β-galactosidase were measured by using *p*-nitrophenyl substrates. The assay being applicable to a large number of soils within hours of sampling was considered suitable for system-level evaluations.

Fluorimetric and spectrophotometric assays: Moscatelli and others by comparing fluorimetric and spectrophotometric assays of β-glucosidase in soils subjected to different management practices tried to find out reasonable answers on the role of β-glucosidase kinetic parameters as reliable indicators in detecting changes due to soil management and the capability of different analytical approaches to give the same results. Both fluorimetric and spectrophotometric approaches highlighted changes occurring under organic and conventional management. Higher V_{max} values were measured under organic cropping system although the whole catalytic behaviour of β-glucosidase was not affected by organic management as indicated by the values of the specificity constant (K_a). This parameter was strongly influenced by the analytical approach. Increases of its values were detected using the fluorogenic substrate, 4-MUF-β-d-glucopyranoside, indicating a higher catalytic efficiency of the enzyme with the fluorimetric approach.

Studies on the same enzyme were also performed by Dick and others to evaluate the reliability of *p*-nitrophenol (pNP)-based and MUF-based microplate methods for estimating β-glucosidase activity in two soils with different characteristics. Evaluation of the two methods was also extended to the kinetics of the enzyme and the effect of different preparation procedure of soil samples (e.g., sample, sonication, soil suspension) on the two methods was studied, as well. While values of enzyme activities measured by the MUF and bench methods were comparable, more variable results in terms of reproducibility, kinetic parameters and less sensitivity were measured with the pNP microplate method, probably because of high background absorbance that affected its reproducibility, sensitivity and accuracy. These results suggest that the pNP microplate method is not as a valid substitute for the standard bench method.

Microplate fluorimetric assay: A better efficiency of the microplate fluorimetric (F) assay was also demonstrated by Trap and others when fluorimetric assay was compared to the standard spectrophotometric (P) method. The activities of five enzymes (cellulase, *N*-acetyl glucosaminidase, β-glucosidase, acid phosphatase and alkaline phosphatase) in contrasting land uses, including woodland, grassland, cultivated and contaminated lands were measured. Although activities measured with the P methods were higher (around 8 times) than those measured with the F methods, significant differences in enzyme activity were revealed by the F methods in different soils. Smith and others concluded that 'the F method improves the effectiveness and the efficiency of measuring universal soil quality indicators using enzymes'.

Standardised and validated enzyme assays: Further studies to define standardised and validated enzyme assays for meaningful data comparison and interpretation were carried out by Deng and others. They compared bench scale and microplated format assays of soil enzyme activities using spectroscopic (pNP) and fluorometric (MUF) based approaches. Investigations were extended to pNP-bench, pNP-microplate and MUF-microplate and three different enzymes were tested in 16 different soils. MUF-based assays were found about 14 times more sensitive and precise than pNP based assays when soil suspensions were used, although standard errors raised by the presence of soil suspensions. Results were different with the three analytical protocols but being in the same order of magnitude and significantly correlated to each other indicated that the different protocols detected the same pool of isoenzymes.

13.4.2 Location and distribution of soil enzymes

Another important and complex aspect of soil enzymology is the identification of enzyme location and/or distribution in the soil matrix, i.e., the spatial variability in a soil profile and localisation in soil structural fractions of different nature and size. Extracellular enzymes extruded outside the cell may diffuse far from the cell in the soil solution or remain still attached to it by operating within the cell periplasm, associated to the cell wall being confined in biofilm or capsule of polysaccharidic nature or contained in structures such as polysomes.

Typical examples of enzymes contained in polysomes or cellulosomes, are polysaccharidases involved in the degradation of polysaccharides and lignocellulose. It appears quite obvious that in the case of cellbound enzymes substrates have to reach the enzymes and diffusional limitations may arise. In this case, their producing cells implement some strategies such as production of signals trough the so-called quorum-sensing system to monitor the environment and/or contact the target substrates by responding via chemotaxis to gradient concentrations. When the enzymatic action, in the majority of case macro-molecule degradation, occurs at the cell-soil solution interface formed monomeric products will be easier up-taken by the cell with evident advantages for the cell metabolism. It is widely accepted that the activities of soil enzymes usually decrease with soil depth accordingly to the decrease of microbial biomass and organic C content. Moreover, the response of enzymes to different treatments and/or disturbance may change along a soil profile and with depth.

Geng and others examined the response of six hydrolytic enzyme activities and two oxidative enzyme activities in forested ecosystems under three different treatments (intact forest controls, canopy tree thinning and canopy tree thinning plus ercicaceous stem removal and soil tilling) at different depths. All enzyme activities except for peroxidase were significantly lower at 10–20 cm than 0–10 cm soil depth. Moreover, in the 0–10 cm soil horizon the activities of protease and arylsulphatase significantly decreased in each treatment compared to controls as cellulase and phenoloxidase activities also did. By contrast, an opposite trend for peroxidase activities was observed and no significant differences between treatments for glucosaminase, glucosidase and acid phosphatase activities were detected.

A different behaviour occurred at 10–20 cm soil depth, where treatments affected only the activities of acid phosphatase and phenol oxidase, whereas all other soil enzyme activities remained unchanged. The effects of management or disturbance on agricultural ecosystems are usually examined in surface soils and no attention is devoted to the changes in microbial activity occurring in deep soils where large amounts of organic carbon are potentially stored with possible consequences on carbon biogeochemical cycle.

Interesting is the behaviour of enzymes in the rhizosphere, an oasis of biological activity where plant properties dominate microbial and enzymatic behaviour, a different gaseous regime is present and mineral solubilisation and competition effects influence nutrient cycling.

The distribution of enzyme activities between bulk and rhizosphere soil and the accumulation of enzymes at soil-plant root interfaces, the contribution of enzymes deriving from enzymatic proteins produced by plant roots and released in their surrounding rhizosphere soil, the effect of plant species composition and root type as well as the contribution of root mycorrhisation on enzyme activities of rhizosphere soil have attracted the attention of several researchers.

Phosphatases are enzymes typically more abundant in the rhizosphere, particularly in rhizosphere soils of mycorrhise-infected plants and their activity is related to phosphorous (P) availability and/or deficiency. In their review on phosphatase activity in natural and mined soil collin and others demonstrated that both acid and alkaline phosphatase activities had good relationships with inorganic P fractions and their activity varied with mycorrhizal association.

Further insights in the distribution of microbial- and root-derived phosphatase activities in the rhizosphere of *Lupinus albus* L and their dependence on P availability and carbon (C) allocation have been recently provided by Spohn and Kuzyakov in their studies on the coupling of soil zymography results with ^{14}C imaging, a technique capable of reveal the distribution of photosynthates after labelling plants with ^{14}C. Larger (5.4 times) amounts of both acid and alkaline phosphatase activity occurred in the rhizosphere than in the bulk soil, being the first mainly associated with roots although produced by roots and micro-organisms. By contrast, alkaline phosphatase activity was produced only by micro-organisms and was more broadly distributed in a larger area as respect to acid phosphatase. The activities of two enzymes were also differently affected by P fertilisation and rhizodeposition of photosynthates, thus indicating that evidently different ecophysiological groups of organisms capable of mineralising organic P exist and their spatial differentiation possible reduces a potential competition between them.

Not only phosphatases are clearly present in the rhizosphere. An interesting study was performed by Wu and others that correlated the activity of protease and β-glucosidase in the rhizosphere of *Citrus unshiu* with the spatial distribution of glomalin-related soil protein (GRSP), root mycorrhisation, soil aggregates and carbohydrates. Positive correlations were found between two-categories (total and easily extractable) of GRSP with β-glucosidase, soil water-stable aggregates water-extractable or hydrolysable carbohydrates and mycorrhisation whereas a negative correlation was found with protease.

Correlations between five enzymes involved in the transformation of C, P and nitrogen (N) substrates and arbuscular mycorrhizal, ectomycorrhizal, dual-colonised (arbuscular and ectomycorrhizal) and ericoid mycorrhizal plants mycorrhizospheres were found by Gartner and others in a fire chronosequence in Alaska. In particular, β-glucosidase and peroxidase were lower in arbuscular mycorrhizospheres and ericoid mycorrhizospheres, respectively, than in bulk soil. Additionally, mycorrhizosphere types influenced the entity of each enzyme activity, thus indicating that 'the community composition of mycorrhizal host plants might mediate enzymatic activity in boreal soils'.

Additional information on rhizosphere enzymes and their relation with soil, plants and microbes was provided by studies of Bell and others in which the stoichiometry approach developed by Sinsabaugh and co-workers (was applied to rhizosphere).

In their pioneering study Sinsabaugh and others assembled a comparative database of soil EEs potential for 40 ecosystems and related soil microbial function, as expressed by EE activities, to global biomass composition, nutrient dynamics and soil organic matter storage. In other words an elemental stoichiometric evaluation was derived between microbial biomass and detrital organic matter and microbial nutrient assimilation, growth and expression. The main result achieved was that a similar stoichiometry likely exists between the more common measured soil enzymatic activities and all microbial communities.

Starting from the hypothesis that 'soil nutrient and microbial stoichiometry would differ among plant species and be correlated within plant rhizospheres' Bell and others studied soil C, N, P, microbial biomass C, N and soil enzyme C, N, P nutrient acquisition activities using rhizospheres of eight different intact species-specific plants in a semiarid grassland in Wyoming, USA. Overall results indicated that strong positive correlations exist between soil and plant tissue stoichiometry, whereas these components and microbial or enzyme stoichiometry weakly correlated to each other. Indeed, contrarily to what expected no negative correlation was observed between soil microbial enzyme activities and microbial biomass stoichiometry. By contrast, plant tissue, microbial and soil nutrient stoichiometry showed similarity among many plant species independently of plant functional type, thus indicating that not a strong conservation occurred for plant tissue and rhizosphere stoichiometry at the plant species level.

Extensive studies have been also devoted to examine the micro-scale distribution of enzyme activities in particle-size soil fractions to evaluate their catalytic potential and kinetics as affected by biotic and abiotic soil constituents as well as their variation in response to different agricultural practices. Useful information may be also achieved on the role exerted by different soil fractions on micro-organisms and enzyme activities protection and in the turnover of organic C, by following the response of microbial processes to a range of long-term and short-term organic C-inputs. Usually soft fractionation methods have been utilised to guarantee the complete recovery of soil enzyme activities.

For instance, Marx and others investigated the distribution of four carbohydrases (β-cellobiohydrolase, *N*-acetyl-β-glucosaminidase, β-glucosidase and β-xylosidase), acid phosphatase and leucine-aminopeptidase in four particle-size fractions (>200, 200–63, 63–2 and 0.1–2 mm) obtained by a combination of wet-sieving and centrifugation, after low-energy ultrasonication. The effect of the location on enzyme distribution and long-term N fertiliser management on their respective kinetics was studied as well. Carbohydrases distributed mainly in the coarser fractions while phosphatase and leucine-aminopeptidase predominated in the clay-size fractions. Moreover, the association with different particle-size fractions influenced the substrate affinity of each enzyme as revealed by Michaelis constant (K_m) measurements whereas no significant effects were observed by N fertiliser management.

Relationships between the specific location of cellulase within the matrix of a paddy soil and soil organic matter (SOM) quality and carbon turnover were found by Yan and others. In particular, a physical fractionation procedure was used to investigate whether soil carbon was spatially isolated from degradative enzymes in the soil. In particulate organic matter fractions (POM) the activity of carboxymethyl cellulase was greatest in coarse fractions and generally decreased from the coarse to the silt-size fraction, concurrently to organic C concentrations did.

The different depth gradients shown by hydrolytic and oxidative enzymes in the study of Kramer and others could indicate that hydrolytic enzymes are very likely linked to POM whereas oxidative enzymes mainly prevail on mineral fractions. Moreover, a relationship between microbial abundance and substrate availability possibly exist only for hydrolytic enzymes.

Limitation to soil organic matter accumulation was observed in desert grassland soils where the higher and stabilised activities of phenol oxidase and peroxidase occurred as compared to temperate soils and uniformly distributed across particles ranging from >1 mm to <38 μm.

13.5 Factors affecting soil enzymes

Inherent: Soil enzymes have varying optimum pH and temperature values at which they function most effectively. For example, the activity of phosphatase, aryl sulphatase and amidase involved in phosphorus, sulphur and nitrogen cycling, respectively, is strongly correlated to variations in soil pH. Since enzyme structure and substrate binding can be altered by heat and extreme cold temperature, enzyme activity decreases above and below the optimum temperature. The activity of many enzymes often correlates with soil moisture content, as well. Drought may suppress enzyme activity.

Soil texture influences enzyme activity and normally enzyme activities are significantly and positively correlated with clay content. Clayey soils have greater ability to store organic matter that promotes microbial communities and clay forms clay-enzyme complexes. In contrast, sandy soils tend to exhibit low rates of enzyme activity because they are naturally low in organic matter and have poor water holding capacity which results in lower microbial biomass and therefore lower enzyme activity.

Dynamic: Addition of organic amendments and adoption of management practices that increase soil organic matter lead to increased enzyme activity. Plant roots stimulate enzyme activity because of their positive effect on microbial activity and production of exudates rich in substrates acted on by enzymes.

Elevated soil concentrations of chemical compounds that are end products of enzymatic reactions can inhibit enzyme activity by feedback inhibition. For example, phosphatase activity increases in phosphorus deficient soil, but its activity decreases in soil with high phosphorus concentration. Similarly, urease activity may be suppressed by ammonia-based nitrogen fertiliser because ammonium is the product of urease activity. Compaction may limit the activity of enzymes involved in nutrient mineralisation because of decreased oxygen in the soil for those reactions or organisms requiring an aerobic environment.

Conversely, anaerobic conditions from compaction or water saturation increase enzymatic reaction rates related to denitrification. Application of materials containing heavy metals can reduce enzyme activity (e.g., amidase) due to their toxic effect on soil organisms and roots or direct inhibition of enzyme reactions.

13.5.1 Relationship to soil function

Enzymes respond to soil management changes long before other soil quality indicator changes are detectable. Soil enzymes play an important role in organic matter decomposition and nutrient cycling. Some enzymes only facilitate the breakdown of organic matter (e.g., hydrolase, glucosidase), while others are involved in nutrient mineralisation (e.g., amidase, urease, phosphatase, sulphates). With the exception of phosphatase activity, there is no strong evidence that directly relates enzyme activity to nutrient availability or crop production.

The relationship may be indirect considering nutrient mineralisation to plant available forms is accomplished with the contribution of enzyme activity.

13.5.2 Problems with poor activity

Absence or suppression of soil enzymes prevents or reduces processes that can affect plant nutrition. Poor enzyme activity (e.g., pesticide degrading enzymes) can result in an accumulation of chemicals that are harmful to the environment, some of these chemicals may further inhibit soil enzyme activity.

13.5.3 Improving enzyme activity

Organic amendment applications, crop rotation and cover crops have been shown to enhance enzyme activity. The positive effect of pasture is associated with the input of animal manure and less soil disturbance. Agricultural methods that modify soil pH (e.g., liming) can also change enzyme activity.

13.5.4 Measuring enzyme activity

Enzymes are measured indirectly by determining their activity in the laboratory using biochemical assays. Enzyme assays reflect potential activity and do not represent true *in situ* activity levels and must be viewed as an index.

13.5.5 Interpretation and assessment

When possible, compare the site of interest to samples taken from an adjacent, undisturbed site on the same soil type. Alternatively, for a newly implemented land management system, track changes from time zero to five or more years with annual sampling to detect temporal changes in activity of soil enzymes.

13.6 Environmental factors affecting production and expression of soil enzyme activities

Soil enzyme activities may be affected by numerous factors of both natural (i.e., physico-geological, geographical or physico-chemical properties of soils, organic, clay or biomass contents, etc.), and anthropogenic (agricultural management, environmental pollution, additives such as fertilisers, pesticides, salts, heavy metals, etc.), nature. These factors may influence the production, activity, catalytic behaviour and persistence of soil enzymes through different mechanisms involving direct, either reversible or irreversible and indirect effects. Reversible or irreversible action on the catalytic active site of soil enzymatic activities as well as alteration of the protein conformation may occur. Measurable changes in size, structure and functionality of the microbial community as well as altered production (induction and/or repression) of enzymes may result as indirect effects on soil microbial growth and activity.

The factors selected and analysed in this section are climate changes (increasing and decreasing temperature, precipitation and associated variation of soil moisture and drought and the possible interference of nitrogen fertilisation) and soil management by biochar addition.

13.6.1 Temperature, precipitation and drought, soil moisture

Temperature strongly affects and controls soil enzyme activities, changing enzyme kinetics and stability, substrate affinity and enzyme production because it can influence the size and activity of microbial biomass. As soil hydrolytic enzymes are the main drivers of soil organic matter (SOM) degradation and litter decomposition, the dependence of these enzymes on global changes including warming, precipitation, drought and associated soil moisture will assist in understanding the relationships among SOM stock, global carbon cycle and microbial nutrient demand. Moreover, the possible interference of nitrogen demand in soil has also to be considered, being nitrogen a fundamental element not only for several metabolic routes but mainly because involved in protein and therefore enzyme synthesis.

To evaluate the role of EEs in soil and ecosystem responses to climate change, Henry reviewed field studies in which the responses and inter-annual variation of soil EE activities were examined to warming and altered precipitation, seasonal variations, elevated atmospheric CO_2, increased atmospheric N deposition and changes in disturbance regimes. Henry underlined that the measurement of potential soil enzyme activities, as afforded by assays performed under laboratory conditions, can be inadequate to provide a real picture of temperature and moisture dependence of enzyme activity *in situ*. Moreover, some criticisms about the methodologies used to simulate warming in field experiments were also evidenced. Controlled environment and field experiments usually involve open top chambers, greenhouses, retractable passive warming curtains, snow removal, heated coils/fluid filled tubes inserted into soil or overhead infrared heaters. Each of these methods, often selected by financial, logistical or spatial constraints rather than scientific reasons, may present advantages and disadvantages and introduce artifacts. In many cases they do not represent real situations occurring *in situ* following warming or precipitation or wind, mainly in cold areas where snow very often covers lands. These methodological aspects can influence the interpretation of enzyme activity responses and have to be taken in duly account. The main conclusion drawn by Smith and others was that in general soil moisture manipulations in field studies have had a much greater influence on potential EE activities than warming treatments, though no high warming effects could have been achieved in the field due to the used methodology.

Opposite conclusions were achieved by Baldrian and others in their studies in a temperate hardwood forest soil with dominant *Quercus petraea*, with large seasonal temperature differences and moderate changes in soil moisture content. The effects of soil temperature and seasonality on the sizes of EE pools and activities along with microbial biomass, soil moisture content and pH were investigated for three years in the litter (L), organic horizon (O) and upper mineral horizon (Ah). Strong increases in enzyme activity as measured *in vitro* were observed up to temperature of 20–25°C and higher activity levels occurred in the warm period of the year, whereas no significant changes were detected in the pools of most EEs and in the content of microbial biomass. Soil moisture did not significantly affect enzyme activities.

Contrasting results (increase or decrease) of specific enzyme activities, involved in C, N and P cycles, were obtained in a temperate grassland of northern China as six-year responses to warming and increased precipitation. The responses were soil depth-dependent and indicated that enzymes differentially responded to warming and increased precipitation at two depths in this region. For instance, the activities of acid phosphatase and β-glucosidase dramatically decreased with warming at 10–20 cm soil depth whereas a strong increase of acid phosphatase and β-*N*-acetylglucosaminidase occurred in surface soils, where the patterns of enzymes were mainly driven by soil pH, ammonium and nitrate.

A different temperature-sensitivity behaviour as well as low activity levels could be expected by enzymes acting in cold areas. Wallenstein and others investigated seasonal variation and temperature sensitivities of six hydrolytic enzymes at 4 and 20°C on four sampling dates in tussock, intertussock, shrub organic and shrub mineral soils in Arctic tundra soils at Toolik Lake, Alaska. Greatest activities of β-*N*-acetylglucosaminidase, β-glucosidase and peptidase were observed at the end of winter, suggesting that enzymes were produced even in frozen soils. Soil enzyme pools generally showed high sensitivity to temperature (as measured by Q_{10}) in the same period of the year and not higher activity levels were observed during Arctic summer. Temperature was found to be the strongest factor driving low *in situ* enzyme activities in the Arctic, as achieved by modelling the potential activity of *in situ* β-glucosidase for tussock and shrub organic soils based on measured enzyme activities, temperature sensitivities and daily soil temperature data. However, limitation in nitrogen supply could have affected enzyme production in summer determining their non-expected lower values.

The influence of nitrogen fertilisation on enzyme responses to climate changes (precipitation variability and relative varied soil drought) was successively confirmed by the findings of Alster and others. Direct and indirect negative effects of drought on litter decomposition and faster decomposition by *N*-adapted microbial communities in *N*-fertilised plots than in non-fertilised plots were shown as measured by the activities of nine EEs in a semi-arid study site in southern California. Smith and others highlighted the hypothesis that changes in fungal biomass, considered dominant in the area and potential EE activities would relate directly to litter decomposition responses. Contrarily to what hypothesised, bacteria and not fungi predominated in the area and they negatively responded to drought treatment. However, fungal biomass and many potential enzyme activities positively responded to drought treatment although their efficiency (measured as the mass loss of chemical substrates per unit of potential enzyme activity) declined with drought, possibly because an increase of enzyme immobilisation and a reduced diffusion rate occurred due to the low water availability. In *N*-fertilised plots, larger efficiencies of some enzymes (β-glucosidase, β-xylosidase and polyphenol oxidase) were observed when microbes were transplanted into originating environments, thus suggesting that microbial enzymes may adapt to their local environment. Thus, Smith and others concluded that a link between enzyme potentials and *in situ* activities is possible if the impact of drought and N addition on the efficiencies of EEs can be predicted.

Nitrogen addition showed significant effects also on the kinetics (V_{max} and K_m parameters) of cellobiohydrolase, β-glucosidase, β-xylosidase, β-glucosidase and β-*N*-acetylglucosaminidase involved in SOM degradation in two forest soils. The values of K_m were considerably decreased by N addition in a soil, while variable K_m was observed in the other soil. By contrast significant increases of V_{max} were detected for all investigated enzymes in the two soils. Both kinetic parameters were temperature sensitive but a greater response from V_{max} (higher Q_{10} values) than K_m was observed. No measurable interaction between *N*-addition and temperature variation was detected. Overall results suggested that a larger degradation of substrates possibly occurred under *N*-fertilisation and higher hydrolytic enzymatic activities could be expected by the simultaneous increase of temperature and *N*-addition.

The role of temperature sensitivities in SOM decomposition on both kinetic parameters of the same enzymes was investigated in five sites crossing from a boreal forest to a tropical rainforest. Q_{10} values ranging from 1.53 to 2.27 for V_{max} and 0.90 to 1.57 for K_m were calculated and a significant ($P = 0.004$) negative relationship with mean annual temperature was observed for the K_m of β-glucosidase, suggesting that temperature can have higher effect on enzymes in cooler climates. When data were parameterised in a mathematical model it was concluded that V_{max} and K_m temperature sensitivities could counterbalance to each other for the SOM losses. Moreover, results confirmed the adaptation of microbial EEs, as measured by their kinetic behaviour, to local environmental temperature.

Recent findings seem to partly contrast this conclusion at least in an alpine grassland ecosystem. The response of five soil EE activities (phenol oxidase, β- and α-glucosidase, β-xylosidase, cellobiohydrolase) and temperature sensitivity (Q_{10}) to experimental warming was investigated by setting up a free air-temperature enhancement system. While extracellular enzymes adapted to seasonal temperature variations showing higher values during the warm period of the year, no acclimation to the field experimental warming was observed.

Similar results were obtained for four peptidases and four glycosidases when exposed to experimental climate manipulation in a long-term experiment in northern Sweden. Experimental soil warming and/or winter snow addition did not show significant effect on either the potential activities or the temperature sensitivity of the two group of enzymes, whereas opposite trends of significantly season-dependent patterns occurred for the two functional enzymatic groups. This behaviour may suggest that warming acts indirectly on soil processes by the seasonality of substrate supply and microbial nutrient demand rather than directly by influencing the production of enzymes. Moreover, the potential divergence between the different seasonal patterns observed for the two groups of enzymes must be considered when enzyme-based models are implemented to describe soil processes.

Detailed studies on the temperature and moisture sensitivity of *in situ* β-glucosidase enzyme activity were performed by Steinweg and others and data incorporated in a mathematical model indicated that beside temperature and moisture β-glucosidase activity was affected also by substrate concentrations and diffusion constraints, thus providing a template for analysing the role of specific abiotic managing *in situ* enzyme activities, facilitating their incorporation in biogeochemical models. The new experimental protocol considered soil samples collected every two weeks over a 10-week period with precipitation inputs manipulated to obtain drought (50% ambient precipitation), ambient and wet (150% ambient precipitation) treatments. Enzyme activity was assayed in soil slurries at three different temperatures (15, 25 and 35 °C) and to different moisture levels in the lab and adding substrate to homogenised dry or moist soils instead of slurries. Stable temperature sensitivity but significant variation of moisture sensitivity was detected among the five sample dates and treatments. As respect to ambient and wet plots, drought plot soils showed strong responses of the enzymatic activity to increases of moisture. When the effect of only temperature or only moisture or their simultaneous action were estimated, results indicated that the highest β-glucosidase activity occurred in ambient plots and that it responded primarily to changes in temperature except in dry soil. These results corroborated what usually retained, i.e., *in situ* enzyme activity in soils can be strongly limited by low soil moisture, thus reducing any increasing effect of warming.

Further studies were extended to evaluate how climate variables (warming, precipitation and interaction between them) affected the activities of β-glucosidase, cellobiohydrolase, xylosidase, acid phosphatase, β-N-acetylglucosaminidase and leucine-aminopeptidase and of microbial biomass carbon (MBC) in different seasons and in soils treated with the same levels of precipitation (ambient, 150% of ambient during growing season and 50% of ambient year-round) and four levels of warming (unwarmed to ~4°C above ambient) over the course of a year. Negligible or little effects by climate manipulations (warming, precipitation, moisture and season) were observed on potential enzyme activity, whereas mass-specific potential enzyme activity (calculated by potential enzyme activity/MCB ratio) raised with temperature, reached its peaking value under medium warming, then declined at the highest warming. Moreover, a stable enzyme pool under drought was observed thus indicating that mass-specific enzyme production increased with temperature and drought or that decreased enzyme turnover rates occurred in dry soils. The stoichiometry of potential enzyme acquisition activities strongly changed with season, with an increase in C-acquiring enzymes and a decline in the potential activity of *N*-acquiring enzymes from summer to winter when a reduction in organic N degradation likely occurred. A stable enzymatic C:P activity ratio instead was measured over the year indicating a consistent, continuous need of P sources. At low soil temperature and in frozen soils slower reaction rates as well as reduced *in situ* activities will be likely present because of the limited diffusion of substrates.

As compared to presence and composition of fungal communities, abiotic factors such as seasonality, nutrient concentrations, moisture, pH, carbon and nitrogen content and mean annual precipitation, showed higher and more significant effects on the activities of - and β-glucosidase, β-xylosidase, cellobiohydrolase, β-*N*-acetylglucosaminidase and acid phosphatase in five sites during the growing season in March and 17 sites during the dry season in July throughout southern California. The correlations between abiotic factors, fungal community composition and soil enzyme activities were examined by analysing the response of enzyme activities to all factors (abiotic vs. fungal composition) separately and concomitantly across a variety of ecosystems. The analysis showed no correlation between fungal community composition and enzyme activities at the species, genus, family or order levels, whereas significant correlations were found between soil carbon, nitrogen and pH and hydrolytic activities and between mean annual precipitation and oxidative activities. In dry season, fungal composition explained only 27.4% of the variation in all enzyme activities but 35.3% of it was accounted for by abiotic factors.

13.6.2 Biochar fertilisation

Soil management is strongly related to global food security and research, then practices and policies influencing soil management are of paramount importance to assure a sustainable agriculture. Soil management by addition of inorganic fertilisers and organic residues is an old and very common agricultural practice to increase soil fertility and crop production.

Recently, a great attention has been devoted to the use of biochar to improve soil fertility by promoting sustainable resource efficiency, mitigate climate changes, increase crop yield per area, increase soil carbon (C) storage by its sequestration, reduce soil acidity and reduce irrigation and fertiliser requirements. In addition the biochar capability to decrease soil emissions of greenhouse gases and reduce nitrous oxide (N_2O) emissions was taken into account as healthy effect.

Biochar is a stable carbon-rich by-product synthesised through pyrolysis/carbonisation of plant- and animal-based biomass that can last in soil for very long periods. Biochar chemical and physical properties depend on the raw material characteristics and pyrolysis conditions as well as its soil ameliorant capacity is dependent on the soil properties and environmental conditions. Another, interesting and valuable property of biochar is its reactivity and capability to reduce and mitigate soil pollution by contributing to the restoration of soils contaminated by both organic and inorganic contaminants when added to contaminated soils. Biochar large surface area and cation exchange capacity, strictly dependent on the originating material and pyrolysis treatment, facilitate and improve sorption of pollutants to its surfaces, reducing not only mobility but also bioavailability and phytoavailability of organic contaminants in soil.

Char materials may, however, contain toxic compounds. Therefore, to avoid undesired effects on soil biota and environmental quality, a detailed characterisation of biochar nature and quality is mandatory before any application either for agricultural or environmental purposes.

It has been well established that amendment of soil with biochar affects not only microbial activity, biomass and community structure, but also other soil biota as fauna and plant roots. Weak or strong correlations might establish among primary properties of biochar, soil processes they influence and soil biota, with resulting effects on soil behaviour.

Usually, increase of enzyme activities have been detected as a result of the increase of soil microbial biomass by biochar addition, although opposite responses have been also obtained. Stable biochar materials obtained from the weed *Parthenium hysterophorus* and used at different amounts in laboratory experiments increased the activities of soil dehydrogenase and catalase while hydrolytic enzyme activities decreased. Increases of acid phosphatase, alkaline phosphatase and fluorescein hydrolase activities as well as those of dehydrogenase and catalase were instead observed with biochar obtained from water hyacinth (*Eichornia crassipes*) applied to a red soil from Dhanbad, India. The increase was dependent on the amount applied and was the highest in 20 g kg^{-1} biochar treatment.

To better enhance soil productivity and biological activity biochar was co-applied with fly ash to the same soil and soil nutrients, biological properties and the yield of *Zea mays* were measured. Dehydrogenase (+60.7%), alkaline phosphatase (+32.2%) and fluorescein hydrolase (+12.3%) activities increased due to co-application of the two materials. Further positive effects were great in increasing of soil nutrients (+110% P and +64% K), microbial biomass (+25.3%) and maize grain yield (28.1%). pH-buffering effect and sorption of SOM to mineral surfaces probably induced more reactive interactions among water, air and nutrient in soil. For a clear understanding of biochar effect on soil enzyme activities, possible interaction with substrate and products should be, however, considered, because real biochar effect on soil enzymes can be hampered and masked by such interactions.

Experiments performed on sandy silt loam soils amended with different amounts of pine wood and barley straw biochar demonstrated that significant decreases (up to 80%) of substrates (pNPP and INT) and respective products (pNP) and iodonitrotetrazolium formazan (INTF) occurred in the assays of dehydrogenase and phosphatase, suggesting that solid-phase adsorption on biochar surfaces did occur. Bioavailability of the substrates decreased and no saturating condition possibly arose. Statistical analysis was conducted to evaluate the effects of soil type, biochar concentration, biochar feedstock and their interactions on the concentration or extractability of enzyme assay substrates and products. Results indicated that the chemical properties and nature of biochar as well as the soil type affected the relationship between sorption and biochar concentrations, such as no unique correlation or correction factor could be developed.

Similar sorption phenomena have been already observed by Bailey and others when effects on the activities of soil β-glucosidase, β-*N*-acetyl-glucosaminidase, lipase and leucine aminopeptidase were investigated in biochar-amended soils or *in vitro* reaction. Indeed, to explain variable effects of biochar on soil enzymatic activities observed with field studies, *in vitro* colorimetric and fluorescent assays were performed in microcosms with three different soils added with biochar. As compared to non-amended soils, the activity of β-glucosidase increased and that of lipase decreased as evidenced by assays performed with fluorescent substrates. When substrates were temporarily put in contact with biochar and successively tested, a reduction of the apparent activity of the enzymes was observed thus confirming that sorption of substrates on biochar surface hampered enzyme activity. By contrast, a stimulating effect (increases by 50–75%) of β-Nacetylglucosaminidase activity was measured when the purified enzyme was previously exposed to biochar. Therefore, when biocharsorption effects are not rightly considered, soil assays can underestimate the increase of soil enzymatic activities due to biochar.

Biochar adsorption of β-glucosidase but not of its substrate cellobiose was found by Lammirato and others when both were exposed to chestnut wood char *in vitro* phosphate buffered system. A reduction by <30 % of the reaction rate occurred, indicating a slight influence of the material on the activity of the enzyme. By contrast, complete inhibition of the reaction rate occurred with activated carbon that adsorbed more than 97% of either β-glucosidase or cellobiose, probably because of its high specific surface area and high porosity.

13.7 Enzyme activities as a component of soil biodiversity

Soil enzyme activities are the direct expression of the soil community to metabolic requirements and available nutrients. While the diversity of soil organisms is important, the capacity of soil microbial communities to maintain functional diversity of those critical soil processes through disturbance, stress or succession could ultimately be more important to ecosystem productivity and stability than taxonomic diversity.

Understanding and maintaining biodiversity has become an increasingly important field of research, as well as a resource management goal. In soil microbial communities, maintaining critical functions may ultimately be more important than maintaining taxonomic diversity. One essential microbial function in soils is the processing and recovery of key nutrients from detrital inputs and accumulated soil organic matter. This often requires the activity of extracellular enzymes to process complex organic compounds into assimilable subunits (sugars, amino acids, NH_4^+, PO_4^{-3}).

Distinct from the physiological or genetic diversity of the soil microbial biomass which assess potential, functional diversity of soil enzymes is related to the actual activities resulting from that potential.

13.7.1 Substrate specificity

Substrate specificity, as either an independent measure of enzyme diversity or as means to distinguish different reaction mechanisms, could resolve those enzyme activities that attack specific detrital components either between or within major nutrient pools (Table 13.1).

Table 13.1: Examples of specific soil enzyme activities and assays available to assess functional diversity between and within nutrient cycles.

Nutrient	*Form*	*Compound*	*Enzyme*
Carbon	Polysaccharide	Cellulose	Endo-cellulase
			β-glucosidase
		β(1–3) glucan	β(1–3) glucanase
		Hemicellulose	Xylanase
		Chitin	Endo-chitinase
			N-acetylglucosaminidase
		Starch	Amylase
	Aromatic	Lignin	Phenoloxidase
			Peroxidase
			Mn-peroxidase
	Aliphatic	Fatty acid esters	Lipase
Nitrogen	Peptide	Protein	Endo-protease
		Peptides	Aminopeptidase
			Carboxypeptidase
	Non-peptide	Primary amine	Amidase
			Urease
			(Adenosine) deaminase
			(Aryl) deaminase
Phosphorus	Diester		Phosphodiesterase
			DNAase, RNAase
	Monoester		Phosphomonoesterase
			Phytase

Within each type of nutrient, there are specific chemical forms based on structure and bonding. The major forms of carbon are polysaccharide, aromatic (lignin) and aliphatic (polymethylene). The bulk of organic nitrogen is thought to be the in amide form, either as peptide or non-peptide C–N bonds. Most organic phosphorus occurs in either a mono- or di-ester form.

Within each of these major nutrient groups, there are specific compounds against which major classes of soil enzymes are active. Keystone to the breakdown of litter are the various cellulolytic activities requiring endo-cellulases, cellobiohydrolases and β-glucosidases and ligninolytic activities requiring a variety of polyphenol oxidases and peroxidases. Within the nitrogen cycle, substrate diversity for proteins and peptides can be based on hydrolysis of different amino acid groups.

Release of ammonium from various non-peptide C–N bonds can also be measured for a variety of different substrates, including the frequently measured urease activity. Mineralisation of phosphate from organic esters can be resolved into phosphodiesterase and phosphomonoesterase activities, reflecting the use of tissue-based and soil organic phosphates pools, respectively.

14

Enzyme catalysis for polymer function and degradation

14.1 Introduction

Enzyme catalysed reactions are green alternative approaches to functionalise polymers compared to conventional methods. This technique is especially advantageous due to the high selectivity, high efficiency, milder reaction conditions and recyclability of enzymes. Selected reactions can be conducted under solventless conditions without the application of metal catalysts. Hence this process is becoming more recognised in the arena of biomedical applications, as the toxicity created by solvents and metal catalyst residues can be completely avoided.

'Green chemistry' is still an emerging field that strives to work at the molecular level to achieve sustainability. The field has received widespread interest in the past decade due to its ability to harness chemical innovation to simultaneously meet environmental and economic goals. Synthesis of polymeric materials with unique properties can be achieved not only by polymerisation reactions but also by modification/functionalisation of existing polymers. Enzymatic catalysis is an alternative strategy to increase the diversity of functional groups in polymeric materials. Specifically, enzyme catalysed polymer functionalisation carried out under solventless conditions is a great advancement in the design of green processes for biomedical applications, where the toxicity of solvents and catalyst residues needs to be considered. The polymers obtained through this method are free from any metal catalyst, which often broadens the application of the products, particularly for biomedical use.

Enzymes are natures catalysts that accelerate specific metabolic reactions in living cells. As an environmentally friendly alternative to conventional chemical catalysts, enzymes offer several advantages, including high selectivity, high efficiency, the ability to operate under mild conditions and catalyst recyclability. New synthetic and catalytic methods are necessary to design new classes of functional polymers for covalent polymer-drug conjugates. Despite all the advantages that enzymatic catalysis offers, the area of quantitative functionalisation of preformed polymers has not fully been developed. Enzymes can catalyse the modification of a polymer through functional groups located in the main chain, in the side chains or at the polymer terminals (synthetic symmetric or asymmetric telechelic polymers). As already discussed enzymes are classified into six main groups according to the International union of biochemistry and molecular biology. Today more than 4000 enzymes are available commercially and some of them are mutated for industrial applications. Generally, oxidoreductases, hydrolases and isomerases are relatively stable and the most widely used catalysts in biotransformations. Among these, some of the isolated enzymes are conveniently used as catalysts in practice. In contrast, lyases and ligases are present in lesser amounts in living cells and are less stable for isolation or separation from living organisms.

Lipases catalyse the hydrolysis of triglyceride substrates into fatty acids and glycerol at the lipid-water interface *in vivo.* They are widely used in esterification, transesterification, aminolysis and Michael addition reactions in organic solvents. These enzymes display almost no activity as long as the substrate is in a monomeric state. However, when the substrate concentration is beyond its critical micelle concentration, a lipophilic phase forms and a sharp increase in lipase activity takes place. This phenomenon is called 'interfacial activation'. Interfacial activation is attributed to a rearrangement process within the enzyme *in vitro*. X-ray structural analysis of lipases usually shows a 'closed' conformation where a 'lid' blocks the active site. This lid opens when the enzyme is exposed to an interface of a biphasic water-oil system and thus the catalytic activity of the lipase increases. Lipases also work in organic solvents.

14.2 Enzymatic degradation of polymers

Nature has always provided, in considerable amounts and variety, polymeric materials with interesting compositions and structures and with great potential to be used in biomedical applications. These natural polymers include polysaccharides, such as cellulose, chitin, starch, alginate, galactan, hyaluronic acid, dextran and gellan, obtained from plants, animals and microbial sources and polyesters, such as poly(β-hydroxybutyrate) and poly(β-hydroxybutyrate-co-β-hydroxyvalerate), produced by numerous bacteria. In biological systems, polysaccharides and their derivatives are found as energy reservoirs, as components of the cell wall of plants, in bacteria and in the connective tissues.

Since the early 1960s, several synthetic degradable polymer systems have been used as medical implant materials, which include various polyesters, polyurethanes, polyanhydrides, polyacrylates, polyphosphoesters and polydiaxanone, among others. Over decades of research, many new biomaterials have been developed, but the prospect for novel biomaterials and novel applications continues to be immense. This chapter discusses the degradation mechanisms of biodegradable polymers with special emphasis on the main parameters affecting the enzymatic degradation of polymeric biomaterials.

Biodegradable polymers are materials with the ability to function for a temporary period and subsequently degrade, under a controlled mechanism, into products easily eliminated in the body's metabolic pathways. In this way, biodegradability not only eliminates the risk of complications associated with the long-term presence of a foreign material and the need for a second surgery for implant removal, but also allows for improved healing, as viable tissue interacts and grows into the degrading construct.

The use of polymeric materials in drug delivery applications also requires that the polymer degrade under physiological conditions and slowly release the encapsulated drug. The polymer should demonstrate, therefore, a continuous mass loss profile to facilitate repeated dosing and to ensure the successful effect of the treatments. The degradation of biomaterials is also important in terms of biocompatibility, since the changes that occur in the physico-chemical properties of the materials during degradation may alter their functionality and the associated biological response. In addition, the nature of the degradation products will, in part, define the ultimate biocompatibility of the materials since it may also induce alterations to cellular function.

Understanding the degradation mechanisms of biomaterials (degradation kinetics, evolution of mechanical properties, identification of degradation products) is, therefore, of crucial importance when selecting and designing materials for specific applications since the degradation process may affect a range of events such as cell growth, tissue regeneration, drug release, host response and the material function.

14.3 Degradation processes in biodegradable polymers

The degradation of a biomaterial can occur at different stages of its preparation, including during its storage. For instance, it was observed that the molecular weight of poly(L-lactide) decreased from 431000 to 202000 Da upon storage.

The conditions used during the processing and fabrication of polymeric materials may also lead to polymer degradation, consequently influencing their degradation behaviour *in vivo*. Melt based techniques (injection molding, extrusion, compression molding) are performed at high temperatures and in the first two cases at high shear rates, which may cause some degradation of the material. The production of samples by injection molding leads to a partial material orientation, which is typically higher in the skin than in the core of the molding. The chain orientation across the sample upon processing may be responsible for a faster degradation in the centre than in the skin. In addition, it is very common to use additives (plasticisers, lubricants, antioxidants, salts, stabilisers) during the processing of polymeric materials, which will leach out after immersion and may enhance or inhibit the degradation process.

Another aspect to be considered is the fact that biomedical materials need sterilisation before being implanted. Sterilisation can be performed using heat, steam, gas (ethylene oxide, EtO), or ionising radiation, mainly γ or β. Each of these sterilisation methods may have an effect on the material degradation, but sterilisation by radiation requires high doses of high-energy radiation, resulting in some cases in polymer cross-linking and degradation. γ sterilisation was shown to reduce significantly the molecular weight of poly(lactide–glycolide) polymers, but cold-cycle EtOH sterilisation did not cause any changes in the molecular weight of polylactides. Mechanical stress may also affect the degradation, either as a result of loading under service or due to residual stress arising during manufacturing, but this type of degradation is more significant on materials subjected to mechanical stress such as sutures, scaffolds for tissue engineering and fixation devices.

Materials exposed to the body fluids may undergo changes in their physico-chemical properties as a result of chemical, physical, mechanical and biological interactions between the material and the surrounding environment.

14.3.1 Chemical and enzymatic oxidation

Polymeric biomaterials may be degraded by chemical and enzymatic oxidation when exposed to the body fluids and tissues. It is well known that during inflammatory response to foreign materials, inflammatory cells, particularly leukocytes and macrophages, are able to produce highly reactive oxygen species such as superoxide (O_2^-), hydrogen peroxide (H_2O_2), nitric oxide (NO) and hypochlorous acid (HOCl). The oxidative effect of these species may cause polymer chain scission and contribute to their degradation.

Several studies have been carried out to assess the effect of oxygen species in the degradation of polymeric biomaterials. Lee and Chu studied the role of superoxide in the degradation of absorbable sutures and found that O_2^- could accelerate the degradation of aliphatic polyesters by the cleavage of ester bonds via nucleophilic attack of O_2^-. The effect of oxidative enzymes, such as horseradish peroxidase, catalase and xanthine oxidase, on the degradation of poly(urethane)s (PUs) was also studied, but these oxidative systems were unable to induce degradation of PUs.

14.3.2 Nonenzymatic hydrolysis

Polymer hydrolytic degradation may be defined as the scission of chemical bonds in the polymer backbone by the attack of water to form oligomers and finally monomers. In the first step, water contacts the water-labile bond, by either direct access to the polymer surface or by imbibition into the polymer matrix followed by bond hydrolysis. Hydrolysis reactions may be catalysed by acids, bases, salts, or enzymes.

After implantation, the biomaterial absorbs water and swells and degradation will progress from the exterior of the material toward its interior.

The hydrophilic and hydrophobic nature of polymeric materials influences their degradation rate and the susceptibility to hydrolysis follows this order: (i) hydrophilic with hydrolysable bonds, (ii) hydrophobic with hydrolysable bonds, (iii) hydrophilic with no hydrolysable bonds and (iv) hydrophobic with no hydrolysable bonds. For instances, *N*-vinylpyrrolidinone (NVP) is capable of absorbing relatively large amounts of water, but it is not prone to hydrolysis.

All biodegradable polymers contain hydrolysable bonds, such as glycosides, esters, orthoesters, anhydrides, carbonates, amides, urethanes, ureas, etc. Polymers with strong covalent bonds in the backbone (like C-C) and with no hydrolysable groups require longer times to degrade.

14.3.3 Enzyme catalysed hydrolysis

Enzymes are biological catalysts, i.e., they accelerate the reaction rates in living organisms without undergoing themselves any permanent change. In fact, in the absence of enzymes, most of the reactions of cellular metabolism would not occur. Hydrolysis reactions may be catalysed by enzymes known as hydrolases, which include proteases, esterases, glycosidases and phosphatases, among others. This class of enzymes comprises cell-derived proteins that are responsible for the catalysis of several reactions in the human body. For example, hydrolytic enzymes are present in the plasma and interstitium, in the brush border membrane and lumen of the gastrointestinal tract and in the tubular epithelium of the kidneys, where they ensure the efficient hydrolysis of different substrates to facilitate absorption of nutrients and solutes.

In this sense, it is expected that some of these enzymes may play an important role in the degradation of biomaterials by catalysing their hydrolysis. It has been shown that the degree of biodegradation of polyurethanes, in the presence of cholesterol esterase enzyme, is about 10 times higher than in the presence of buffer alone. This fact may explain the higher *in vivo* degradation rates of some biomaterials when compared with *in vitro* experiments.

While some enzymes catalyse only one reaction involving only certain substrates, others are not very specific. This indicates that the degradation of synthetic polymers may also occur by enzymatic hydrolysis.

Factors affecting enzymatic hydrolysis

The enzymatic hydrolysis of polymeric biomaterials is a heterogeneous process that is affected by the mode of interaction between the enzymes and the polymeric chains and involves typically four steps: (i) diffusion of the enzyme from the bulk solution to the solid surface, (ii) adsorption of the enzyme on the substrate, resulting in the formation of the enzyme–substrate complex,

(iii) catalysis of the hydrolysis reaction and (iv) diffusion of the soluble degradation products from the solid substrate to the solution. The rate of the global reaction is controlled by the slowest step.

The adsorption and rate of hydrolysis reaction is affected by the physico-chemical properties of the substrate (molecular weight, chemical composition, crystallinity, surface area) and also by the inherent characteristics of a specific enzyme (activity, stability, local concentration, amino acid composition and 3-D conformation). It is also very important to take into account the medium conditions such as pH and temperature, since they influence both the properties of the substrate and of the enzyme. The presence of stabilisers, activators, or inhibitory products in the medium, resulting from material degradation or leaching out of processing additives, may affect the enzyme catalysed reactions by influencing enzyme adsorption and activity.

The enzymatic hydrolysis of solid substrates is normally characterised by an enzyme saturation point, at this enzyme concentration, no further increase in the degradation rate is observed when more enzyme is added. This has been attributed to a decrease in the exposed polymer surface as the enzyme molecules saturate the surface and appears to be limiting the progress of degradation of some biomaterials.

The chemical modification of polymers (cross-linking, removal, or introduction of chemical groups in the polymer chain) also affects the enzymatic degradation rates since, depending on the degree of chemical modification, it may compromise the ability of the enzyme to recognise the modified substrate. This seems to be the case of lysozyme (enzyme responsible for the degradation of peptidoglycan and also chitin materials), which exhibited low activity toward chitosans with high degrees of deacetylation or cross-linked chitosan.

It was shown that different hard segment distributions at the surfaces of polyether-urea-urethanes could influence the manner in which the enzyme cholesterol esterase adsorbs, binds and expresses its activity on the surface of the polymers. It was found, for instance, that as hard segment increased, the enzyme hydrolytic activity was reduced.

The complex chemical nature of body fluids and inflamed tissues around the implant, the variability in the polymer material and the variability associated with biological systems (e.g., rate of metabolism of the host) make the overall degradation of biomaterials a rather complicated process. The first interaction of an implanted biomaterial with the host tissue is the adsorption of proteins. The subsequent interactions are determined by the nature of the proteins adsorbed. Blood plasma contains over 150 proteins and any of these may adsorb to the biomaterial, depending on the binding potential of the particular protein. Proteins may adsorb to a biomaterial surface with low affinity and may be replaced by other proteins with higher binding affinities. The deposition of proteins on the surface of biomaterials is followed by adherence of certain type of cells. The adsorption of other proteins to the polymer surface may influence the enzymatic degradation of biomaterials. It was shown, for example, that the preadsorption of fibrinogen (Fg) onto the modified and nonmodified surfaces of poly(ether-urethane) (PEU) polymers provided a temporary protective effect against the hydrolytic action of cholesterol esterase (CE) enzyme. This effect may be explained by the fact that the prior adsorption of Fg on the polymer surface might have occupied sites for CE adsorption, consequently limiting the access of the enzyme to the susceptible hydrolysable bonds. Another explanation may be related to the inability of the degradation products to release from the polymer surface due to the adsorbed layer of Fg in the immediate area.

Potential enzymes involved in the degradation of biodegradable polymers, their activities and half-lives in human plasma

The enzymes present in serum can be divided into two categories: (i) plasma-specific enzymes and (ii) non-plasma-specific enzymes.

The former are enzymes whose normal function is related to blood coagulation, complement activation and lipoprotein metabolism. The latter are enzymes that have no physiological function in the plasma, whose cofactors or even substrates are not normally present in the plasma. This category includes enzymes that are secreted by tissues, e.g., amylase, lipase, phosphatases and also enzymes associated with cellular metabolism.

Their presence in the serum at low levels may be due to the turnover of cells within the tissue causing release of the enzyme. The determination of enzyme activities in plasma has been used for clinical diagnosis since high levels of certain enzyme activities in serum may indicate tissue damage or malfunction. It should be noted, however, that increases in enzyme activities might also be due to other factors such as increased cell turnover, cellular proliferation, or decreased clearance by the kidney.

The initial stages of certain diseases and the presence of internal injuries give rise to elevated levels of enzyme concentrations in body fluids (lymph, blood and urine). Inflammation usually occurs at the biomaterial–tissue interface and reflects surface adsorption of plasma proteins, complement activation, neutrophil and macrophage infiltration, hyperplasia and release of inflammatory mediators, free radicals and proteolytic enzymes.

Lipase is an enzyme that is able to catalyse the hydrolysis of ester bonds in polyesters when in the presence of an aqueous media. It was found that certain lipases enhanced the degradation of polycaprolactone (PCL) when compared with incubation in buffer only. The enzymatic hydrolysis of PCL occurs mainly at the polymer surface since it is difficult for a hydrophilic enzyme to diffuse into a hydrophobic polymer like PCL. The surface erosion mechanism of PCL demonstrated by lipase enzyme may be useful to deliver bioactive agents by an enzymatically controlled process.

14.3.4 Degradation-monitoring techniques

After implantation, biodegradable polymers may undergo a variety of changes in their physico-chemical properties as a result of degradation. These alterations may compromise their desired function and evoke undesirable tissue response. It is important, therefore, to characterise and quantify the changes that occur in the biomaterial at different stages of degradation and to evaluate the corresponding tissue response, since the tissue reaction may change during the course of the degradation process.

The main techniques used to evaluate the degradation of biomaterials can be divided into surface analysis (infrared spectroscopy, X-ray photoelectron spectroscopy, contact angle measurements), which are more appropriate to monitor the changes occurring in the first stages of degradation and bulk analysis (determination of changes in molecular weight, weight loss, temperature transitions, mechanical properties) for characterising the later stages of degradation. Some of the most common properties and techniques used to evaluate the course of degradation of biodegradable polymers are described in the following section.

Water absorption and weight loss

The main factors influencing the wettability of polymers are related to their degree of crystallinity, chemical composition and aqueous media. Marois and others found different water absorption behaviours for polyhydroxyoctanoate (PHO) films in water and in PBS solution. It was observed that incubation in PBS led to lower water absorption than when incubated in water and this difference was attributed to osmotic phenomena and to the ionic strength of the PBS. Water uptake measurements can give some indication of the hydrophilicity/hydrophobicity nature of the materials and therefore of their tendency to be degraded by hydrolysis. The water uptake values are normally obtained after equilibrium of the materials in solution. In some cases, the equilibrium cannot be reached because the material is degrading at the same time. In this case, the level of water uptake usually increases with the degradation time due to an increase in the permeability of the materials (a porous structure is normally obtained as a result of the release of degradation products).

During degradation, the mass of the material may undergo changes and these changes can be monitored by comparing the mass before and after the degradation period. The material should be dried to a constant mass before measuring the initial weight of the sample to avoid residual moisture in the samples. The drying temperature should not exceed the temperature at which the materials undergo irreversible changes (e.g., melting temperature). After degradation, the sample should be washed thoroughly with distilled or deionised water to remove traces of soluble degradation products, enzymes, salts, or other impurities and dried under vacuum conditions until constant weight. The extent of degradation is commonly determined by calculating the percentage of weight loss.

Molecular weight

The determination of the polymer molecular weight (M_w) during the degradation process is one of the most important analyses to be made when studying the degradation mechanisms of polymeric biomaterials. Many other properties are affected by changes to the M_w, including mechanical properties, crystallinity, weight loss and morphology. The evolution of M_w during degradation can be determined by gel permeation chromatography (GPC) or by viscometry techniques. The main useful parameters to be analysed are the number-average (M_n) and the weight-average (M_w) molecular weights. The ratio M_w to M_n is known as the polydispersity index, which measures the breadth of the molecular weight distribution.

Crystallinity

In semicrystalline polymers, an increase in polymer crystallinity has been frequently observed at early stages of degradation and this is likely related to the plasticising effect of the absorbed water that makes chain movements possible, allowing crystallisation to proceed toward thermodynamic equilibrium. The generation of crystallised monomers and oligomers was also indicated as a source for increased crystallinity after degradation. As chain scission occurs in the crystalline regions, during later stages of degradation, a decrease in the overall crystallinity of the polymer can be expected.

Morphology and dimensional changes

Changes in the surface morphology of biomaterials after degradation, like roughness and the appearance of cracks or micro/macropores, may be examined by microscopy methods such as light microscopy, scanning electron microscopy (SEM) and atomic force microscopy (AFM). In addition, AFM gives information about the sample topography at low magnifications and allows for roughness calculations.

Surface chemistry

The surface chemistry of biodegradable polymers, before and after degradation, can be analysed by several techniques that include Fourier transform infrared

(FTIR) spectroscopy, X-ray photoelectron spectroscopy (XPS) and contact angle measurements. Surface chemistry characterising techniques may give indications about the type of degradation mechanism.

Mechanical properties

Certain biomedical applications, such as hard tissue replacement, require biodegradable materials with mechanical properties typically on the same order of magnitude as those of the tissue that it replaces. In addition, the material should degrade while maintaining a specified minimum mechanical strength to support the formation of new tissue. Thus, in these cases, it is critical to evaluate the mechanical performance of biomaterials during the course of their degradation.

Isolation and identification of degradation products

The development of materials for biomedical applications always requires extensive biological testing to demonstrate the safety of both the material and its degradation components. A potential disadvantage of using biodegradable polymers in biomedical applications is the eventual toxicity of the degradation products. If the biomaterial degrades, either spontaneously or due to biological activity, components can leach into surrounding tissues and may enter the circulation and be easily metabolised via normal pathways, causing toxic effects systemically and in distant sites. It is necessary, therefore, to identify the major species produced at different stages of degradation and the kinetics of their formation.

14.3.5 Mechanisms of degradation

In general, the first degradation event after contact with water molecules is the hydrolytic scission of the polymer chains leading to a decrease in the molecular weight. At this initial stage, the first degradation products are not small enough to become soluble and no significant change in the material weight is detected. With increasing time, the molecular weight of degradation products is reduced by further hydrolysis, which allows them to diffuse from the bulk material to the surface and then to the solution, causing significant weight loss.

Degradation by surface erosion mechanism is characterised by loss of the material from the surface only, resulting in very predictable mass loss profiles. The materials get smaller but keep their original geometric shape. This feature may be beneficial for delivering molecules at constant rate and maintaining the mechanical and structural integrity of the material with degradation. Poly(anhydrides) and poly(orthoester)s are examples of surface-eroding polymers.

Enzyme catalysed degradation of polymeric biomaterials may follow a surface erosion mechanism, especially for highly crystalline and hydrophobic homopolymers. Due to their relatively large size, the enzyme molecules cannot penetrate the tightly packed structure of certain polymers, so that the enzymatic catalysis occurs at the polymer–enzyme interface. As the degrading surface becomes roughened or fragmented, enzymatic action may be enhanced as a result of increased surface area. Enzymatic degradation mechanisms of polymeric materials depend, however, on many other factors such as chemical composition, degree of homogeneity and processing technique.

14.3.6 Strategies for controlling the degradation rate of biodegradable polymers

Depending on the application (long-term implants or temporary matrices for tissue engineering and drug delivery), three main strategies have been observed to alter the degradation rate of biodegradable polymers. Some strategies intend to increase the degradation rate of biomaterials and this is normally obtained by blending or copolymerisation techniques or by adjusting the polymer physico-chemical properties. The reduction of the degradation is mainly achieved by selecting the polymer chemical composition (high degree of crystallinity, hydrophobicity, elevated molecular weight, nonhydrolysable bonds) and by using surface modification methods. The third approach may be obtained by selecting the right combination of the other two strategies and may take advantage of using smart systems, which means that polymers will degrade only in response to a certain stimulus.

The degradation of some biomaterials involves nonenzymatic hydrolysis, which is largely nonregulated. In such situations, only the calendar may determine the rate of resorption of the material and in these cases, the degradation of biomaterials is controlled by manipulating their physico-chemical properties. Blending and copolymerisation have been used to change the degradation rate of polymeric materials, since, in general, blends and copolymers degrade faster than homopolymers from the same family.

Other approaches have been used, such as including enzymatic recognition sites in polymer backbones (e.g., amino acids) to enzymatically modulate the material degradation or making the material sensitive to the feedback provided by the cells involved in the healing response, as cells enzymatically degrade the extracellular matrix around them.

Section IV

Industrial applications of enzymes

15

Applications of oxidoreductase enzymes

15.1 Introduction

Oxidoreductases are widely distributed among microbial, plant and animal organisms. They catalyse the exchange of electrons or redox equivalents between donor and acceptor molecules, in reactions involving electron transfer, proton abstraction, hydrogen extraction, hydride transfer, oxygen insertion, or other key steps. In general, two half reactions (one oxidative and one reductive) take place and at least two substrates (one reducing and one oxidising) are activated/transformed. To accomplish their physiological function, oxidoreductases employ various redox-active centers. The centers are shielded by the polypeptide backbone of oxidoreductases, which can modulate their selectivity, redox potency, reactivity, stability and inhibition resistance.

Common redox centers include amino acid residues (e.g., tyrosine or cysteine), metal ions or complexes (e.g., Cu, Fe, Mo, Fe-S cluster, or heme) and coenzymes (e.g., FMN, FAD, pterin, or pyrroloquinoline quinone, or PPQ). Oxidoreductases can be classified according to their sequence or three dimensional structure, which is very informative for the study of structure function relationship, enzyme evolution, functional genomics and in silico new enzyme discovery.

15.2 Applications of oxidoreductase

A few oxidoreductases are at present in market for textile, food and other industries and more candidates are being actively developed for future commercialisation. The application of oxidoreductase may be divided into industrial-technical, specialty chemical synthesis, environmental, food, medicinal and personal care fields. Being specific, energy-saving and biodegradable, oxidoreductase-based biocatalysts fit well with the development of highly efficient, sustainable and environment-friendly industries.

15.2.1 Industrial-technical applications

In this field, oxidoreductases can be applied as advantageous biocatalysts to replace hazardous/expensive chemicals, save on energy/resources consumption, create novel functionalities, or reduce detrimental impacts on the environment.

Lignocellulosics transformation

Conventional pulp delignification relies on either chlorine- or oxygen-based chemical oxidants. Although very effective, these agents can cause serious problems in by-products disposal or cellulose fibre-strength loss. Enzymatic delignification systems are attractive alternatives. Laccase, peroxidase and other oxidoreductases participate in the natural delignification by lignolytic white-rot fungi. Various laccases have been shown capable of degrading both natural and synthetic lignin. They oxidise either directly the phenolic components of lignin or in the presence of a proper redox mediator, indirectly, the heterogeneous phenolic and nonphenolic (especially methoxybenzene) components. As a result, radicals could be generated in lignin, which can lead to aliphatic or aromatic C–C bond cleavage and depolymerisation.

Lignin peroxidase is also a powerful delignifying agent. Its highvalent ferryl-oxo species can abstract electrons or protons from the nonphenolic component of lignin, thus generating radicals that can disintegrate the heterogeneous polymer. Like lignin peroxidase, Mn peroxidase is also employed by white-rot fungi to degrade lignin. Mn peroxidase is of particular interest, because its oxidative agent, Mn(III), can be stabilised by small chelators (e.g., oxalate) and diffuse onto the sites in lignin normally inaccessible to an enzyme.

Currently, there are a few major hurdles in commercialising an enzymatic delignification system. For laccase, the need of a relatively high dosage (~0.1 to 1%) of a redox mediator not only significantly increases the overall cost, but also leads to a waste-disposal problem. For lignin peroxidase and Mn peroxidase, economically viable, large-scale production is yet to be realised.

Polymerised lignocellulosics are useful for the production of composites such as particle board and liner board. However, most chemical adhesives used in the conventional polymerisation, such as urea, formaldehyde, isocyanate and petrochemical resins, are hazardous. Based on its lignin-oxidation property, laccase could be used to replace these chemicals and serve as a bioadhesive. To initiate or enhance the cross-linking efficiency, laccase could be used in three ways: to directly oxidise wood particles/pulp to generate radicals for cross-linking, to functionalise wood particles or pulp with small compounds (such as aromatic, carboxyl, isocyanate, or acrylamide substances) which act as cross-linking agents and to transform isolated lignin (often a by-product from pulping), starch, phenolic polysaccharide, or protein into radical-rich and nontoxic adhesives. Such applications of laccase could not only replace toxic or expensive chemical adhesives, but also transform wastes such as lignin from paper industry into value added product.

To commercialise such enzymatic bioadhesives for particle-/linerboard manufacturing, further optimisation is needed so that they can be competitive against the inexpensive chemical counterparts.

Carbohydrates derivatisation

Carbohydrates can serve as bio-based, renewable and inexpensive raw materials, precursors, building blocks, or additives for many industry products. Historically, useful organic acids, such as lactate, have been produced from sugars by whole-cell fermentation. Cell-free oxidoreductases may also be used for the organic acid production.

Commodity sugars, such as glucose and sucrose, can be modified to produce value-added sugars or other substances. In one example, D-glucose is converted to 2-keto-D-glucose by glucose 2-oxidase and the keto glucose is then converted to D-fructose by chemical hydrogenation. In another example, the cheese-making by-product lactose is converted by a carbohydrate oxidase to lactobionic acid, a valuable food additive, acidulant, chelator, drug formulant and polymer precursor. Glucose and other carbohydrates are rich in (pro)chiral sites, which can be functionalised by specific carbohydrate oxidoreductases. The derivatised carbohydrates may become useful synthons for valuable industrial or pharmaceutical products.

Oligo and polymeric carbohydrates may also be oxidatively modified to gain new properties. In one example, galactomannan is treated with galactose oxidase, which converts the C_6–OH into highly reactive aldehyde, making the polymer amenable for various functionalisations, including use as a paper additive. Adding reactive groups onto sugars, via oxidoreductase catalysis, can lead to the synthesis of new materials, such as glycolipid surfactants. Because of the fragmentation of the carbohydrate-derived specialty chemical field, the use of carbohydrate-specific oxidoreductases would be relatively low-volume and of high specialty.

Textile applications

Potential oxidoreductase applications for the textile industry include cotton fibre whitening, dye finishing and waste treatment.

A recent report describes an enzymatic cotton fibre-whitening method. A significant effect of applying laccase, in combination with peroxide, to bleach cotton is observed. Potential benefits of the application include chemical, energy and water saving.

Laccase-catalysed textile dye-bleaching is useful in finishing dyed cotton fabric. Replacing conventional chemical oxidants (e.g., hypochlorite), a laccase-based system has been commercialised, which is used to bleach the indigo dye in denim to achieve desirable bleached appearance on the fabric.

Laccase can be used to convert dye precursors *in situ* for better, more efficient fabric dyeing. In the application, fibres are soaked with dye precursors first. Addition of laccase then promotes their conversion on the fibre, leading to stronger binding, more even dyeing and less usage of dye/fixation chemicals.

Catalase has been used to degrade the remaining peroxide after the bleaching treatment of fabric.

Detergent applications

Oxidoreductases have potential for cleansing applications, such as cloth- and dishwashing. The majority of the stains in cloth, as well as all the cleansing targets on dish, are of biological origin. Laccase, peroxidase and other enzymes are being studied for bleaching stains that are tough to be removed under normal washing conditions. An enzymatic bleaching could be better than the conventional chemical bleaching because the former may selectively target the stains over the fabric dye.

There is another detergent application for oxidoreductases. A heme peroxidase has been developed to inhibit so-called 'dye-transfer,' so that cloths of different colour may be mixed for washing. Oxidoreductase may be included in a cleansing formulation to eliminate the odour on fabrics, including cloth, sofa surface and curtain, or in a detergent to eliminate the odour generated during cloth washing.

Other technical applications

Other miscellaneous technical applications exist for oxidoreductase. One example is laccase-based oxidative transformation and subsequent coupling of dye precursors to the collagen matrix in hides. Under laccase catalysis, soluble dye precursors could be adsorbed, oxidised and polymerised to give a desired tanning effect. The process could enhance dyeing efficiency, reduce cost (by using inexpensive precursors), or provide improved hide characteristics. Another example is the use of catalase to clean up the residual H_2O_2 during semiconductor/circuit chip manufacturing.

15.2.2 Food applications

Many oxidoreductase substrates, such as carbohydrates, unsaturated fatty acids, phenolics and thiol-containing proteins, are important components of various foods and beverages. Their modification by oxidoreductase may lead to new functionality, quality improvement, or cost reduction. Sometimes O_2 is detrimental to the quality or storage of food/beverage because of unwanted oxidation. Oxidases may be used as O_2-scavengers for better food packaging.

Food improvement

Glucose oxidase has been commercialised for bread making applications. Adding the enzyme to dough can lead to various physico-chemical changes including cross-linking of wheat albumin, globulin and to lesser extent, glutenin. Consequently the dough shows better visco-elastic/rheological characteristics and the baked bread has improved crumb, larger volume, or other properties. The effect is likely caused by the H_2O_2 produced by the enzyme. However, the performance of the enzyme is not superior to that of chemical oxidative additives such as bromate and azodicarbonamide. Thus there is an interest to develop other carbohydrate oxidases for this application.

Lipoxygenase is a promising candidate for baking applications. Modifying the endogenous lipids/unsaturated fatty acids (and their emulsification property) and generating oxidative peroxide, the enzyme may provide dough-strengthening and hitening effects. However, adding the enzyme to certain foods may cause off-flavour or loss of endogenous antioxidants.

The application of oxidases is also being extended to other wheat products, such as noodle, pasta and cake. Similar to the case of bread, these enzymes modify flour components either directly or indirectly through produced active oxygen species.

Laccase may be applied to certain processings that enhance or modify the colour appearance of food or beverage. One interesting case involves the ripe-olive processing in which laccase replaces conventional lye solution and oxidatively polymerises various phenolics (such as oleuropein) in olive, resulting in colour darkening and debittering.

Some oxidoreductases, such as glucose oxidase and catalase, have been shown able to improve the freshness preservation of shrimp and fish. It is suggested that the active oxygen species generated by the enzymes can act as bactericides. Disulphide-reducing enzymes may be useful in recovering proteins from low-value sources (such as fishery by-products) for reconstructed meat products.

Wine, juice and brewery applications

Browning, haze formation and turbidity development during the processing or storage of clear fruit juice, beer and wine can be major problems for the industry. It is believed that phenolic compounds are involved in this process. Conventionally, undesirable phenolics are adsorbed and removed by various fining agents (e.g., gelatin, bentonite) that usually have low specificity, may affect colour or aroma and can pose disposal problems. Laccase and other oxidases may be used to remove or modify problematic phenolic saccharides and improve the clarity, colour appearance, flavour, aroma, taste, or stability in fruit juice or fermented alcohol beverages. Following treatment by laccase, oxidised and polymerised (or precipitated) unwanted phenolic substances could be removed by silicate fining or filtration.

A laccase has recently been commercialised for preparing cork stoppers for wine bottles. The enzyme oxidatively reduces the characteristic cork taint/astringency which is frequently imparted to the bottled wine.

Dairy application

One recent example of applying oxidoreductase to dairy is the use of a carbohydrate oxidase to convert lactose during cheese-making. Lactose, found in the whey fraction, is currently discarded as a cheese-making by-product. Lactobionic acid, the product of the enzymatic oxidation, is a valuable chemical widely used as a food additive, acidulant, chelator, drug formulant and polymer precursor. Converting the lactose *in situ* to lactobionic acid may have many benefits in terms of added value and improved quality.

15.2.3 Environment protection

Oxidoreductases have high potential in environment-protection applications, along with hydrolases, lyases and transferases. Many microbes can use polycyclic aromatic hydrocarbons, chlorinated compounds and other pollutants as their carbon, nitrogen, or energy source. A significant part of their metabolism is carried out by oxidoreductases. The degradation reactions catalysed by these enzymes include electron transfer, H extraction and O insertion, which can induce ring opening, depolymerisation, mineralisation and other transformations.

Biodegradation

Oxidoreductases may be applied to degrade various substances such as undesirable contaminants, by-products, or discarded materials. Some of these applications have been described above.

Recently, laccase has been shown capable of oxidising and degrading lipids such as trilinolein and methyl linoleate. These unsaturated fatty compounds are not typical laccase substrates. The products include hydroperoxides and epoxides. It seems that laccase promotes the initial pentadienyl and subsequent peroxy radical formation. The reaction is of interest because of the occurrence of the fatty compounds in wood and food, which may get involved in laccase catalysed delignification and food modification, respectively.

A heme peroxidase has been developed for transforming discarded dyes from the effluent of dyehouse before its discharge into the environment. The oxidation of the dyes by peroxidase, or laccase, may lead to bleaching, degradation, or polymerisation to facilitate downstream treatments (e.g., physical adsorption). These enzymes may also be used to treat effluents from pulp/cotton mills, food/fruit processing plants, or breweries.

Laccase may be applied to degrade plastic waste having olefin units. Likely, an oxidation of the olefin units by the enzyme, preferably in the presence of small redox mediators, could initiate a radical chain reaction, leading to the disintegration of the plastic. Laccase, peroxidase and oxygenase are being studied as biocatalysts for degrading hazardous coal substances, particularly the sulphur-containing components. The study is of interest in terms of reducing the pollution around coal mines and emission of acid rain causing agents from power plants.

Oxidoreductase may also be used to eliminate odour emitted from places such as garbage disposal sites, livestock farms, or pulp mills. Various oxidases can oxidatively degrade sulphide, thiol, ammonia, amine, short-chain fatty acids, or other volatile organic compounds that cause malodour.

15.2.4 Biodetoxification and biodecontamination

Many pesticides, xenobiotics, coal substances and industrial products derived from polycyclic, aromatic, halogenated hydrocarbons and other organic compounds are hazardous environment pollutants. Using oxidoreductases to detoxify and remove them is attracting active research efforts.

Laccase and peroxidase have been used to transform (often in the presence of redox mediators) various xenobiotics, polycyclic aromatic hydrocarbons and other pollutants found in industrial waste and contaminated soil or water. In general, the redox potential of these compounds is too high for laccase to directly oxidise them via electron transfer. The use of redox mediator allows other reactions, such as H extraction, to take place. The laccase catalysis could result in either direct degradation or polymerisation/immobilisation. Reported examples of direct degradation by laccase include dechlorination of chlorophenols, cleavage of aromatic rings and mineralisation of polycyclic aromatic hydrocarbons. The processes include polymerisation among pollutants themselves or copolymerisation with other nontoxic substances (such as humic materials). Polymerised pollutants often become insoluble or immobilised, thus facilitating easy removal by means such as adsorption, sedimentation, or filtration.

One of the major biodetoxification targets is chlorinated compounds, such as tetrachloroethene and polychlorinated biphenyl, which contaminate underground water. Oxidoreductases, particularly mono and dioxygenases, are powerful dehalogenating agents. Enzymes such as ethene mono-oxygenase and biphenyl dioxygenase are promising candidates for large-scale industrial application in the field.

Recent advances in the sequencing of the genomes of environmentally relevant microbes and the direct DNA sequencing of soil microbes are exposing new metabolic enzymes/pathways potentially beneficial for bioremediation. Protein engineering is also creating new enzymes/functions for biodetoxification. For instance, a dioxygenase that oxidises arene to vicinal arene diols has been engineered to extend its reactivity from toluene, naphthalene and biphenyl to polychlorinated biphenyls, trichloroethylene and dioxin.

15.2.5 Biosensors and bioreportors

Many oxidoreductases have co-substrates that are either chromogenic, fluorogenic, chemiluminescent, or electroactive. They may produce products that are reactants/substrates for coupled chemical or enzymatic reactions suitable for optic, electric, or other physical measurements. Thus many oxidoreductases can be applied as biosensors or bioreporters.

Portable and online biosensors

Examples of the oxidase applied in the field include laccase, horseradish peroxidase and glucose oxidase. Laccase catalysis, coupled with various physical transducers, could be useful as biosensors for detecting O_2 and a wide variety of reducing substrates, especially phenols or anilines. One type of biosensor can monitor electric current or voltage changes from a modified oxygen electrode on which O_2 reduction is enhanced under the electrocatalysis of immobilised laccase. For detecting phenols or anilines, three types of biosensors can be made: one may detect the photometric change resulting from the oxidation of a chromogenic substrate, one may monitor the O_2 concentration change which is coupled to the substrate oxidation and one may use an electrode to accept the electrons surrendered by the substrate. For these applications, laccase is preferably immobilised onto the biosensor surface and the coupled signal transducer is either of optic, amperometric and piezoeffect nature, or a field-effect transistor or thermister. A horseradish peroxidase, glucose oxidase, or other oxidoreductase-based biosensor may be constructed similarly. The targeted substances may be biogenic, such as blood sugar and cholesterol, environment-related, such as phenol and peroxide, or food related, such as ascorbate. Stereo-selective oxidoreductases, such as anomer-specific carbohydrate oxidase, may be used for sensing chiral compounds.

The biosensors described above contain one enzyme responsible for both target sensing and initial signal transduction. Other biosensors may be constructed by multiple enzymes, with separated primary sensing and secondary signal-relay/transduction tasks. These biosensors are of interest for quality control, environment monitoring, health care and other fields.

Reporters

Oxidoreductase catalysis can be used to assay other enzymes. In these assays, either the enzyme of interest catalyses the production of a compound whose subsequent transformation by an oxidoreductase generates detectable physical change, or a product from the oxidoreductase catalysis (whose production is accompanied by a detectable physical change) is quenched by the activity of the enzyme of interest. The strategy has been applied to use laccase to assay various enzymes including amylase, alanine-, cysteine-, or leucine-specific aminopeptidases, alkaline phosphatase, angiotensin I converting enzyme, arylamidase, cellobiose oxidase, chymotrypsin, glucosidase, γ-glutamyl transpeptidase, kallikrein, plasmin, thrombin and coagulant X factor.

Oxidoreductase covalently conjugated to a bio-binding molecule can be used as a reporter for immunochemical (ELISA, Western blotting), istochemical, cytochemical, or nucleic acid-detection assays. The bio-binding molecules include antibody, antigen, DNA, RNA, biotin, or streptavidin. An example is a linked laccase antibody. In this application, binding of the antibody to its antigen target is detected by localised laccase activity on a gel or a blot membrane, much like the conventional peroxidase or phosphatase assisted immunochemical assays.

Under certain conditions, the antibody-antigen binding impairs the function of conjugated laccase, thus allowing immunochemical detection through modulation of laccase activity. For DNA detection, an oxidoreductase, such as horseradish peroxidase, can be linked to a DNA probe for direct hybridisation, or to biotin for binding to a streptavidin-linked DNA probe for indirect hybridisation. The bioreporter applications are of interest for the high-sensitivity diagnostic field.

15.2.6 Organic synthesis

Oxidoreductase-based biocatalysts have high potential for efficient, asymmetric synthesis applications. This is mainly because of their broad versatility, high specificity, high stereo-selectivity, low demand for stringent/extreme conditions and environment-friendliness. Oxidoreductases can transform a wide range of aliphatic/aromatic molecules, functionalise inert hydrocarbons (by hydroxylation, sulphoxidation, epoxidation, etc.), carry out regio-, enantio- (on racemic substrates), enantiotopo – (on prochiral substrates) and chemo-selective reactions, build useful synthons from inexpensive and renewable biomaterials, save on energy and other resources and relieve negative environment impact. Although few oxidoreductases have been commercialised in the field for now, significant growth in using oxidoreductase-based biocatalysis is projected for the coming decades.

Asymmetric/chiral synthesis

NAD(P)H-dependent dehydrogenases are very useful for asymmetric synthesis. Catalysing reversed dehydrogenation reactions, the enzymes can transform prochiral aldehydes/ketones to chiral alcohol, hydroxyl acids, amino acids, etc., with high regio- or enantio-selectivity (with enantio excess >90 to 95%).

Alcohol dehydrogenases can catalyse the synthesis of chiral alcohols from aldehydes/ketones, although the yield of hydrophobic alcohols is relatively low.

Hydroxy acid dehydrogenases can catalyse the synthesis of chiral hydroxyl acids from aliphatic, linear/branched, or aromatic 2-oxo acids, which are valuable synthons.

Amino acid dehydrogenases can catalyse the reductive amidation of 2-keto acids to either L- or D-amino acids, useful for making unnatural/rare amino acids with selected aliphatic, aromatic, linear, or branched side chains. Such synthesis is almost unattainable by conventional fermentive amino acid-production methods.

In general, dehydrogenase-catalysed chiral synthesis has higher yield than that of hydrolase, but is more difficult or expensive due to its NAD(P)-dependency. Oxygenases/hydroxylases are also good biocatalysts for asymmetric synthesis. They can functionalise inert molecules, such as hydrocarbons, by introducing active O atoms.

Flavin-containing monoxygenases/hydroxylases, such as phydroxybenzoate 3-hydroxylase, can carry out highly regio- and stereo-selective arene hydroxylation. Baeyer-Villiger mono-oxygenase is of particular interest, because of its ability to catalyse the nucleophilic oxygenation of ketone (the Baeyer-Villiger reaction). It can also catalyse the nucleophilic oxygenation of boron, as well as the electrophilic oxygenation of sulphur, selenium, nitrogen, or phosphorus, all with high enantio- or enantiotopo-selectivity. The enzymatic way is more advantageous than the traditional chemical ways (using oxidants such as *m*-chloroperbenzoic acid or trifluoroperacetic acid), because the latter are hazardous and poorly enantioselective.

Iron-containing dioxygenases can transform arenes into vicinal, *cis* arene diols, a task not attainable by conventional chemical synthesis. More than 300 vicinal arene diols are identified from microbial oxidation of aromatic hydrocarbons, which provides a useful pool for selecting desirable synthons for various industrial and medicinal substances, such as postaglandins and other hypertensive agents.

Having high regio- and stereo-selectivity towards its unsaturated fatty acid substrates, lipoxygenase can be used to synthesise many natural products and valuable build blocks. Nonaqueous media, immobilisation and protein engineering have been applied to optimise its biocatalysis for asymmetric synthesis.

Chloroperoxidase and cytochrome P450 can be applied for asymmetric olefin epoxidation, allylic hydroxylation and sulphoxidation, due to their high-valent ferryl-oxo species. One major hurdle of commercialising chloroperoxidase is its instability in the presence of a medium to high level H_2O_2, the oxidising substrate. This may be solved by *in situ* H_2O_2 generation with either a secondary enzyme (e.g., glucose oxidase) or an electrode. As for cytochrome P450, large-scale production of the enzyme remains a challenge.

Nanomaterial and polymer synthesis

Oxidoreductases capable of generating radicals are potential biocatalysts for polymer synthesis. Phenol oxidases (such as tyrosinase) are used to make polyphenolic polymers from monomeric phenols (such as catechol). Carbohydrate oxidases (e.g., galactose oxidase) are used to modify oligo-polymeric carbohydrates to promote cross-linking.

Amino acid-based polymers are desirable due to their biocompatibility and biodegradability. Horseradish peroxidase is able to catalyse the polymerisation of tyrosine. When nano-patterned on a defined surface, the enzymatic polymerisation generates well-organised nanofeatures potentially useful for specialty membrane and circuitry. Immobilised horseradish peroxidase can direct the synthesis of polyaniline film of controlled thickness, a property very useful for coating the surface of biosensors. Such a property is unattainable by chemically synthesised polyaniline because it is insoluble and uncompressible. Horseradish peroxidase is also used to synthesise phenol/aniline polymer via oxidative dehydrogenation.

Immobilised peroxygenases may be used to catalyse the formation of fatty epoxides, important polymer precursors, plasticisers and plastic stabilisers. The enzymatic process may avoid the shortcomings of the current chemical process whose strong acidity causes significant oxirane ring opening and other undesirable side activities.

Medicinal and other synthesis

Laccase can be used to synthesis several complex medicinal agents including triazolo(benzo)cycloalkyl thiadiazines, vinblastine, penicillin X dimer, cephalosporin antibiotics and dimerised vindoline. Laccase can also be used to synthesise various functional organic compounds including polymers with specific mechanical/electrical/optical properties, textile dyes, cosmetic pigments, flavour agents and pesticides.

Applying oxidoreductases may lead to new industrial synthetic methods. For instance, Baeyer-Villiger mono-oxygenase can catalyse a useful ring-expansion reaction by transforming a cyclic ketone to a corresponding lactone. Macrophomate synthase can catalyse Diels-Alder reaction. Sometimes, the action of an oxidoreductase on its substrate may induce a secondary reaction on other parts of the substrate, leading to a new type of biocatalysis.

Applying oxidoreductases may lead to the usage of materials that are otherwise inactive or difficult chemical reactants. For instance, chloroperoxidase and cytochrome P450 can functionalise inert hydrocarbons via hydroxylation. Horseradish peroxidase can carry out sulphoxidation on alkyl aryl sulphides, *N*-oxidation of nitrosoand hydroxylamino-compounds, or epoxidation on styrene. Enone reductase can hydrogenate unsaturated bonds to transform ketones to hydrocarbons. The so-called 'old yellow enzyme' from yeast, an FMN-containing enzyme, can catalyse the reduction (by NADPH) of the olefinic >C=C<, not carbonyl >C=O, the site of cyclohexenone.

Applying oxidoreductases may lead to new intermediates/synthons. Methane mono-oxygenase can transform methane to methanol, a useful synthon. Xylose reductase can produce xylitol, a natural sweetener, from glucose.

15.2.7 Medical and personal-care applications

Many products generated by oxidoreductases are antimicrobial, detoxifying, or active personal-care agents. Due to their specificity and bio-based nature, potential applications of oxidoreductases in the field are attracting active research efforts.

Disinfection

One potential application is laccase-based *in situ* generation of iodine, a reagent widely used as disinfectant. A laccase-iodide salt binary iodine-generating system (for sterilisation) may have several advantages over the direct iodine application. First, the iodide salt is more stable and much safer than I_2 in terms of storage, transport and handling. Second, the release of iodine from a laccase-iodide system can be easily controlled (by means such as adjusting laccase concentration). The system may be used in various industrial, medical, domestic and personal-care applications such as sterilisation of drinking water and swimming pools, as well as disinfection of minor wounds. Peroxidases may replace laccase for the application, although they would need H_2O_2 as cosubstrate.

The ClO- and Mn(III) species produced by haloperoxidase and Mn peroxidase are highly potent oxidants and antimicrobial agents. Coupled with an oxidase as an *in situ* H_2O_2 provider, a peroxidase system may be applied to various processes where a disinfection is involved.

Skin care

Poison ivy dermatitis is caused mainly by urushiol, a catecholderived toxin. Oxidised urushiol (an *o*-quinone derivative), however, is nontoxic. Laccase has been shown to oxidise, polymerise and detoxify urushiol, thus reducing the effect of poison ivy dermatitis. Lysyl oxidase, whose physiological activities include extracellular matrix formation, can be applied to accelerate wound-healing. Peroxidase may also be used to cross-link collagen, beneficial to the recovery of damaged skin. An NO-generating enzyme, such as nitrite reductase or NO synthase, may too find a topical use for skin tissue regeneration. Glucose oxidase may be applied as an *in situ* H_2O_2 generator in a skin-care lotion/

foam, useful for skin whitening and disinfecting and catalase and peroxidase may be applied as an oxygenator for skin rejuvenation.

Hair care

Current hair-dyeing or waving processes often involve oxidative or additive chemicals that are either malodourous, irritants, or difficult to handle. A laccase-based system may overcome these drawbacks by replacing harsh chemicals and operating at milder conditions (in terms of pH and solvent). Laccase-catalysed oxidation, transformation and cross-linking of various precursors (mostly phenols and anilines) have been reported to result in satisfactory hair dyeing or waving. In addition to providing an easier-to-handle hair-care procedure, a laccase-based system may also improve or complement the cosmetic effect achieved by conventional chemical methods.

Other health/hygiene applications

A glucose oxidase, lactoperoxidase and iodide system has been tested for dental care. The oxidase produces H_2O_2 (from glucose oxidation) to feed the peroxidase, so that it can generate iodine that can kill plaque-causing bacteria.

Oxidoreductases may find use as deodorants for personal-hygiene products, including toothpaste, mouthwash, chewing gum, detergent, soap and diapers. For instance, laccase can oxidise many thiol, sulphide, ammonia and amine compounds that cause the malodour in halitosis, bromhidrosis and hyperhidrosis. Rather than simply masking the malodour with fragrances as conventional deodorants do, an enzymatic system can degrade the offensive molecules, or even kill the microbes that generate them.

There are many other applications in the field for oxidoreductases. For instance, haloperoxidase may be used to oxidatively modify rubber latex surfaces, making them less allergenic. Catalase may be applied as part of contact lens cleansing solution. A secreted oxidoreductase may even be developed as a vaccine against secretor microbes. For instance, *Aspergillus oryzae* catalase A protein has been studied as a potential aspergillosis vaccine.

15.3 Future prospects

Multidisciplinary efforts are being made to discover better oxidoreductase biocatalysts. New genes are being found from nature's repertoire, by means of classic screening, direct cloning ('metagenomic'), or in silico mining. New genes are also being constructed by rational design or directed evolution, a rapidly developing, powerful way of engineering proteins and enzymes.

Selected oxidoreductases are preferably produced in fermentative microbial hosts, although transgenic plants and animals (through their milk or eggs) are emerging as promising alternative hosts.

As industrial catalysts, enzymes (including oxidoreductases) may have many advantages over chemical- or living cell-based catalysts. Again, as catalysts for synthesis, oxidoreductases may be preferred over other enzymes (e.g., hydrolases, lyases, or transferases). Oxidoreductases may have higher stereo-selectivity, be able to act on substances (such as hydrocarbons) that are often inert to other enzymes and use bio-based/renewable starting materials. However, many oxidoreductases suffer from lower specific activity and the need for cofactors. The established technical and food application fields, as well as the emerging synthesis and biodegradation fields, should continue to grow for oxidoreductases. Today the growth is mainly fuelled by performance improvement and cost reduction. In the future, however, it might be driven by new, unique functionalities of oxidoreductase biocatalysis that could lead to high-value bio-based materials (polymers, plastics, adhesives, or pigments), bio-enhanced energy sources (bio-fuel cell), or protein pharmaceuticals.

15.3.1 Potential new applications

There are many new and classic ideas, as well as laboratory-scale trials, of applying oxidoreductases as biocatalysts.

Biofuel cells

Fuel cells are very attractive energy sources, particularly at micro-, mini-, portable-, or mobile-scale, that potentially have higher energy conversion/usage efficiency and lower pollution effect than any of the existing/emerging energy sources. For instance, fuel cells may be more desirable than batteries because they can refuel almost instantly and have fewer disposability problems. Fuel cells may tap on the chemical energy in renewable resources or even waste/pollutants, converting it into electricity with high efficiency.

A fuel cell uses two electrodes (in two compartments) to reroute the electron flow in a redox reaction: an anode, to take electrons from a reducing substance (e.g., carbohydrate, alcohol, hydrocarbon, or H_2) and a cathode, to give the electrons (after carrying out work in the circuit) to an oxidising substance (e.g., O_2). A membrane, permeable only to the solvent, H^+, OH^-, or other supporting electrolytes), separates the two compartments to prevent the reducing and oxidising substrates from reacting directly with each other.

Oxidoreductase may be applied as a biocatalyst for the electrode reactions. For instance, a laccase may be adsorbed, entrapped, or wired onto the cathode to catalyse the O_2 reduction. A carbohydrate oxidase may be applied to the anode to catalyse the oxidation of a sugar, when it serves as the reducing substance. An H^+-pumping oxidoreductase, such as NADH-quinone oxidoreductase or cytochrome *c* oxidase, may be embedded onto the membrane to facilitate the H^+ movement or fuel cell's internal conductance.

Fossil fuels upgrading

Heteroatomic compounds, especially sulphur-containing aromatics, are detrimental for the usage of fossil fuels. They impede the refining of crude oil, downgrade the quality of coal, cause the emission of acid rain agents and damage the catalytic converter in motor vehicles. Removal of these compounds can correct these problems and allow low-value oil to be used for making chemicals, plastics, or surfactants, however, this is very challenging. Conventional chemical and physical desulphurisation methods (such as oxidising and solubilising inorganic pyrite sulphur by ferric salt or chlorine) have low efficiency towards organic sulphur compounds and require extreme conditions (high temperature, pressure and corrosion-resistant equipment). Oxidoreductase-based desulphurisation could be an alternative. At laboratory-scale, inorganic and organic sulphur compounds in coal slurry can be oxidised by laccase, solubilised in water and separated from solid (desulphurised) coal particles.

Another potential oxidoreductase application is in the oil field. To drill a well, a drill-in fluid containing polymers like cellulose, starch, or xanthan is used to deposit a low-permeable filter cake on the borehole wall to prevent leakage. The polymeric material needs to be removed after drilling, which is carried out at present by acids (e.g., HCl) or strong oxidants (e.g., Na-perborate). Because of their hazardous nature, it would be of interest to replace these chemicals with oxidoreductase and/or hydrolase.

15.4 Potential new oxidoreductase-based biocatalysts

For future biocatalysis, 'smart' oxidoreductases and novel reaction media may be applied to improve performance or economy.

Smart oxidoreductases

Target-specific and -binding molecules may be used to guide oxidoreductases for improved specificity and reactivity. For industrial applications, Llama heavy-chain antibodies are of particular interest. Because they are made from heavy-chain dimers only, their expression and production in heterologous hosts are relatively straightforward. In one report, the variable domains of the Llama antibodies are raised against a major plaque-causing bacterium, cloned from the B lymphocytes of immunised Llama, screened for their affinity towards the antigen and fused with glucose oxidase. The fusion protein is expressed and added to a mixture with lactoperoxidase and iodide. The antibody enhances the specificity of the oxidase towards the bacterium, leading to improved overall performance of the enzymatic system in killing the bacterium. The antibody-peroxidase fusion useful for antitumour activity is another example.

Naturally occurring binding motifs, such as carbohydrate-binding motifs, as well as peptide-targeting segments, such as the peroxisome-targeting signal, may be fused with an oxidoreductase biocatalyst to enhance its searching, binding and attacking function. For instance, a carbohydrate-binding motif may bring a detergent oxidoreductase to close vicinity of cellulose-adsorbed stains for better bleaching and a peroxisome-targeting signal may insert a lipoxygenase to fatty particulates for better oxidation.

Artificial binding peptides, specific for a target substance, may be constructed by either organic synthesis (e.g., combinatorial chemistry) or genetic engineering (e.g., phage display) combined with high-throughput screening. Mono-polyclonal antibodies can be raised for a target by established techniques. Selected binding peptides and antibodies may then be fused onto an oxidoreductase to create a 'smart' biocatalyst.

New catalysis media

Oxidoreductases can be applied in nonaqueous solution or multiphasic systems. For water-immiscible organic solvents, an oxidoreductase may be entrapped in reverse micelles or immobilised onto a carrier. Newer media for biocatalysis include ionic liquid and supercritical CO_2.

Ionic liquids, especially the room-temperature ones, are emerging as the 'green' and high-tech media for industry. They have low volatility, high thermal stability, as well as tunable polarity, hydrophilicity/phobicity, or miscibility with other liquids. Such properties make ionic liquids particularly suited for catalyst recycling and product recovery. A few oxidoreductase-based biocatalyses in ionic liquid have been reported. Further application in the field may lead to expanded biocatalysis and improved performance.

16

Enzymes in dairy, meat, fish and wine industry

16.1 Introduction

The food manufacturing and ingredients industry makes widespread use of enzymes in both traditional sectors such as baking, brewing and cheese making, but also in new areas such as fat modification and sweetener technology. A degree of care and ingenuity is often needed to adapt these fragile biological catalysts to industrial processes, but a combination of basic biochemical knowledge and modern biotechnology is opening up new areas of application, especially for enzymes of microbial origin and animal enzymes produced in microbes by genetic engineering technology.

Dairy enzymes are enzymes used for the production of cheese and yoghurt as well as other milk products. The function of these enzymes varies widely from coagulants, which are used to make cheese, to bioprotective enzymes to enhance the shelf life and safety of dairy products. For improving the quality of milk and milk products, a number of different enzymes from microbial as well as from non-microbial sources have potential applications in dairy processing.

16.2 Enzymes used in dairy industries

The application of enzymes (proteases, lipases, esterases, lactase and catalase) in dairy technology is well established. Rennets (rennin, a mixture of chymosin and pepsin obtained mainly from animal and microbial sources) are used for coagulation of milk in the first stage of cheese production. Proteases of various kinds are used for acceleration of cheese ripening, for modification of functional properties and for modification of milk proteins to reduce the allergenic properties of cow milk products for infants. Lipases are used mainly in cheese ripening for development of lipolytic flavours. Lactase (galactosidase, EC 3.2.1.23) is used to hydrolyse lactose to glucose and galactose as a digestive aid and to improve the solubility and sweetness in various dairy products. Many people do not have sufficient lactase to digest milk sugar. Lactose hydrolysis helps these lactoseintolerant people to drink milk and eat various dairy products. Galactosidases from *K. fragilis*, *A. niger*, or *A. oryzae* are inhibited by galactose. Immobilised enzyme systems are used to overcome this inhibition problem and to lower the cost of lactase use. The cheese manufacturing industry produces large quantities of whey as a by-product of which lactose represents 70–75% of the whey solids. The hydrolysis of lactose by lactase converts whey into more useful food ingredients. Lactases have also been used in the processing of dairy wastes and as a digestive aid taken by humans in tablet form when consuming dairy products. Hydrogen peroxide is used as an effective chemical sterilant for the treatment of raw milk as an alternative to pasteurisation with heat. Catalase (EC 1.11.1.6), which catalyses the decomposition of hydrogen peroxide, is used at the end of the process to remove the remaining peroxide. Bitter off-flavours that develop in ripened cheese as it matures are due to the formation of bitter flavoured peptides from milk proteins. Peptidases can break down the bitter peptides as they are formed and thus help maintain the traditional flavour of cheese.

16.2.1 Microbial rennets in dairy applications

Animal rennet (bovine chymosin) is conventionally used as a milk-clotting agent in dairy industry for the manufacture of quality cheeses with good flavour and texture. Owing to an increase in demand for cheese production worldwide coupled with reduced supply of calf rennet, has therefore led to a search for rennet substitutes, such as microbial rennets. Rennin acts on the milk protein in two stages, by enzymatic and by non-enzymatic action, resulting in coagulation of milk. In the enzymatic phase, the resultant milk becomes a gel due to the influence of calcium ions and the temperature used in the process. Many micro-organisms are known to produce rennet-like proteinases which can substitute the calf rennet. Micro-organisms like *Rhizomucor pusillus*, *R. miehei*, *Endothia parasitica*, *Aspergillus oryzae* and *Irpex lactis* are used extensively for rennet production in cheese manufacture.

Different strains of species of Mucor are often used for the production of microbial rennets. Whereas best yields of the milk-clotting protease from *Rhizomucor pusillus* are obtained from semisolid cultures containing 50% wheat bran, *R. miehei* and *Endothia parasitica* are well suited for submerged cultivation. Using the former, good yields of milk-clotting protease may be obtained in a medium containing 4% potato starch, 3% soyabean meal and 10% barley. During growth, lipase is secreted together with the protease. Therefore, the lipase activity has to be destroyed by reducing the pH, before the preparation can be used as cheese rennet.

16.2.2 Lactase enzyme

Popularly known as lactase, beta-galactosidases are enzymes classified as hydrolases. They catalyse the terminal residue of β-lactose galactopiranosil (*Galb1 - 4Glc)* and produce glucose and galactose. Lactose, the sugar found in milk and whey and its corresponding hydrolase, lactase or β-galactosidase, have been extensively researched during the past decade. This is because of the enzyme immobilisation technique which has given new and interesting possibilities for the utilisation of this sugar. Because of intestinal enzyme insufficiency, some individuals even a population, show lactose intolerance and difficulty in consuming milk and dairy products. Hence, low-lactose or lactose-free food aid programme is essential for lactose-intolerant people to prevent severe tissue dehydration, diarrhea and at times, even death.

Another advantage of lactase-treated milk is the increased sweetness of the resultant milk, thereby avoiding the requirement for addition of sugars in the manufacture of flavoured milk drinks. Manufacturers of ice cream, yoghurt and frozen desserts use lactase to improve scoop and creaminess, sweetness and digestibility and to reduce sandiness due to crystallisation of lactose in concentrated preparations. Cheese manufactured from hydrolysed milk ripens more quickly than the cheese manufactured from normal milk.

Technologically, lactose crystallises easily which sets limits to certain processes in the dairy industry and the use of lactase to overcome this problem has not reached its fullest potential because of the associated high costs. Moreover, the main problem associated with discharging large quantities of cheese Whey is that it pollutes the environment. But, the discharged whey could be exploited as an alternate cheap source of lactose for the production of lactic acid by fermentation. The Whey permeate, which is a by-product in the manufacture of whey protein concentrates, by ultrafiltration could be fermented efficiently by *Lactobacillus bulgaricus*.

Lactose can be obtained from various sources like plants, animal organs, bacteria, yeasts (intracellular enzyme), or molds. Some of these sources are used for commercial enzyme preparations. Lactase preparations from *A. niger*, *A. oryzae* and *Kluyveromyces lactis* are considered safe because these sources already have a history of safe use and have been subjected to numerous safety tests. The most investigated *E. coli* lactase is not used in food processing because of its cost and toxicity problems. The remarkable interest in *Aspergillus niger*, a species of great commercial interest with a highly promising future and already widely applied in modern biotechnology, is due to its several and diverse reactions. Moreover, *A. niger* not only produces various enzymes but it is one of the few species of the fungus kingdom classified as Generally Recognised as Safe (GRAS) by the Food and Drug Administration (FDA). The species is used in the production of enzymes, its cell mass is used as a component in animal feed and its fermentation produces organic acids and other compounds of high economic value.

16.2.3 Microbial enzymes in accelerated cheese ripening

Cheese ripening is a complex process mediated by biochemical and biophysical changes during which a bland curd is developed into a mature cheese with characteristic flavour, texture and aroma. The desirable attributes are produced by the partial and gradual breakdown of carbohydrates, lipids and proteins during ripening, mediated by several agents, viz. (i) residual coagulants, (ii) starter bacteria and their enzymes, (iii) nonstarter bacteria and their enzymes, (iv) indegenous milk enzymes, especially proteinases and (v) secondary inocula with their enzymes.

Proteolysis occurs in all the cheese varieties and is a prerequisite for characteristic flavour development that can be regulated by proper use of the above agents. Cheese ripening is essentially an enzymatic process which can be accelerated by augmenting activity of the key enzymes. This has the advantage of initiating more specific action for flavour development compared to use of elevated temperatures that can result in accelerating undesirable nonspecific reactions and consequently off flavour development.

Enzymes may be added to develop specific flavours in cheeses, for example lipase addition for the development of Parmesan or Blue-type cheese flavours. Attempts to accelerate the multiple secondary flavour-forming-reactions, e.g., Strecker degradation, have been scarce.

The pathways leading to the formation of flavour compounds are largely unknown and therefore the use of exogenous enzymes to accelerate ripening is mostly an empirical process.

16.2.4 Proteolytic enzymes of lactic acid bacteria in fermented milk products

The proteolytic system of lactic acid bacteria is essential for their growth in milk and contributes significantly to flavour development in fermented milk products. The proteolytic system is composed of proteinases which initially cleaves the milk protein to peptides, peptidases which cleave the peptides to small peptides and amino acids and transport system responsible for cellular uptake of small peptides and amino acids. Lactic acid bacteria have a complex proteolytic system capable of converting milk casein to the free amino acids and peptides necessary for their growth. These proteinases include extracellular proteinases, endopeptidases, aminopeptidases, tripeptidases and proline-specific peptidases, which are all serine proteases. Apart from lactic *streptococcal* proteinases, several other proteinases from non-lacto-streptococcal origin have been reported.

There are also serine type of proteinases, e.g., proteinases from *Lactobacillus acidophillus*, *L. plantarum*, *L. delbrueckii* sp., *bulgaricus*, *L. lactis* and *L. helveticus*. Aminopeptidases are important for the development of flavour in fermented milk products, since they are capable of releasing single amino acid residues from oligopeptides formed by extracellular proteinase activity.

16.2.5 Other dairy enzymes

Other enzymes used for dairy food application include: Proteases to reduce allergic properties of cow milk products for infants and lipases for development of lipolytic flavours in speciality cheeses.

The functional properties of milk proteins may be improved by limited proteolysis through the enzymatic modification of milk proteins. An acid-soluble casein, free of off flavour and suitable for incorporation into beverages and other acid foods, has been prepared by limited proteolysis. The antigenicity of casein is destroyed by proteolysis and the hydrolysate is suitable for use in milk-protein-based foods for infants allergic to cow milk. Recently proteases represent 60% of industrial enzymes on the market, whereas microbial proteases, particularly fungal infections, are advantageous because they are easy to obtain and to recover. An enzyme extract, which coagulates milk and which is derived from the fungus *Aspergillus niger* var. awamori, is already produced industrially.

Lipolysis makes an important contribution to swiss cheese flavours, due mainly to the lipolytic enzymes of the starter cultures. The characteristic peppery flavour of Blue cheese is due to short-chain fatty acids and methyl ketones. Most of the lipolysis in Blue cheese is catalysed by *Penicillium roqueforti* lipase, with a lesser contribution from indigenous milk lipase. Lipolytic enzymes such as lipases and esterases are an important group of enzymes associated with the metabolism of lipid degradation. Lipase-producing micro-organisms such as *Penicillium restrictum* may be found in soil and various oil residues. The industries Novozymes, Amano and Gist Brocades already employ microbial lipases.

The NOVO process for production of enzyme modified cheese (EMC) uses medium-aged cheese which is emulsified, homogenised and pasteurised, after which 'palatase' (α-lipase from *R. miehei*) is added, with or without a proteinase and the blend is ripened at a high temperature for one to four days. The mixture is reheated, a paste results which is suitable for inclusion in soups, dips, dressings, or snack foods. EMC technology has been developed to produce a range of characteristic cheese flavours and flavour intensities, for example swiss, blue, cheddar, provolo-nemor or romano, suitable for inclusion at low levels in many products. The claims that exogenous enzymes are effective in accelerating ripening have not led to their wide-spread use, possibly due to their high cost, difficulties in distributing them uniformly in the curd and the possible danger of overripening the cheese.

The other minor enzymes having limited applications in dairy processing include glucose oxidase, catalase, superoxide dismutase, sulphydryl oxidase, lactoperoxidase and lysozymes. Glucose oxidase and catalase are often used together in selected foods for preservation. Superoxide dismutase is an antioxidant for foods and generates H_2O_2, but is more effective when catalase is present. Thermally induced generation of volatile sulphydryl groups is thought to be responsible for the cooked off-flavour in ultra-high temperature (UHT) processed milk. Use of sulphydryl oxidase under aseptic conditions can eliminate this defect. The natural inhibitory mechanism in raw milk is due to the presence of low levels of lactoperoxidase (LP), which can be activated by the external addition of traces of H_2O_2 and thiocyanate.

It has been reported that the potential of LP-system and its activation enhances the keeping quality of milk. Cow milk can be provided with protective factors by the addition of lysozyme, making it suitable as an infant milk. Lysozyme acts as a preservative by reducing bacterial counts in milk without affecting the *L. bifidus* activity. The scope of application of minor enzymes to milk and milk products has been recently reviewed.

16.3 Enzymes in baking industry

For decades, enzymes such as malt and fungal alpha-amylases have been used in bread-making. Rapid advances in biotechnology have made a number of exciting new enzymes available for the baking industry. The importance of enzymes is likely to increase as consumers demand more natural products free of chemical additives. For example, enzymes can be used to replace potassium bromate, a chemical additive that has been banned in a number of countries. The dough for bread, rolls, buns and similar products consists of flour, water, yeast, salt and possibly other ingredients such as sugar and fat. Flour consists of gluten, starch, non-starch polysaccharides, lipids and trace amounts of minerals. As soon as the dough is made, the yeast starts to work on the fermentable sugars, transforming them into alcohol and carbon dioxide, which makes the dough rise.

The main component of wheat flour is starch. Amylases can degrade starch and produce small dextrins for the yeast to act upon. There is also a special type of amylase that modifies starch during baking to give a significant anti-staling effect

Gluten is a combination of proteins that forms a large network during dough formation. This network holds the gas in during dough proofing and baking. The strength of this gluten network is therefore extremely important for the quality of all bread raised using yeast. Enzymes such as hemicellulases, xylanases, lipases and oxidases can directly or indirectly improve the strength of the gluten network and so improve the quality of the finished bread.

Table 16.1 lists some of the bread properties that can be improved using industrial enzymes.

Table 16.1: Some of the bread properties that can be improved using industrial enzymes.

Enzyme	*Effect*
Amylase	Maximises the fermentation process to obtain an even crumb structure and a high loaf volume
Maltogenic alpha-amylases	Improves shelf-life of bread and cakes
Glucose oxidase	Oxidative reaction with gluten to make weak doughs stronger, drier and more elastic
Lipase	Modifies the natural lipids in flour to strengthen the dough
Lipoxygenase	Bleaching and strengthening dough
Xylanase	Dough conditioning. Easier dough handling and improved crumb structure
Protease	Weakens the gluten to give plastic properties required in doughs for biscuits
Asparginase	Reduces the amount of acrylamid formed during baking

16.3.1 Flour supplementation

Malt flour and malt extract can be used as enzyme supplements because malt is rich in alpha-amylases. Commercial malt preparations can differ widely in their enzyme activity, whereas an industrial enzyme is supplied with a standardised activity.

The alpha-amylases degrade the damaged starch in wheat flour into small dextrins, which allows yeast to work continuously during dough fermentation, proofing and the early stage of baking. The result is improved bread volume and crumb texture. In addition, the small oligosaccharides and sugars such as glucose and maltose produced by these enzymes enhance the Maillard reactions responsible for the browning of the crust and the development of an attractive baked flavour.

Staling is believed to be due to changes in starch structure during storage. When the starch granules revert from a soluble to an insoluble form, they lose their flexibility, the crumb becomes hard and brittle. For decades, emulsifiers have been used as anti-staling agents. However, they actually have a limited anti-staling effect and are subject to special labelling rules.

By contrast, Novozymes' bacterial maltogenic alpha-amylases has been found to have a significant anti-staling effect. It modifies the starch during baking at the temperature when most of the starch starts to gelatinise. The resulting modified starch granules remain more flexible during storage. Bread produced with maltogenic alpha-amylase has a far softer and more elastic crumb than bread produced with distilled monoglycerides as emulsifiers.

Dough conditioning

Flour contains 2.5-3.5% non-starch polysaccharides, which are large polymers (mainly pentosans) that play an important role in bread quality due to their water absorption capability and interactions with gluten. Although the true mechanism of hemicellulase, pentosanase or xylanase in bread-making has not been clearly demonstrated, it is well known that the addition of certain types of pentosanases or xylanases at the correct dosage can improve dough machinability yielding a more flexible, easier-to-handle dough. Consequently, the dough is more stable and gives better ovenspring during baking, resulting in a larger volume and improved crumb texture. Normal wheat flour contains 1–1.5% lipids, both polar and non-polar. Some of these lipids, especially the polar lipids such as phosphorlipids and galactolipids are able to stabilise the air bubbles in the gluten matrix. The addition of functional lipases modifies the natural flour lipids so they become better at stabilising the dough.

This ensures a more stable dough in case of over-fermentation, a larger loaf volume and significantly improved crumb structure. Because of the more uniform and smaller crumb cells, the crumb texture is silkier and the crumb colour appears to be whiter. It also reduces the need for addition of emulsifiers like DATEM and SSL that otherwise are commonly added to dough in order to stabilise it. This in turn means that emulsifiers can be removed from the label. Chemical oxidants such as bromates, azodicarbonamide and ascorbic acid have been widely used to strengthen the gluten when making bread. As an alternative, oxidases such as glucose oxidase can partially replace the use of these chemical oxidants and achieve better bread quality.

Synergistic effects of enzymes

Each of the enzymes mentioned above has its own specific substrate in wheat flour dough. For example, lipases work on the lipids, xylanase works on the pentosans and amylases work on the starch. Because the interaction of these substrates in dough and bread is rather complex, the use of enzyme combinations can have synergistic effects that are not seen if only one enzyme is used, not even at high dosages. Quite often an overdosage of enzymes will have a detrimental effect on either the dough or the bread. For instance, an overdose of fungal alpha-amylase or hemicellulase/xylanase may result in a dough that is too sticky to be handled by the baker or baking equipment. It is therefore beneficial for some types of bread formulation to use a combination of lower dosages of alpha-amylase and xylanase with low dosages of lipase or glucose oxidase to achieve optimum dough consistency stability and bread quality. Another example is to use maltogenic alpha-amylase in combination with fungal alpha-amylases and xylanase or lipase to secure optimum crumb softness as well as optimum bread quality in terms of crumb structure, bread volume, etc.

16.3.2 Reduction of acrylamide content in food products

During recent years it has been shown that the amount of the potentially carcinogenic substance acrylamide is relatively high in a number of cereal and potato based products like biscuits, crackers, crisp bread, French fries and potato crisps. This is a substance that is formed at high temperatures when the amino acid asparagine reacts with a reducing sugar like glucose. To meet this issue the enzyme asparaginase have been developed in order to reduce the formation of acrylamide. Asparaginase converts asparagine to aspartic acid which does not take part in the formation of acrylamide. The use of an asparginase is able to reduce the formation of acrylamide with up to 90%.

16.4 Enzymes in poultry, meat and fish processing

The biggest single expense in any system of poultry production is feed accounting for up to 70% of total production cost per bird. Poultry naturally produces enzymes to aid the digestion of feed nutrients. However, they do not have enzyme to break down fibre completely and need exogenous enzymes in feed to aid digestion. This section covers the information on enzymes and their uses in poultry nutrition.

Feeding enzymes to poultry is one of the major nutritional advances in the last sixty years. It is the culmination of something that nutritionists realised for a long time but until 1980's it remained beyond their reach. Indeed, the theory of feed enzymes is simple. Plants contain some compounds that either the animal cannot digest or which hinder its digestive system, often because the animal cannot produce the necessary enzyme to degrade them. Nutritionists can help the animal by identifying these indigestible compounds and feeding a suitable enzyme. These enzymes come from micro-organisms that are carefully selected for the task and grown under controlled conditions.

The poultry industry readily accepts enzymes as a standard dietary component, especially in wheat and barley-based rations. But still many questions are partially answered. For example, how do enzymes work? Do growth rates reflect differences in the potency of different enzyme preparations? What is the link between gut viscosity, enzyme action and growth rates? and are enzymes necessary in all poultry rations?

Since, enzymes are one of the many types of protein in biological systems. Their essential characteristic is to catalyse the rate of a reaction but is not themselves altered by it. They are involved in all anabolic and catabolic pathways of digestion and metabolism. Enzymes tend to be very specific catalysts that act on one or, at most, a limited group of compounds known as substrates. Enzymes are not living organisms and are not concerned about viability or cross infection. They are stable at 80–85°C for short time.

Another important feature of enzymes is that the rate of an enzyme catalysed reaction increases with increasing substrate concentration, to the point where there is no further response and the enzyme is said to be saturated. Therefore, we need to match the amount of enzyme with the quantity of substrate. It is common practice to name enzymes by adding the suffix-ase to the name of the principal substrate. For example, β-glucanase is an enzyme that splits β-glucans and proteases break protein. We may also broadly categorise the digestive enzymes as endogenous or exogenous - referring to those produced by the animal and those administered from outside, respectively. For example, pancreatic lipase, which splits fat or lipid into glycerol and fatty acids, is an endogenous enzyme. Those enzymes added to feed as a supplement are exogenous.

16.5 Enzymes in poultry nutrition

The use of enzymes in animal feed is of great importance. Consistent increase in the price of feed ingredients has been a major constraint in most of the developing countries. As a consequence cheaper and nonconventional feed ingredients have to be used which contain higher percentage of non-starch polysaccharides (soluble and insoluble/crude fibre) along with starch. Non-starch polysaccharides (NSPs) are polymeric carbohydrates which differ in composition and structure from starch and possess chemical cross-linking among them therefore, are not well digested by poultry. A part of these NSPs is water-soluble which is notorious for forming a gel like viscous consistency in the intestinal tract thus by reducing gut performance. Predominantly water soluble and viscous arabinoxylans, which belong to pentosan group, are assumed to be the factor responsible. These pentosans also greatly increase the water intake by the birds, which lead to unmanageable litter problems caused by wet and sticky droppings. This deteriorates the hygienic conditions and carcass quality. On the other hand, β-glucans adversely affect all nutrients, especially protein and starch utilisation and are known to give rise highly viscous conditions in the small intestine of the chicks. Poultry do not produce enzymes for the hydrolysis of non-starch polysaccharide present in the cell wall of the grains and they remain un-hydrolysed.

This results in low feed efficiency. Research work has suggested that the negative effects of NSPs can be overcome by dietary modifications including supplementation of diets with suitable exogenous enzyme preparations. Enzymes break down the NSPs, decreases intestinal viscosity and eventually improve the digestibility of nutrients by improving gut performance.

16.5.1 Types of enzyme available for poultry

Some of the enzymes that have been used over the past several years or have potential for use in the feed industry include cellulase (β-glucanases), xylanases and associated enzymes, phytases, proteases, lipases and galactosidases (Table 16.2).

Table 16.2: Enzymes used in poultry feeds.

Enzymes	*Substrate*
β-glucanases	Barley oats
Xylanases	Wheat, rye, triticale rice bran
β-galactosidases	Grain legumes lupins
Phytases	Plant feedstuffs
Proteases	Proteins
Lipases	Lipids
Amylases	Starch

Enzymes in the feed industry have mostly been used for poultry to neutralise the effects of the viscous, nonstarch polysaccharides in cereals such as barley, wheat, rye and triticale. These antinutritive carbohydrates are undesirable, as they reduce digestion and absorption of all nutrients in the diet, especially fat and protein. Recently, considerable interest has been shown in the use of phytase as a feed additive, as it not only increases the availability of phosphate in plants but also reduces environmental pollution. Several other enzyme products are currently being evaluated in the feed industry, including protease to enhance protein digestion, lipases to enhance lipid digestion, β-galactosidases to neutralise certain antinutritive factors in noncereal feedstuffs and amylase to assist in the digestion of starch in early-weaned animals.

Benefits of enzymes

Benefits of using feed enzymes to poultry diets include, reduction in digesta viscosity, enhanced digestion and absorption of nutrients especially fat and protein, improved Apparent Metabolizable Energy (AME) value of the diet, increased feed intake, weight gain and feed–gain ratio, reduced beak impaction and vent plugging, decreased size of gastrointestinal tract, altered population of micro-organisms in gastrointestinal tract, reduced water intake, reduced water content of excreta, reduced production of ammonia from excreta, reduced output of excreta, including reduced N and P.

Reduction in digesta viscosity

Enzymes added to poultry diets, especially diets containing cereals rich in NSP such as wheat, barley and rye, reduced viscosity in the diet and digesta. Morgan and others found that enzyme supplementation of wheat based diets significantly reduced foregut digesta viscosity of birds. The reduction in foregut digesta viscosity was achieved primarily by reducing the molecular weight through hydrolysis of xylan backbone by endo-xylanase into smaller compounds and thus reduction in viscous effects of the feed because foregut digesta viscosity is directly proportional to the molecular weight of wheat arabinoxylans. As a result of endo-xylanase and β-glucanase supplementation, the long backbones of the arabinoxylans and β-glucans are cleaved into shorter fragments, thereby reducing their viscosity. Similar findings on digesta viscosity were also reported by Bedford and others who inferred that the high viscosity in the gut contents caused by the pentosans led to increased water intake of the birds, which resulted in the wet and sticky droppings.

Increase in available energy

One of the main reasons for supplementing wheat- and barley-based poultry diets with enzymes is to increase the available energy content of the diet. Increased availability of carbohydrates for energy utilisation is associated with increased energy digestibility. The apparent metabolisable energy (AME) of wheat has been extensively studied and found to have a considerable range, i.e., 9500–16640 kJ/kg. Enzyme supplementation improves this range by enhancing carbohydrate digestibility, reducing gut viscosity and improving fat utilisation. The improvements in AME resulting from enzyme supplementation are variable because of the variability in the NSP content of wheat. Classen and others, Schutte and van der Klis and other reported improvements of 5–16, 3.1–4.5 and 4.5–12.4%, respectively. The increase in AME with the use of enzymes is difficult to predict, as nutrient ratios, such as energy–protein and other factors also play an important part in poultry-feed formulations. The AME value of wheat has been correlated with its content of water-soluble NSPs, which in turn affects gut viscosity. Unfortunately, NSP analyses are relatively lengthy processes and in a commercial situation rapid testing of incoming grains is required. No chemical test or detectable physical characteristic can be used to rapidly predict the AME value of wheat or to estimate the improvements to be expected from the use of enzymes.

This is part of the difficulty in trying to accurately estimate the energy content of wheat or barley in poultry feeds and compensate for the deficiency by adding enzymes. Adding adequate activity levels of α-amylase, β-glucanase and xylanase to broiler starter and grower corn-soyabean diets with a 3% reduction in dietary metabolisable energy (ME) allowed full restoration of growth performance of broilers comparable to those fed the adequate energy.

Improvement in nutrient digestibilities

Enzymes have been shown to improve performance and nutrient digestibility when added to poultry diets containing cereals, such as barley, maize, oats, rye and wheat and to those containing pulses, such as lupins. The effect of enzyme supplementation on Dry Matter Digestibilities (DMD) in pigs and poultry depends on the type of diet and the type of animal: increases in DMD range from 0.9 to 17% in poultry.

The enzymes currently used in monogastric diets are predominantly glycanases, which cleave NSPs into smaller polymers, thereby removing their ability to form viscous digesta and enhancing nutrient digestibilities. The effects of glycanases are generally nonspecific, except for their effect on fat (greater effect on saturated fat than on unsaturated fat). Another enzyme used in feed is phytase, which increases the utilisation of phytate phosphorus. The ability of phytase to improve the digestion of phytate phosphorus and subsequently to reduce the output of organic phosphorus to the environment has attracted a great deal of scientific and commercial interest. In poultry use of phytase was reported to reduce phosphorus excretion by as much as 40% for broilers. When phytase was added to layer diets, increased egg production and positive effects on egg weight and tibia ash were also noted.

Reduction in excreta moisture: A reduction in the moisture content of poultry excreta is often noted when glycanases are included in the diet. Supplementing the NSP-enriched diet with three different commercial glycanase products improved performance, but their effectiveness in reducing the moisture levels of the excreta differed from 10 to 29%. This supports the view that different glycanases have similar performance-enhancement effects in monogastric animals but the site of the breakdown of the NSPs in the gut and the molecular sizes of the released products differ. Graham, also reported an increased water uptake and excretion in broilers given diets containing higher levels of viscous cereal grains.

Health improvement

Morgan and Bedford reported that coccidiosis problems could be prevented by using enzymes. Birds fed a wheat-based diet with and without glycanase supplementation showed vastly different responses to coccidiosis challenge. Growth was depressed by 52.5% in the control group but by only 30.5% in the enzyme group, which also had a much better lesion score. An increase in digesta passage rate and a reduction in excreta moisture are often noted when glycanases are added to poultry diets, which may be detrimental to the life cycle of the organism.

Precision and flexibility in least cost feed formulation: Enzymes provide greater flexibility in feed formulation and allow the use of a wide range of ingredients without compromising bird performance and hence provide great flexibility in least-cost feed formulation. The nutritive value of cereal grains for poultry varies greatly and no suitable assays are currently available for rapid in-mill testing. For instance, the variability in the AME of wheat for poultry can be as great as 4 MJ/kg DM. This problem can be largely overcome by using glycanases to bring the AME of different wheats to comparable levels.

Impact on environment

Enzymes have been approved for use in poultry feed because they are natural products of fermentation and therefore pose no threat to the animal or the consumer. Enzymes not only will enable livestock and poultry producers to economically use new feedstuffs, but will also prove to be environmentally friendly, as they reduce the pollution associated with animal production. As well as contributing to improved poultry production, feed enzymes can have a positive impact on the environment. In areas with intensive poultry production, the phosphorus output is often very high, resulting in environmental problems such as eutrophication. This happens because most of the phosphorus contained in typical feedstuffs exists as the plant storage form phytate, which is indigestible for poultry. The phytase enzyme frees the phosphorus in feedstuffs and also achieves the release of other minerals (e.g., Ca, Mg), as well as proteins and amino acids bound to phytate. Thus, by releasing bound phosphorus in feed ingredients, phytase reduces the quantity of inorganic phosphorus needed in diets, makes more phosphorus available for the bird and decreases the amount excreted into the environment.

16.5.2 Factors affecting the benefits of enzyme

The degree of improvement obtained by adding enzymes to the diet depends on many factors, including the type and amount of cereal in the diet, the level of antinutritive factor in the cereal, which can vary within a given cereal (for example, low- versus high-β-glucan barley), the spectrum and concentration of enzymes used, the type of animal (poultry tend to be more responsive to enzyme treatment than pigs) and the age of the animal (young animals tend to respond better to enzymes than older animals), type of gut micro flora present and the physiology of the bird. Older birds, because of the enhanced fermentation capacity of the micro flora in their intestines, have a greater capacity to deal with negative viscosity effects.

Use of enzymes in layers: Although the majority of research trials were conducted on broilers. However, the responses of laying hens to enzyme-supplemented feeds are also well documented. Typically, enzymes added to layer feed appear to have little effect on egg mass but improve feed efficiency, energy utilisation. Wyatt and Goodman reported that corn-fed layers exhibited better feed efficiency than those fed enzyme-supplemented barley-based diets. Nevertheless, enzyme supplementation improved the utilisation of barley diets. Increased energy utilisation in laying hens appears to be due to microbial fermentation of solubilised NSPs and the subsequently higher absorption of volatile fatty acids. Wet litter arising from the use of barley and newly harvested wheat can result in an increased incidence of dirty egg shells and in ammonia buildup in poultry barns. Adding enzymes to both wheat- and barley-based diets has been shown to reduce the moisture content of fecal matter in layers.

To sum up, the use of enzymes as a feed additive has rapidly expanded. In the last decade, extensive studies have been conducted to study the effects of feeding exogenous enzymes on the performance of poultry. By compiling these studies into a single focused work, this review provides evidence that enzyme is a significant instrument for the use in poultry feed.

Although the economic and social benefits of enzymes have been well established and the future of feed enzymes is a bright one. However, further research is required if enzymes are to reach their full potential in the industry and to answer some of the questions that this section raises, particularly those regarding the mode of action of enzymes, how best to match the levels of enzyme and substrate and how enzymes counter the variable environments in the animal's gut. Ultimately, *in vitro* tests will tell us not only whether we need exogenous enzymes, but also the correct enzyme cocktail for a specific batch of a feed ingredient. This will allow the formulation of better rations using a wider range of ingredients. Any advances in this field must ultimately improve the welfare of chickens, reduce the production of wastes and conserve resources.

16.6 Enzymes in meat industry

Beef palatability is affected by many factors and tenderness is cited as one of the most important. Consumers are willing to pay a premium for a guaranteed tender product. Consequently, the meat industry is in a continual search for methods to improve the tenderness of the middle meats and upgrade other cuts and muscles to steak quality. This search for tender product not only adds value to the industry but can help to supply a more consistent and uniform product to consumers while increasing overall beef satisfaction.

Tenderness is a complex trait. Generally, the two primary structural features of muscle that influence tenderness are integrity of the myofibrils (termed the actomyosin effect) and the connective tissue contribution (termed a background effect). Fat also plays a minor role. Smith and others identified six characteristics of perceived tenderness.

Three characteristics relate to the myofibrillar portion, two relate to product adhesion and the remaining one relates to connective tissue components. Myofibrillar proteins are located intracellularly while connective tissue proteins (primarily collagen) are located extracellularly. This difference in distribution of key proteins within the muscle increases the challenge of improving tenderness. Stated simply, some muscles may lack palatability because of myofibrillar proteins while others may lack palatability because of the connective tissue proteins. A plausible tenderisation strategy should target one or both of the main structural constraints to be successful.

16.6.1 Type of enzymes used in meat industry

Inherent proteolytic enzymes

The primary mechanism of postmortem improvement in tenderness is through the disruption of the native structure of the muscle. Of the methods identified as having a positive effect on tenderness, the easiest and oldest is postmortem cooler ageing. The calpain system, the primary enzyme system responsible for the ageing process, is comprised of three primary components, μ-calpain, *m*-calpain and calpastatin. Calpains and their inhibitor calpastatin are calcium-dependent. Calpain proteolysis of myofibrillar proteins occurs through the degradation of the *z*-line, costameric proteins and titan. Negligible degradation occurs for collagen primarily due to its extracellular nature. Postmortem proteolysis is self limiting because the calpains undergo autolysis. As a result of this self-degradation, calpains do not over-tenderise meat.

Exogenous enzymes

Research into the use of exogenous proteolytic enzymes has been conducted for over 60 years and has investigated countless enzymes from plant, bacteria and fungal sources. Exogenous enzymes added to meat to enhance tenderness react differently to the myofibrillar and connective tissue portions of the meat. Currently, just five of the many exogenous enzymes that have been studied have been classified as 'Generally Recognised as Safe' (GRAS) by USDA's Food Safety Inspection Service (FSIS) and come from varying plant, bacterial and fungal sources (Table 16.3).

Table 16.3: Enzymes classified as generally recognised as safe (GRAS).

Enzyme	*Type*	*Source*	*Protease class*
Papain	Vegetable	Papaya	Cysteine
Bromelain	Vegetable	Pineapple	Cysteine
Ficin	Vegetable	Figs	Cysteine
Bacillus protease	Bacterial	*Bacillus subtilis*	Serine
Aspartic protease	Fungal	*Aspergillus oryzae*	Aspartic

Consequently, only these five can be added to meat for the purposes of enhancing tenderness. Additional, promising sources of exogenous enzymes have been identified. In particular, enzymes have been isolated from kiwi fruit (actinidin) and ginger that show potential for future inclusion in meat systems.

Temperature is probably the most controllable factor influencing enzyme activity. Most of the exogenous enzymes used to tenderise meat have an optimal activity in the 50–70°C range (Table 16.4). As a result, most of the activity takes place during the cooking process and to a lesser degree during cooler storage. Product pH will also influence enzymatic activity. Most of the enzymes have an optimal pH in the normal range of meat, but some function best at a more acidic or alkaline pH for optimal protein degradation. The addition of salts can have positive or negative effects on enzymatic activity. Many of the enzymes do not penetrate meat, causing the method of application to play an important role in uniform distribution into the meat product. The proper identification of enzyme, time, temperature and pH will be dependant upon the processing system and desired product. The optimal pH and range of activity is shown in Table 16.5.

Table 16.4: Temperature effects on activity.

Enzyme	*Active range*	*Optimal temperature*
Papain	50–80	65–75
Bromelain	50–80	65–75
Ficin	45–75	60–70
Bacillus	50–65	55–60
Aspergillus	40–60	55–60

Table 16.5: pH optimal and range of activity.

Enzyme	*Active pH*	*Optimal pH*
Papain	4.0–9.0	4.0–6.0
Bromelain	4.0–7.0	5.0–6.0
Ficin	5.0–9.0	7.0
Bacillus	5.0–9.0	7.0
Aspergillus	2.5–7.0	<6.5

Enzymes are often sold on the basis of activity, however, the assay depends upon which enzyme is being tested. Papain, bromelain and ficin activity is often measured in milk clot units. Enzymes from *Aspergillus oryzae* are measured in haemoglobin units of tyrosine. Enzymes from *Bacillus subtilis* are measured by a proteolytic test on casein. The relationships among assays are not strong, making it difficult to directly compare the activities of enzymes. The same enzyme can show differing results in activity levels and optimal pH and temperature ranges depending upon the substrate. The milk clot assay relies on a subjective determinate of clotting time while the others provide results using a spectrophotometer. The assay used is partly based on which one conforms to the optimal activity level for each enzyme, depending on pH, temperature and substrate.

GRAS enzymes

Papain: Papain, a mixture of enzymes from the papaya fruit, has been studied for its tenderising effects since the 1940's. Papain is a highly aggressive, indiscriminate enzyme causing significant degradation to both myofibrillar and collagen proteins, yielding protein fragments of several sizes. Even though papain shows massive disruption of the Z disc, as observed in aged meat and collagen, it is much more effective when injected into the product due to its poor ability to penetrate surfaces. The optimal temperature range for papain is 65–80°C but Ashie and others did find a significant increase in tenderness after one and two weeks of storage. Tappel and others showed that a majority of the tenderising activity occurs during the cooking process. Although active over a wide range of pH, the highest activity occurred in the pH range of 4.0–6.0. Studies disagree on the ability of papain to solubilise collagen with some showing little to no degradation and some a major increase in collagen solubility. Caution must be taken when using papain in order to prevent over-tenderisation creating a mushy or mealy product.

Bromelain

An enzyme mixture coming from pineapple, bromelain (a cystein protease), has been studied since the 1950's. Bromelain first degrades 40% of the collagen in the sarcolemma followed by degradation of myosin in the myofibrillar component. This enzyme has low, but significant, activity at 0°C which dramatically increases at 50–70°C. The activity is sustained without a marked decrease up to 80°C. Kim and Taub found that 5.0 is the optimal pH for bromelain activity. McKeith and others found a significant increase in tenderness when an enzyme solution was injected into muscle versus dipping or tumbling in brine.

Ficin

Ficin, a vegetable-based enzyme, is derived from figs. The enzyme exhibits less activity against all types of proteins when compared to papain and bromelain. This thiol protease, although capable of minimal degradation of collagen and elastin, will preferentially degrade myofibrillar protein. Ficin is able to degrade elastin at temperatures as low as 20°C whereas it has little activity against collagen and myofibrillar proteins below 40°C. Optimal activity occurs between 60–70°C. El-Gharbawi and Whitaker also showed the optimal pH for activity is near 7 for collagen and myofibrillar proteins and approximately 5.0–5.5 for elastin degradation. Sensory evaluations tend to show only a marginal increase in tenderness when compared to untreated samples.

Bacillus subtilis proteases

A blend of proteolytic enzymes (alkaline elastase and neutral protease) derived from *Bacillus subtilis* were approved as GRAS in 1999. Takagi and others show alkaline elastase specifically degrades collagen and elastin with little degradation of myofibrillar protein resulting in less tenderising effect but also decreasing the possibility of over-tenderising. *Bacillus subtilis* proteases' optimal and activity ranges vary depending upon the organism from which they are produced. McConn and others found neutral protease activity to rapidly increase at 50°C with a dramatic drop above 65°C. The same authors found a broad pH activity from 5.0–9.0 pH with an optimal pH found at 7.0. Enzymatic activity occurs during storage for one, two and three days at 4°C as measured by a decrease in relative hardness. This bacterial protease was developed to be an inexpensive alternative for ficin.

Aspergillus oryzae proteases

Aspergillus oryzae produces an aspartic protease that shows a self-limiting proteolytic activity in meat systems. Proteolytic activity of *Aspergillus* has been known and studied since the 1950's. Myofibrillar proteins are the main substrate for activity while little, if any, collagen breakdown occurs. This aspartic protease shows minimal activity during refrigerated storage through 14 days of storage but increases to an optimum at 55°C before dropping dramatically at 60°C. The enzyme from *Aspergillus* is active in acidic conditions and activity is maintained up to pH 7.0 before rapidly declining. As a self-limiting enzyme, some limited degradation can improve tenderness without the risk of a mushy or mealy texture.

16.6.2 Characteristics of enzymes

The general characteristics of the two plant derived enzymes vary somewhat since they all have different temperatures of inactivation and operate with different kinetics when applied. Papain is the most temperature stable and can require a temperature as high as 170–185°F to completely inactivate it. This has certain advantages and certain disadvantages. The main disadvantage is that a piece of meat cooked to what we call 'medium rare' will not reach a temperature high enough to inactivate the papain. Thus, subsequent storage of the meat will allow the enzyme to continue to tenderise and if extended over too long a period will produce a mushy unpalatable texture.

Papain should be used in very controlled processes where each step and cut of meat is under controlled time and temperature and served properly to the consumer. This is the best process for large scale highly organised restaurant chains where the process is thoroughly outlined and adhered to. The pH optimum of papain is typically similar to that of meat itself.

Bromelain has a lower temperature of inactivation and a slightly different mode of operation. The temperature of inactivation of Bromelain is around 160°F which, again, will not be high enough for inactivation in a medium rare piece of beef. The rate of action for both papain and bromelain are similar and, therefore, timing for processing would be similar.

16.6.3 Activity-potency

The most important consideration in selecting a tenderising enzyme is the activity of the enzyme. Further considerations are that the material be of food grade quality, that it have a low microbial count and that it meets all incidental government specifications. Activity is a measure of the enzyme's ability to react with a specific substrate chosen by the supplier. Enzymes are sold on the basis of activity or potency. One of the most common assays for Papain and Bromelain is the milk clot assay.

The milk clot assay is a very accurate and yet simple test procedure which measures the amount of time required to form clotted milk in the presence of the proteolytic enzyme under specified and controlled conditions, i.e., temperature, etc. Using this number, whether it is 100 units per mg. or 500 units per mg., the buyer can immediately assign formulations that will consistently yield the same quality of tenderisation during the application.

16.7 Enzymes in fish processing

Fish processing generates valuable by-products that are high in protein and lipids, such as viscera, skin, tails, heads and frames. The disposal of these nutrient-rich by-products can often increase costs. By using enzymes to valorise by-products, processors can potentially obtain additional revenue from nutritional supplements, ingredients for pet food, biodiesels, fertilisers, agricultural products, animal and fish feeds.

Many processors do not capture the full value of these by-products. This may be due to inefficient secondary processing and over-reliance on mechanical, heat or chemical methods. Enzymes can enhance the quality and value of fish processing by-products, optimising time, energy and resources while also increasing revenue. Some of the enzymes and their benefits are shown in Table 16.6.

Table 16.6: Enzymes and their benefits in fish processing.

Enzyme	*Benefits*
Protease	Improved oil separation
	Lower temperatures
	Higher oil yields energy savings

(Cont'd...)

Enzyme	*Benefits*
Protease	Reduced water binding of proteins
	Increased water removal in evaporators
	Lower need for water removal in spray-drying
	Energy savings
	Less evaporator down-time
Protease	Improved protein value, flavour and digestibility
	Enhanced protein functionality: emulsifying properties, solubility, fat- and water-binding
	Generation of high-value peptides and proteins
Protease	Improved fish sauce flavour
	Faster fermentation
	Reduced fermentation cost
	Increased fish sauce yield
Protease	Conversion of waste material to high value dietary supplement
	Production of value-added products: hydrolysed fish protein
Protease	Improved removal of residual meat
	'Clean' fish frames for use as calcium source
	Increased recovery of hydrolysed mollusk protein
Glucose oxidase	Removal of oxygen
	Reduced potential for oxidative rancidity
	Preservation of colour
Catalase	Fast, efficient removal of hydrogen peroxide

16.8 Enzymes in winemaking

Enzymes are nature's tools. They are needed now-a-days winemaking processes for mastering the production steps, creating value and most important protecting the quality of the wine and the environment. The grapes themselves have many enzymes, acting during the fruit maturation and along the winemaking process. Most of the enzymes are hydrolases, meaning they hydrolyse substrates and liberate products. Enzymes need to be purified as the micro-organisms used in fermentation–commonly *Aspergillus sp.,* - and the fermentation conditions favour the production of negative enzyme activities such as the cinnamoyl esterase (CE) activity.

16.8.1 Pectinases for extraction and clarification

Under the term pectinases, many enzyme activities are present. In fact to hydrolyse pectins various enzymes are needed such as polygalacturonase, pectin esterase, pectin lyase, arabinanase, galactanase, rhamno-galacturonase.

Commercial enzymatic preparations contain a mixture of these enzymes needed in the application. The enzymes are obtained by fermentation of the fungi *Aspergillus sp.*, either in a fermenter (submerged fermentation) or on plates (surface fermentation). The enzyme concentration is a key factor in the applications, therefore the dose and the contact time are detrimental to get the best results. These pectinases are preferably free of cinnamoyl esterase (FCE) activity - for use in winemaking applications.

16.8.2 Glycosidases for aroma liberation

These enzymes are also produced by *Aspergillus sp.*, Under the term glycosidases, many enzyme activities are present like the glucosidase, arabinosidase, rhamnosidase, apiosidase. However to liberate the aroma molecules (R) present in the glycosylated aroma precursors mostly under the form of glycosylated terpens, glucosidase is the main activity liberating the aroma in the free form and aroma can be smelt only if they are in their free form.

16.8.3 Glucanases for ageing on lees

In winemaking applications, the glucanases involved in the hydrolysis of glucans - originating either from the yeast or from *Botrytis cinerea* (moldy grapes) - are composed of two major enzyme activities: the beta 1,3-glucanase and the beta 1,6-glucanase. These enzymes favour the release of mannoproteins by hydrolysing the cell wall yeast's glucan. Acid proteases might also be present in the preparation liberating small peptides. These glucanases differ from the glucanases needed to hydrolyse cellulose. It is important to note that the main activity involved in the hydrolysis of the cellulose is the beta 1,4-glucanase. These enzymes are produced by strains of *Trichoderma sp*. The preparation also contains an acid protease very similar to the yeast protease, thus explaining the release of savoury peptides during the ageing on lees.

16.8.4 Pectinases/glucanases blend for filtration

In this application, pectinases are needed to hydrolyse pectins, in particular the ones released from the pomace in red winemaking in the press wines and glucanases to hydrolyse the glucan produced by the yeast during the alcoholic fermentation. Little is known about the impact of yeast strains on the release of glucan and its effect on wine filterability.

16.8.5 Lysozyme or muramidase for microbial control

Lysozyme is a food preservative. This enzyme degrades the cell walls of gram-positive bacteria such as lactic bacteria in wine: *Oenococcus*, *Pediococcus* and *Lactobacillus*. Its use can greatly reduce the need for SO_2, which poses a health hazard to individuals allergic to sulphites.

16.8.6 Mastering the processes

Getting best out of grapes using enzymes

The major applications concern the use of pectinases either on grapes or on the juice (exogenous enzymes). On grapes, pectinases are used for direct pressing, skin contact and red maceration. On juice, enzymes are used for clarification in various processes: static clarification, flotation, centrifuge and filtration (rotary vacuum filter for example).

The parameters affecting enzyme activity are the pH, the temperature and the time. In order to get good results, the enzyme dose is very important because the other parameters are given and cannot be changed easily. The dose is linked to the enzyme concentration in the product. Among the oenological parameters, sulphites (SO_2) are not affecting the enzyme activity up to 500 ppm. Enzymes should not be used in presence of bentonite, the major enzyme inhibitor. In all applications, it is advisable to use the enzymes prior bentonite addition. If bentonite has been used, a racking should be done before adding any enzymes.

Treatments	*Impact on enzymes*
SO_2 addition	Bearable up to 500 ppm
Bentonite	Non-bearable, enzyme activity lost even at low dosage
Gelatin	Bearable, more effective if use after enzyme
Silice gel	Bearable, more effective if used after enzyme addition

Enzymes are available both in solid and liquid forms. Solid enzymes should be diluted 1 to 10 (100 g for 1000 ml of juice or water). The enzyme solution can be prepared for one day of operation in the winery, as it remains stable. Typical enzyme doses vary from 3–4 g/100 kg grapes in extraction (dose can be higher with thick and slippery grape cultivars up to 5–6 g/100 kg) and 1–3 g/hL in clarification. The dose depends on the enzyme concentration. Large producers favour the enzymes in liquid form for convenience of use. Specific liquid purified pectinases exits also for wine like the newly introduced Vinozym Ultra FCE.

Extraction and clarification

The major effect of pectinases is on the viscosity reduction of the mash or the juice by hydrolysing the larger polysaccharide molecules into smaller fragments. As a result, the juice is extracted easily, at lower pressure, the juice yield is increased and the process runs smoothly. In red extraction, pumping overs are facilitated as more juice is available from the mash, colour extracted quicker and stabilised earlier with tannins. Settling, in respect to clarification, is speeded up because when pectins are hydrolysed, agglomeration and sedimentation starts rapidly. The enzyme should be added as soon as possible in the process: at the crusher for white and rosé pressing, when filling the maceration tank in red production, at the bottom of the vat when filing it for clarification. The addition can be done with a dosing pump, a drop-by-drop system, or using a watering can.

The use of pectinases on grapes, for white or rosé wine production, at the crusher or during the filling of the maceration tank will help decreasing the viscosity of the mash. Enzymes remain active as long as pectins are present. The dose is the key parameter and need to be monitored according to the grape variety and its ripeness. Grapes varieties with thick and slippery berries need addition of higher enzyme dose. Unripe grapes will also need higher dose. Typical dosage varies from 3 to 5 g/100 kg grapes. For hard to press or extract varieties like Muscat or unripe grapes, the dose needs to be set at 5–7 g/100 kg grapes for getting the best results.

In red winemaking, the use of enzymes on grapes has the same effect: quicker and higher juice is released and therefore milder and more efficient extraction through mechanical actions using either pumping over, rack and return or punching down.

To allow flocculation, pectins need to be hydrolysed. The speed of this first step in the clarification process is given by the enzyme concentration used and therefore the enzyme dose to get rapidly a negative pectin test. As soon as a negative pectin test is achieved, agglomeration and sedimentation can start. Then the total time for settling will depend on tank geometry and temperature. A negative pectin test is very important in flotation to achieve a quick and performing clarification. If not achieved winemakers can increase the enzyme dose to get it. If this depectinisation phase is well mastered, reduced dose of flocculent can applied.

16.8.7 Controlling enzyme efficiency

Trials can be run at the start of harvest in order to define the adequate enzyme dose rate. Unripe grapes, thick grape skins and slippery grape varieties like Muscat or Semillon among other cultivars need a higher enzyme dose. Clarification trials can help define the dose by measuring the turbidity reduction over time and performing the pectin test every 30 min. A negative pectin test indicates that pectin is hydrolysed and agglomeration can start. The time needed for sedimentation will then depend on the temperature and the geometry of the tank. Pectin test or alcohol test is a simple tool to monitor enzyme performance.

This test is used in flotation to control if enzyme addition has been efficient enough to obtain a negative pectin test. This is mandatory to be able to produce the flocculation. Addition of flotation agent can be used to help in the process but might be reduced in case a good depectinisation is achieved.

Aroma liberation

Glycosidases are added into the young wine right after the alcoholic fermentation. They are reversibly inhibited by glucose meaning that when glucose is consumed the activity of the enzyme is back. Fructose has no inhibitory effect on glycosidases. Trials conducted at various addition points: beginning, middle and end of alcoholic fermentation have shown that this enzyme gives best sensory results when added into the young wine. The presence of high level of CE in the glycosidase product available into the market can be one of the reasons. At the end of alcoholic fermentation, glycosidases can benefit from higher temperature and activation by the ethanol. The contact time is around 2 to 3 weeks. Results can be monitored by sensory evaluation. The level of free SO_2 has to be checked during the treatment. The enzyme activity can be stopped by bentonite addition (20 g/hL).

Glycosidases help release the terpene aroma molecules on Muscat and related cultivars. They can be used to treat a portion of wine and produce aromatic base wines for blending purposes.

Ageing on lees

In this application, enzyme blends contain beta (1,3–1,6) glucanases to help release the mannoproteins linked to the yeasts cell-wall glucan, often associated with pectinases to hydrolyse the pectins released during pressing in red winemaking. Glucanases from *Trichoderma harzianum* are specific for the hydrolysis of beta (1,3–1,6) yeast and *Botrytis* glucan. This high molecular weight polysaccharide is responsible for clarification and filtration problems. The only way to get rid of this polysaccharide is to hydrolyse it with enzymes. As for pectinases, it is advisable to use the glucanase or the pectinase/glucanase blends at the end of the alcoholic fermentation.

In red winemaking, these enzymes are added under the cap towards the end of alcoholic fermentation or just right after devatting in tanks or barrels. In white winemaking, they are preferably added after racking at the end of alcoholic fermentation in tanks or barrels for ageing on fine or gross lees. As the optimum temperature for glucanases is higher than for pectinases, the enzyme will benefit from higher temperature occurring during the malolactic fermentation. The presence of acid protease in the enzyme preparations will favour the release of small peptides.

Filtration

Enzyme preparations used in the pre-fermentation steps – extraction and clarification processes – will have an impact on filterability improvement of the young wines. Also the use of pectinases/glucanases blend at the end of alcoholic fermentation will help to clarify, stabilise and filter the young wine.

Any fining procedure will be more efficient if done after hydrolysing the long chain polysaccharides. In case no enzymes have been used in none of the processing steps, it is advisable to add enzyme as soon as possible in the young wine in order to hydrolyse the large chains of pectin particularly in red wines and the glucan released by the yeast. The performance can be check by running a filterability test using the CFLA (Critère de Filtration Lamothe-Abiet) measurement. This test is equivalent to the filterability test used prior bottling using membranes of higher porosity (5 and 1,2 micron).

16.8.8 Creating value

The enzyme treatment cost can easily be compensated by the yield increase in the extraction and clarification processes thanks to quicker sediment compaction in the latter, the reduction of fining agent use and wine losses during fining and filtration steps to size reduction of the larger polysaccharide molecules. Also time can be reduced, performances of equipment like presses, flotation, centrifuges and rotary vacuum filter can be improved. In large wineries operation time is valuable as large quantities of grapes are coming in every day. All the savings can be documented by running dedicated trials at the start of the harvest. The yield increase averaged from 4–6% in direct pressing in white and rosés production and 3–4% in red wines with very short or short maceration time.

16.8.9 Protecting and revealing quality

Thanks to milder extraction with the use of enzymes, juice can be released at lower pressure avoiding extracting harsh tannins in particular if the grapes are not enough ripe. In red grape varieties, colour is easily released and stabilised. Clarification of both must and young wine is better performing allowing clarifying quicker and ensuring better microbial stability. Also less classical chemicals are needed and the use of SO_2 can be better rationalised. The enzymes, well-mastered processes have an impact of final quality in all steps of winemaking.

Enzymes are very valuable tools for the winemakers. These biocatalysts help in mastering the production processes. Today time is an important parameter for large production operations and thanks to their properties enzymes have a role to play. They are convenient to use and they are biodegradable. Mastering their use – in particular the dose and the addition point – is important to get the best results in the applications.

17

Enzymes in textile industry

17.1 Introduction

The use of enzymes in the textile industry is an example of white/industrial biotechnology, which allows the development of environmentally friendly technologies in fibre processing and strategies to improve the final product quality. The consumption of energy and raw-materials, as well as increased awareness of environmental concerns related to the use and disposal of chemicals into landfills, water or release into the air during chemical processing of textiles are the principal reasons for the application of enzymes in finishing of textile materials.

17.2 Production of enzymes for textile industry

Commercial sources of enzymes are obtained from any biological source animal, plants and microbes. These naturally occurring enzymes are quite often not readily available in sufficient quantities for industrial use, but the number of proteins being produced using recombinant techniques is exponentially increasing. Screening approaches are being performed to rapidly identify enzymes with potential industrial application. For this purpose, different expression hosts (*Escherichia coli*, *Bacillus* sp., *Saccharomyces cerevisiae*, *Pichia pastoris*, *filamentous fungi*, insect and mammalian cell lines) have been developed to express heterologous proteins.

Among the many systems available for heterologous protein production, the enteric Gram-negative bacterium *E. coli* remains one of the most attractive. Compared with other established and emerging expression systems, *E. coli*, offers several advantages including its ability to grow rapidly and at high density on inexpensive carbon sources, simple scale-up process, its well characterised genetics and the availability of an increasingly large number of cloning vectors and mutant host strains.

However, the use of *E. coli* is not always suitable because it lacks some auxiliary biochemical pathways that are essential for the phenotypic expression of certain functions, so there is no guarantee that a recombinant gene product will accumulate in *E. coli* at high levels in a full-length and biologically active form. In such circumstances, the genes have to be cloned back into species similar to those from which they were derived. In these cases bacteria from the unrelated genera *Bacillus*, *Clostridium Staphylococcus* and the lactic acid bacteria *Streptococcus Lactococcus* and *Lactobacillus* can be used. If heterologous proteins require complex post-translational modifications and are not expressed in the soluble form using prokaryotic expression systems, yeasts can be an efficient alternative once they provide several advantages over bacteria for the production of eukaryotic proteins. Among yeast species, the methylotrophic yeast *Pichia pastoris* is a particularly well suited host for this purpose.

The use of this organism for expression offers a number of important benefits:

1. High levels of recombinant protein expression are reached under the alcohol oxidase1 gene (*aox1*) promoter.
2. This organism grows to high cell densities.
3. Scaled-up fermentation methods without loss of yield have been developed.
4. Efficient secretion of the recombinant product together with a very low level of endogenous protein secretion represents a very simple and convenient pre-purification step.
5. Some post-translational modifications are feasible (such as proteolytic processing and glycosylation).

Furthermore, the existence of efficient methods to integrate several copies of the expression cassette carrying the recombinant DNA into the genome, eliminating problems associated with expression from plasmids, is making this yeast the micro-organism of choice for an increasing number of biotechnologists.

17.2.1 Role of enzymes in textile industry

Textile processing has benefited greatly in both environmental impact and product quality through the use of enzymes.

The principal enzymes applied in textile industry are hydrolases and oxidoreductases. The group of hydrolases includes amylases, cellulases, proteases, pectinases and lipases/esterases. Amylases were the only enzymes applied in textile processing until the 1980s. These enzymes are still used to remove starch-based sizes from fabrics after weaving. Cellulases have been employed to enzymatically remove fibrils and fuzz fibres and have also successfully been introduced to the cotton textile industry. Further applications have been found for these enzymes to produce the aged look of denim and other garments. The potential of proteolytic enzymes was assessed for the removal of wool fibre scales, resulting in improved anti-felting behaviour. However, an industrial process has yet to be realised. Esterases have been successfully studied for the partial hydrolysis of synthetic fibre surfaces, improving their hydrophilicity and aiding further finishing steps. Besides hydrolytic enzymes, oxidoreductases have also been used as powerful tools in various textile-processing steps. Catalases have been used to remove H_2O_2 after bleaching, reducing water consumption. A more detailed description of the most common groups of enzymes applied in the textile industry and the processes where they are applied is given below.

Amylases

Amylases hydrolyse starch molecules to give diverse products including dextrins and progressively smaller polymers composed of glucose units. Starch hydrolysing enzymes are classified according to the type of sugars produced: α-amylases and β-amylases. α-amylases are produced by a variety fungi, yeasts and bacteria, but enzymes from filamentous fungal and bacterial sources are the most commonly used in industrial sectors. Microbial α-amylases range from 50 to 60 KDa, with a few exceptions, like the 10 KDa α-amylase from *Bacillus caldolyticus* and a 210 KDa α-amylase from *Chloroflexus aurantiacus*. α-amylases from most bacteria and fungi are quite stable over a wide range of pH from 4 to 11. *Alicyclobacillus acidocaldarius* α-amylase has a pH optimum of 3, while those from alkalophilic and extremely alkalophilic *Bacillus* sp., have pH optima of 910.5 and 1112, respectively. Optimum temperature for the activity of α-amylases is usually related to growth of the producer micro-organism. Temperatures from 25 to 30°C were reported for *Fusarium oxysporum* α-amylase and temperatures of 100 and 130°C for *Pyrococcus furiosus* and *Pyrococcus woesei*, respectively. Addition of Ca^{2+} can, in some cases, enhance thermostability. They are severely inhibited by heavy metal ions, sulphydryl group reagents, EDTA and EGTA.

In general, microbial α-amylases display the highest specificity towards starch followed by amylase, amylopectin, cyclodextrin, glycogen and maltotriose.

Textile desizing: For fabrics made from cotton or blends, the warp threads are coated with an adhesive substance know as 'size' to lubricate and protect the yarn from abrasion preventing the threads to break during weaving. Although many different compounds have been used to size fabrics, starch and its derivatives are the most common because of their excellent film forming capacity, availability and relatively low cost. After weaving, the sizing agent and natural non-cellulosic materials present in the cotton must be removed in order to prepare the fabric for dyeing and finishing. Before the discovery of amylases, desizing used to be carried out by treating the fabric with acid, alkali or oxidising agents at high temperatures. The chemical treatment was not totally effective in removing the starch, leading to imperfections in dyeing and also resulted in a degradation of the cotton fibre destroying the natural, soft feel of the cotton. Now a days amylases are commercialised and preferred for desizing due to their high efficiency and specificity, completely removing the size without any harmful effects on the fabric. The starch is randomly cleaved into water soluble dextrins that can be then removed by washing. This also reduced the discharge of waste chemicals to the environment and improved working conditions.

Pectinases

Pectin and other pectic substances are complex polysaccharides present in plant cell walls as a part of the middle lamella. Pectinases are a complex group of enzymes involved in the degradation of pectic substances. They are primarily produced in nature by saprophytes and plant pathogens (bacteria and fungi) for degradation of plant cell walls. There are three major classes of pectin degrading enzymes: pectin esterases (PEs), polygalacturonases (PGs) and polygalacturonate lyases (PGLs).

Pectin esterases are mainly produced in plants such as banana, citrus fruits and tomato, but also by bacteria and fungi. They catalyse hydrolysis of pectin methyl esters, forming pectic acid. The enzyme acts preferentially on a methyl ester group of a galacturonate unit next to a non-esterifed galacturonate unit. The molecular weight of most microbial and plant PEs varies between 30–50 kDa. The optimum pH for activity varies between 4.0 and 7.0. The exception is PE from *Erwinia* with an optimum pH in the alkaline region. The optimum temperature ranges between 40 and 60°C and pI between 4.0 and 8.0.

Polygalacturonases are a group of enzymes that hydrolyse α-1,4 glycosidic linkages in pectin using both exo- and endo-splitting mechanisms. Endo-PGs are widely distributed among fungi, bacteria and yeast. These enzymes often occur in different forms having molecular weights in the range of 3080 kDa and pI between 3.8 and 7.6.

Their optimum pH is in the acidic range of 2.5–6.0 and the optimum temperature between 30 and 50°C. Exo PGs are found in *Aspergilus niger*, *Erwinia* sp., and some plants, such as carrots, peaches, citrus and apples. The molecular weight of exo-PGs vary between 30 and 50 kDa and their pI ranges between 4.0 and 6.0.

Polygalacturonate lyase cleaves polygalacturonate or pectin chains via β-elimination that results in the formation of a double bond between C4 and C5 at the non-reducing end and elimination of CO_2. Endo-polygalacturonate lyase cleaves polygalacturonate chains arbitrarily and exo-polygalacturonate lyase splits at the chain end of polygalacturonate which yields unsaturated galacturonic acid. The molecular weight of PGLs varies between 30 and 50 kDa except in the case of PGL from *Bacteroides* and *Pseudoalteromonas*. The optimum pH ranges between 8.0 and 10.0, although PGL from *Erwinia* and *Bacillus licheniformis* were still active at pH 6.0 and 11.0, respectively. The optimum temperature for PGL activity is typically between 30 and 40°C, although PGL from thermophiles have an optima between 50 and 75°C. The potential of some pectate lyases for bioscouring has been exploited.

Enzymatic scouring: Greige or untreated cotton contains various non-cellulosic impurities, such as waxes, pectins, hemicelluloses and mineral salts, present in the cuticle and primary cell wall of the fibre. These are responsible for the hydrophobic properties of raw cotton and interfere with aqueous chemical processes on cotton, like dyeing and finishing. Therefore, before cotton yarn or fabric can be dyed, it needs to be pre-treated to remove materials that inhibit dye binding. This step, named scouring, improves the wetability of the fabric and normally uses alkalis, such as sodium hydroxide. However, these chemicals also attack the cellulose, leading to reduction in strength and loss of fabric weight. Furthermore, the resulting wastewater has a high COD (chemical oxygen demand), BOD (biological oxygen demand) and salt content. Enzymatic or bioscouring, leaves the cellulose structure almost intact, preventing cellulose weight and strength loss. Bioscouring has a number of potential advantages over traditional scouring. It is performed at neutral pH, which reduces total water consumption, the treated yarn/fabrics retain their strength properties, the weight loss is reduced or limited compared with processing in traditional ways and it increases cotton fibre softness.

Cellulases

Cellulases are hydrolytic enzymes that catalyse the break down of cellulose to smaller oligosaccharides and finally glucose. Cellulase activity refers to a multicomponent enzyme system combining at least three types of cellulase working synergistically. Endoglucanases or endocellulases cleave bonds along the length of cellulose chains in the middle of the amorphous region. Cellobiohydrolases or exo-cellulases start their action from the crystalline ends of cellulose chains, producing primarily cellobiose. Cellobiohydrolases act synergistically with each other and with endoglucanases, thus mixtures of all these types of enzymes have greater activity than the sum of activities of each individual enzyme alone. Cellobiose and soluble oligosaccharides, produced by exo-cellulases, are finally converted to glucose by β-4-glucosidase. These enzymes are commonly produced by soildwelling fungi and bacteria, the most important being *Trichoderma*, *Penicillium* and *Fusarium*. Many of the fungal cellulases are modular proteins consisting of a catalytic domain, a carbohydrate-binding domain (CBD) and a connecting linker. The role of CBD is to mediate the binding of the enzyme to the insoluble cellulose substrate. Cellulases are active in a temperature range from 30 to 60°C. Based on their sensitivity to pH, they are classified as acid stable (pH 4.5–5.5), neutral (pH 6.67) or alkali stable (pH 9–10). The application of cellulases in textile processing started in the late 1980s with denim finishing. Currently, in addition to biostoning, cellulases are also used to process cotton and other cellulose-based fibres.

Denim finishing: Many garments are subjected to a wash treatment to give them a slightly worn look, e.g., stonewashing of denim jean, in which the blue denim is faded by the abrasive action of pumice stones on the garment surface. Thanks to the introduction of cellulases, the jeans industry can reduce or even eliminate the use of stones, resulting in less damage to the garment and machine and less pumice dust in the laundry environment. Productivity can also be increased because laundry machines contain fewer stones or none at all and more garments. Denim garments are dyed with indigo, which adheres to the surface of the yarn. The cellulase hydrolyses exposed fibrils on the surface of the yarn in a process known as 'Bio-Stonewashing', leaving the interior part of the cotton fibre intact. Partial hydrolysis of the surface of the fibre removes some of the indigo is creating light areas. There are a number of cellulases available, each with their own special properties. These can be used either alone or in combination in order to obtain a specific look.

Pilling and fuzz fibre removal: Besides the 'biostoning' process, cotton and other natural and man-made cellulosic fibres can be improved by an enzymatic treatment called 'biopolishing'. The main advantage of this process is the prevention of pilling. A ball of fuzz is called a pill in the textile trade. These affect garment quality since they result in an unattractive, knotty fabric appearance. Cellulases hydrolyse the microfibrils (hairs or fuzz) protruding from the surface of yarn because they are most susceptible to enzymatic attack. This weakens the microfibrils, which tend to break off from the main body of the fibre and leave a smoother yarn surface. After treatment, the fabric shows a much lower pilling tendency. Other benefits of removing fuzz are a softer, smoother feel and superior colour brightness. Unlike conventional softeners, which tend to be washed out and often result in a greasy feel, the softness-enhancing effects of cellulases are washproof and non-greasy.

Optimisation of biofinishing processes has been an important area of research. Azevedo and others studied the desorption of cellulases from cotton, for recovering and recycling of cellulases. Lenting and Warmoeskerken came up with guidelines to minimise and prevent loss of tensile strength that can result from cellulase application. The choice of enzyme, enzyme concentration and incubation time, as well as application of immobilised enzymes, use of liquids with different viscosities, use of foam ingredients and hydrophobic agents to impregnate clothes can minimise the drawbacks of cellulases action. Yamada and others reported the action of cellulases on cotton dyed with reactive dyes, which have an inhibitory effect on cellulase activity. The use of ultrasound has been shown to be an efficient way to improve enzymatic action in the bioprocessing of cotton. For cotton fabrics, polishing is optional for upgrading the fabric. However, this step is essential for the fibre lyocell. It is made from wood pulp and is characterised by a tendency to fibrillate easily when wet (fibrils on the surface of the fibre peel up). If they are not removed, finished garments made from lyocell will end up covered with pills. Lyocell fabric is treated with cellulases during finishing, not only to avoid fibrillation, but also to enhance its silky appearance. There are several reports describing lyocell treatment with cellulases and elucidation of their mechanism of action.

Serine proteases: subtilisins

Subtilisins are a family of alkaline serine proteases, generally secreted by a variety of *Bacillus* species. They catalyse the hydrolysis of peptide and ester bonds through the formation of an acyl-enzyme intermediate. Subtilisins are made as preproprotein precursors. The NH_2-terminal prepeptide, of 29 amino acid residues is the signal peptide required for secretion of prosubtilisin across the plasma membrane. The propeptide of 77 amino acids, located between the prepeptide and mature sequence, acts as an intramolecular chaperone required for the correct folding of mature enzyme in active form. Subtilisins are characterised by a common three-layer $\alpha/\beta/\alpha$ tertiary structure.

Enzymatic treatment of wool: Raw wool is hydrophobic due to the epicutical surface membranes containing fatty acids and hydrophobic impurities like wax and grease. Harsh chemicals are commonly used for their removal of alkaline scouring using sodium carbonate, pre-treatment using potassium permanganate, sodium sulphite or hydrogen peroxide. Wool fabric has the tendency to felt and shrink on wet processing. The shrinkage behaviour of wool can be regulated by various chemical means.

The most successful commercial shrink-resistant process available is the chlorine-Hercosett process developed more than 30 years ago. Although this is a beneficial method (good antifelt effect, low damage and low weight loss) there are some important drawbacks (limited durability, poor handling quality, yellowing of fibres, difficulties in dyeing and environmental impact of the release of absorbable organic halogens). Several authors have suggested the use of benign chemical processes such as low temperature plasma to treat wool. Plasma treatment is a dry process, in which the treatment of wool fibre is performed by electric gas discharges (plasma). It is regarded as an environmentally friendly process, as no chemicals are used and it can modify the surface properties of wool without much alteration of the interior part of the fibre. However, costs, compatibility and capacity are obstacles to commercialisation of a plasma treatment process and the shrink-resist properties obtained do not impart a machine-washable finish, which is one of the main objectives. The subsequent application of a natural polymer, such as chitosan, has been investigated to improve wool shrink-resistance or antifelting properties. More recently and mainly for environmental reasons, proteases of the subtilisin type have been studied as an alternative for chemical pre-treatment of wool. Several studies reported that pre-treatment of wool fibres with proteases improved antishrinkage properties, removed impurities and increased subsequent dyeing affinity.

However, due to its small size, the enzyme is able to penetrate into the fibre cortex, which causes destruction of the inner parts of the wool structure. Several reports show that increasing enzyme size by chemical cross-linking with glutaraldehyde or by the attachment of synthetic polymers like polyethylene glycol, can reduce enzyme penetration and the consequent reduction of strength and weight loss. Some of these processes have been tested on industrial process scale. Pre-treatment of wool fibres with hydrogen peroxide, at alkaline pH in the presence of high concentrations of salts, also targets enzymatic activity to the outer surface of wool, by improving the susceptibility of the cuticle to proteolytic degradation.

Cysteine proteases: papain

Cysteine proteases (CP's) catalyse the hydrolysis of peptide, amide, ester, thiol ester and thiono ester bonds. More than 20 families of cysteine proteases have been described. The CP family can be subdivided into exopeptidases (e.g., cathepsin X, carboxypeptidase B) and endopeptidases (papain, bromelain, ficain, cathepsins). Exopeptidases cleave the peptide bond proximal to the amino or carboxy termini of the substrate, whereas endopeptidases cleave peptide bonds distant from the *N*- or C-termini. CPs have molecular masses in the range of 2130 kDa. They are synthesised as inactive precursors with an *N*-terminal propeptide and a signal peptide. Activation requires proteolytic cleavage of the *N*-terminal propeptide that also functions as an inhibitor of the enzyme.

Papain is the best known cysteine protease. It was isolated in 1879 from the fruits of Carica papaya and was the first protease with a crystallographic structure. Papain has 212 amino acids with a molecular mass of 23.4 kDa. The enzyme has three internal disulphide bridges and an isoelectric point of 8.75. The optimal activity of papain occurs at pH 5.8–7.0 and at temperature 50–57°C, when casein is used as the substrate. The general mechanism of action has been very well studied. The catalytic triad is formed by Cys25, His159 and Asn175 residues. Asn175 is important for orientation of the imidazolium ring of the histidine in the catalytic cleft. The reactive thiol group of the enzyme has to be in the reduced form for catalytic activity. Thus, the cysteine proteases require a rather reducing and acidic environment to be active. Generally, papain can cleave various peptide bonds and therefore, have fairly broad specificity.

Degumming of silk: Papain is used for boiling off cocoons and degumming of silk. Raw silk must be degummed to remove sericin, a proteinaceous substance that covers the fibre. Degumming is typically performed in an alkaline solution containing soap, a harsh treatment that also attacks fibrin structure. Several alkaline, acidic and neutral proteases have been studied as degumming agents since they can dissolve sericin, but are unable to affect silk fibre protein. Alkaline proteases seem to be the best for removing sericin and improving silk surface properties like handle, shine and smoothness, although this is not in commercial use.

In the past, papain was also used to 'shrink-proof' wool. A successful method involved the partial hydrolysis of the scale tips. This method also gave wool a silky lustre and added to its value. The method was abandoned a few years ago for economic reasons.

Transglutaminases (TGs)

Transglutaminases are a group of thiol enzymes that catalyse the post-translational modification of proteins mainly by protein to protein cross-linking, but also through the covalent conjugation of polyamines, lipid esterification or the deamidation of glutamine residues. Transglutaminases are widely distributed among bacteria, plants and animals. The first characterised microbial transglutaminase (MTG) was that of the bacterium *Streptomyces mobaraensis*. This enzyme is secreted as a zymogen that is sequentially processed by two endogenous enzymes to yield the mature form. The mature enzyme is a monomeric protein with a molecular weight of 38 kDa. It contains a single catalytic cysteine residue (Cys-64) and has an isoelectric point (pI) of 9. The optimum pH for MTGase activity is between 5 and 8. However, MTGase showed some activity at pH 4 or 9 and was thus considered to be stable over a wide pH range. The optimum temperature for enzymatic activity is 55°C, it maintained full activity for 10 min at 40°C, but lost activity within a few minutes at 70°C. It was active at 10°C and retained some activity at near-freezing temperatures. MTG does not require calcium for activity, shows broad substrate specificity and can be produced at relatively low cost. These properties are advantageous for industrial applications.

Treatment of wool and leather: The use of TGs for the treatment of wool textiles has been shown to improve shrink resistance, tensile strength retention, handle, softness, wetability and consequently dye uptake, as well as reduction of felting tendency and protection from damage caused by the use of common detergents. Treatment of leather with TG, together with keratin or casein, has a beneficial effect on the subsequent dyeing and colour properties of leather. The application of TG for leather and wool treatment seems to be a promising strategy, but is still at the research level.

Lipases/esterases: cutinase: Esterases represent a diverse group of hydrolases that catalyse the cleavage and formation of ester bonds. They are widely distributed in animals, plants and micro-organisms. These enzymes show a wide substrate tolerance, high regio- and stereospecificity, which make them attractive biocatalysts for the production of optically pure compounds in fine-chemicals synthesis. They do not require cofactors, are usually rather stable and are even active in organic solvents. Two major classes of hydrolases are of most importance: lipases (triacylglycerol hydrolases) and 'true' esterases (carboxyl ester hydrolases). Both classes of enzymes have a three-dimensional structure with the characteristic α/β-hydrolase fold.

The catalytic triad is composed of Ser-Asp-His (Glu instead of Asp for some lipases) and usually also a consensus sequence (Gly-x-Ser-x-Gly) is found around the active site serine.

The mechanism for ester hydrolysis or formation is essentially the same for lipases and esterases and is composed of four steps: first, the substrate is bound to the active serine, yielding a tetrahedral intermediate stabilised by the catalytic His and Asp residues. Next, the alcohol is released and an acyl-enzyme complex is formed. Attack of a nucleophile (water in hydrolysis, alcohol or ester in transesterification) reforms a tetrahedral intermediate, which after resolution yields the product (an acid or an ester) and free enzyme. Lipases can be distinguished from esterases by the phenomenon of interfacial activation (which is only observed for lipases). Esterases obey classical MichaelisMenten kinetics, lipases need a minimum substrate concentration before high activity is observed. Structure elucidation revealed that this interfacial activation is due to a hydrophobic domain (lid) covering the lipase active site and only in the presence of a minimum substrate concentration, (a triglyceride phase or a hydrophobic organic solvent) will the lid open, making the active site accessible. Furthermore, lipases prefer water-insoluble substrates, typically triglycerides composed of long-chain fatty acids, whereas esterases preferentially hydrolyse 'simple' esters. Lipases and esterases were among the first enzymes tested and found to be stable and active in organic solvents, but this characteristic is more apparent with lipases.

A comparison of the amino acid sequences and 3D-structures of both enzymes showed that the active site of lipases displays a negative potential in the pH-range associated with their maximum activity (typically at pH 8), esterases show a similar pattern, but at pH values around 6, which correlates with their usually lower pH-activity optimum.

Cutinases are extracellular esterases secreted by several phytopathogenic fungi and pollen that catalyse the hydrolysis of ester bonds in cutin, the structural polyester of plant cuticles. Cutinases are also able to hydrolyse a wide variety of synthetic esters and triacylglycerols, as efficiently as lipases, without displaying interfacial activation. Therefore, cutinases are suitable for application in the laundry industry, dishwashing detergents for removal of fats, in the synthesis of structured triglycerides, polymers and agrochemicals and in the degradation of plastics.

Surface modification of synthetic fibres: Synthetic fibres represent almost 50% of the worldwide textile fibre market. Polyethyleneterephthalate (PET), polyamide (PA) and polyacrylonitrile (PAN) fibres show excellent features like good strength, high chemical resistance, low abrasion and shrinkage properties. However, synthetic fibres share common disadvantages, such as high hydrophobicity and crystallinity, which affect not only wearing comfort (making these fibres less suitable to be in contact with human skin), but also processing of fibres, impeding the application of finishing compounds and colouring agents. Most of the finishing processes/agents are water-dependent, which require an increase in hydrophilicity of fibre surface. Currently, chemical treatments with sodium hydroxide are used to increase hydrophilicity and improve flexibility of fibres. However, chemical treatment is hard to control, leading to unacceptable losses of weight and strength and to irreversible yellowing in the case PAN and PA fibres. Besides, this is not an environmentally appealing process since it requires large amounts of energy and chemicals. A recently identified alternative is the use of enzymes for the surface modification of synthetic fibres.

Laccases

Laccases are extracellular, multicopper enzymes that use molecular oxygen to oxidise phenols and various aromatic and non-aromatic compounds by a radical-catalysed reaction mechanism. They belong to a larger group of enzymes termed the blue-multicopper oxidase family. Laccases have been found in plants, insect, bacteria, but are most predominant in fungi. Laccase activity has been demonstrated in more than 60 fungal strains. Typical fungal laccase is a protein of approximately 60–70 KDa with a pH optimum in the acidic pH range and optimal temperature range between 50 and 70°C. Few enzymes with optima temperature below 35°C have been described, an example being the laccase from *Ganoderma lucidum* with its highest activity at 25°C. The range of substrates with which laccases can react is very broad, showing a remarkable lack of specificity towards their reducing substrate.

Decolourisation of dyes and textile bleaching: Laccases are widely researched for the decolourisation of textile effluents. Due to their ability to degrade dyes of diverse structures, including synthetic dyes, laccases are an environmentally friendly tool to treat dye wastewater. They have also been studied for textile bleaching. Bleaching of cotton is achieved by the decolourisation of natural pigments giving cotton fibres a white appearance.

The most common industrial bleaching agent is hydrogen peroxide usually applied at temperatures close to boiling. However, high temperatures and alkaline pH can cause severe damage to the fibres and large amounts of water are needed to subsequently remove the hydrogen peroxide from fabrics. Laccases can improve whiteness of cotton by oxidation of flavonoids. The substitution or combination of chemical bleaching with an enzymatic bleaching system leads not only to less fibre damage, but also to significant water economy. Pereira and collaborators isolated a new strain of *Trametes hirsuta* for cotton bleaching. Laccases of this organism were responsible for oxidation of the flavonoids morin, luteolin, rutin and quercetin. The authors reported that pre-treatment of cotton with *T. hirsuta* laccases resulted in an increase of whiteness. Later, ultrasound was used to intensify the efficiency of enzymatic bleaching. The authors found that low intensity ultrasound improved diffusion of the enzyme from the liquid phase to the fibre surface, acting synergistically with the enzyme in the oxidation of natural pigments. In denim finishing, there are already some successful industrial applications of laccases like DeniLite® commercialised by Novozyme (Novo Nordisk, Denmark) and Zylite from the company Zytex (Zytex Pvt. Ltd, Mumbai, India).

The application of laccases for the coating of natural and synthetic fibres is under study. Tzanov and others developed a laccase-assisted dyeing process for wool, using low temperatures without dyeing auxiliaries, which permits saving water and energy. More recently, Kim and collaborators described the use of natural flavonoids to dye cotton by an enzymatic process catalysed by laccases.

Catalases

Catalases (CATs), more correctly hydroperoxidases, catalyse the degradation of H_2O_2 to H_2O and O_2. They are produced by a variety of different micro-organisms including bacteria and fungi and most have optima at moderate temperatures (20–50°C) and neutral pH. CATs from animal sources (bovine liver) are generally cheap, therefore, the production of microbial CATs will only be economically advantageous when recombinant strains and cheap technology is used, or for CATs with special properties such as thermostability or operation at alkaline or acidic pH.

Treatment of bleach liquor: In the textile industry, bleaching with H_2O_2 is performed after desizing and scouring, but before dyeing. Historically, a reducing agent was used to destroy the hydrogen peroxide, or water to rinse out the hydrogen peroxide bleach, but CAT can now be used to decompose excess H_2O_2. This eliminates the need for a reducing agent and minimises the need for rinse water, resulting in less polluted wastewater and lower water consumption. The cost of enzyme for degradation of hydrogen peroxide in bleaching effluents could be reduced by the introduction of immobilised enzymes, allowing not only the recovery of enzyme, but also the reuse of treated bleaching effluents for dyeing.

Thus, enzymes can be used in order to develop environmentally friendly alternatives to chemical processes in almost all steps of textile fibre processing. There are already some commercially successful applications, such as amylases for desizing, cellulases and laccases for denim finishing and proteases incorporated in detergent formulations. Further research is required for the implementation of commercial enzyme based processes for the biomodification of synthetic and natural fibres.

18

Enzymes in detergent industry

18.1 Introduction

Microbial lipases are an important group of biotechnologically valuable enzymes, because of the versatility of their applied properties and ease of mass production. Lipases of microbial origin are widely diversified in their enzymatic properties and substrate specificity, which make them very attractive for industrial applications.

Lipases are added to detergents such as household and industrial laundry and in household dishwashers, where their function is in the removal of fatty residues and cleaning clogged drains. The cleaning power of lipase (or other enzyme containing) detergents increases markedly. The enzymes such as proteases, amylases, cellulases and lipases are added to the detergents to improve their efficiency.

Laundry detergents are becoming more and more popular because of their increasing use in washing machine, where it impart softness, resiliency to fabrics, antistaticness, dispersible in water and mild to eyes and skins. There are many different brands or types of laundry detergents and mostly they claim some special qualities. The detergents industry, regulated by the European Commission (EC) 648/2004, detergent directive, is considered as the largest enzyme consuming sector. Concerns over persistence of detergent chemicals in the environment and its possible contamination of ground water, other fresh sources and their subsequent health related issues has raised speculation over biodegradability. This has acted as an impetus for greater consumption of enzymes as Europe is witnessing increasing preference for greener detergents.

The need for greater control over micro-organisms and enzymes for industrial purposes has led to a greater focus on genetic engineering and recombinant DNA technology. This technology, allows genetic modification of micro-organisms to produce the desired enzyme under specific conditions. This helps either to produce a particular type of enzyme or enhance the quantity of enzyme produced from the single recombinant micro-organism. Understanding the structure of enzymes and modifying them to extract benefits is categorised as protein engineering. Studies are being conducted on ways to improve or modify protein structure and its function thus finally the enzyme. Biotechnology is therefore, being increasingly viewed as a possible solution against traditional chemicals processes. The production of enzymes from natural sources and their environment friendly characteristics has led the industry to believe that enzymes are indeed a sustainable alternative to chemicals in industrial processes. The factors that aid their increased application is the availability of wide enzyme types, environmental friendly behaviour, minimised energy consumption, easily controlled processes, possible modification of enzyme characteristics, environmentally friendly by-products, minimal greenhouse gas emissions, decreased use of non-renewable sources and proliferation of chemicals in environment. The multi faceted benefits available through using enzymes is bound to convert into a greener world.

18.2 Enzymes used in detergents

Biotechnology based cleaning agents such as enzyme-based agents, are widely used in industries. The biotechnology based cleaning agents are cheaper and less harmful to the environment. They have specific cleaning action and can also be used at lower temperatures. They produce effluents with lower COD and non-corrosive nature. Enzyme-based cleaners are becoming increasingly popular in the food industry as compared to caustic or acid cleaning regimes.

The enzyme based detergents have better cleaning properties as compared to synthetic detergents. They are active at low washing temperatures and environment friendly also. The enzymes in the detergents do not lose their activity after removing stain. The enzyme containing detergents also improve the fabric quality and keeping colour bright.

The enzyme based detergents are used in small quantity as compared synthetic chemicals. They can work at very low temperature, environment friendly and completely biodegradable. The four classes of enzymes used in detergents are used in Table 18.1.

Table 18.1: Four classes of enzymes are generally used in detergents.

Proteases	Most widely used enzymes in the detergent industry remove protein stains such as grass, blood, egg and human sweat which have a tendency to adhere strongly to textile fibres
Amylases	Used to remove residues of starch-based foods like potatoes, spaghetti, custards, gravies and chocolate
Lipases	Decompose fatty material. Lipase is capable of removing fatty stains such as fats, butter, salad oil, sauces and the tough stains on collars and cuffs
Cellulases	Modify the structure of cellulose fibre on cotton and cotton blends. When it is added to a detergent, it results in, colour brightening, softening and soil removal

Removal of proteins by non-enzymatic detergents can result in permanent stains due to oxidation and denaturing caused by bleaching and drying. Proteases hydrolyse proteins and break them down into more soluble polypeptides or free amino acids. As a result of the combined effect of surfactants and enzymes, such hard to remove stains can be removed from fibres.

An alkaline extracellular enzyme obtained from alkaliphilic *Bacillus* strains is used in detergents. Other enzymes like proteases, alpha amylases and cellulases are produced on large scale and have been used in detergents. The domestic dishwashing detergents contain different chemicals like chelating agents, surfactant, polycarboxylates (PCA), phosphonates, active chlorine compounds, etc. These substances have high environmental risks, particularly via sewage biosolids, poor biodegradability, increased organic load to sewage works, toxicity.

Sodium hydroxide or caustic soda (NaOH) is the most widely used alkaline detergent in the food and beverage industry due to its low price and high cleaning efficiency for fatty-type and protein soils. Nitric acid and phosphoric acid are most commonly used by the food and beverage industries which lead to the increased level of TDS and sodium discharge in effluent. To avoid any damage to soil and vegetation, it is important to reduce the use of traditional chemicals in food and beverage industries.

Enzymes are used in detergents to increase the cleaning ability of detergents. Enzymes can be used instead of chlorine bleach for removing stains on cloth. The enzyme protease was produced from alkaliphilic *Bacillus clausii* KSM-K16 and strain KP-43 and *Bacillus* sp., strain KSM-KP43 and have been incorporated into laundry detergents. Subtilisin-like serine proteases belonging to family A of subtilase super family has been used in laundry and dishwashing detergents. A number of alkaliphilic *Bacillus* produce alkaline cellulase (carboxymethylcellulase) that is used as an additive for improving the cleaning effect of detergents. Enzymatic properties of some cellulases fulfilled the essential requirements for enzymes to be used practically in laundry detergents. Among other enzymes used in some detergents are Guardzyme (α- peroxidase) which inhibits dye transfer and Carezyme removing the fuzz that builds up on cotton clothes. Enzymes used in detergents must be effective at low levels, compatible with various detergent components and to be active at wide range of temperatures.

A number of detergent formulations such as anionic or nonionic surfactants and the powdered lipase are used for the removal of fatty soils from fabrics. These include sodium nitrilotriacetate, sodium tripolyphosphate, sodium silicate, sodium citrate and potassium diphosphate. Optional ingredients include a detergent builder such as, foam boosters, alkalies (e.g., sodium carbonate), optical brighteners (e.g., stilbene derivatives), stabilisers (e.g., triethanolamine), fabric softeners (e.g., quaternary ammonium salts) together with bleaching agents and systems (such as sodium perborate and ethylene diaminetetra acetate). Different fragrances, dyes, foam depressors and corrosion agents are also used additionally. Use of liquid laundry detergents has been increasing over the years. For more than forty years, enzymes in encapsulated form have been used in detergent products, such as laundry formulations. They have increasing importance as they are biodegradable and function at low temperatures and offer environmental benefits. A very small amount of the enzymes is used in detergent preparations and it must withstand the other components of the detergents, such as surfactants, oxidants, optical brighteners and also stable at pH values between 8 and 10.5 (Table 18.2).

Table 18.2: Compositions of an enzyme detergent.

Constituent	*Composition (%)*
Sodium tripolyphosphate (water softener, loosens dirt)	38.0
Sodium alkane sulphonate (surfactant) (cationic/anionic/non-inonic/amphoteric)	25.0
Sodium perborate tetrahydrate, sodium percarbonate (oxidising agent, oxygen bleach)	25.0
Soap (sodium alkane carboxylates)	3.0
Sodium sulphate (filler, water softener)	2.5
Sodium carboxymethyl cellulose/sodium polyacrylate/polyethylene glycol (dirt-suspending agent/(Antiredeposition agents)	1.6
Sodium metasilicate (binder, loosens dirt)	1.0
Bacillus protease (3% active) (any other enzyme)	0.8
Fluorescent brighteners	0.3
Foam-controlling agents	Trace
Perfume	Trace
Water	100%

Alkaline yeast lipases are preferred because they can work at lower temperatures as compared to bacterial and fungal lipases. Cold active lipase detergent formulation is used for cold washing which reduces the energy consumption and wear and tear of textile fibres. Enzymes can reduce the environmental load of detergent products because they save energy by enabling a lower wash temperature to be used. Addition of cold active lipase in detergent become biodegradable, leaves no harmful residues, have no negative impact on sewage treatment processes and do not have risk for aquatic life.

Various factors affect the performance of enzymes in detergents, its composition, type of stains to be removed, washing temperature, washing procedure and water hardness. Some of the additives of the detergents include, builders (Sequestering builders- polyphosphates, citrate, precipitating builders- sodium carbonate, sodium silicate), zeolite, alkaline agents- sodium carbonate, sodium silicates, corrosion inhibitor- sodium silicate, processing aids- sodium sulphate, water, alcohol, xylene sulphonate, colorants, suda control agents, opacifiers, bleaching agents-sodium percarbonate, sodium perborate, hydrogen peroxide.

Enzymes in detergent formulations are stable at high pH and temperature and remove protein and lipid stains. Enzymatic detergent requires low workable temperature, low mechanical energy and is less toxic and non-corrosive. Thermostable enzymes are usually stable in solvents and detergents, therefore, they also have considerable potential for many biotechnological and industrial applications.

The first industrial extremozymes used in textile detergents, were the alkaline cellulases produced by bacteria isolated from an east African soda lake, which was the first industrial extremozymes for use in textile detergents and were commercialised by Genencor. Enzymes used in household detergents have stability at extreme temperatures and pH. The enzymes must fulfil different requirements such as activity and stability, substrate specificity and enantioselectivity, for their application in industry.

The cold active lipases will have a larger share of enzymes of industrial importance in the coming years with the increasing interest in psychrophiles. These enzymes offer a great industrial and biotechnological potential due to their capability to catalyse reactions at low temperature.

18.3 Lipases in detergents

There has been a tremendous increase in the significance of the biotechnological applications of lipases since the last two decades, due to the versatile catalytic behaviour of lipases. Lipases form an integral part of the industries ranging from food, pharmaceuticals and detergents to oleo-chemicals, tea industries agriculture, cosmetics, leather and in several bioremediation processes. Because of the vast applications, newer microbes are to be screened for production of lipases having desirable properties. Molecular analysis of the lipase gene helps to understand the structure and activity relationship of the enzyme, which will help the researchers to construct new lipases for biotechnological applications according to the demand and need of industry.

Detergents are common commercial products used in dish washing, bleaching compositions, in dry cleaning solvents, leather cleaners, cleaning of contact lens, clogged drain clearing in domestic/industrial effluent or food processing treatment plants, in degradation of organic wastes from exhaust pipes, toilet bowls, etc. Lipases and cellulases in certain detergents remove dirt/cattle manure from domestic animals. Lipases are also used in degreasing and water reconditioning in combination with oxidoreductases that require lesser amount of surfactants and work at low temperatures.

18.4 Bacterial lipases in detergents

Lipases for detergents are specifically selected on the basis of low substrate specificity, stability under alkaline (pH 10–11, 30–60°C) conditions, in the presence of surfactants (linear alkyl benzene sulphonates) and enzymes (proteases), constituents of many detergent formulations. Continuous screening and protein engineering leads to the discovery of lipases with the desired properties. Detergent enzymes should be cost-effective and safer to use. Bacterial lipases added to household detergents reduces or replaces synthetic detergents, which have considerable environmental problems.

Normally fat stains are not easy to remove at low temperatures using conventional detergents, therefore lipases are required which are active at lower temperatures and can be used in detergent formulations. The use of lipase active at low temperature in detergent formulation, reduces the energy consumption as well as the wear and tear of textile fibres and maintain the texture and quality of fabrics. Lipases from *Candida* and *Chromobacterium* are also used in detergents.

Lipases which are stable and work at alkaline pH (8 to 11), suitable wash conditions for enzymated-detergent powders and liquids have good potential for use in detergent industry. Lipase produced by *Acinetobacter radioresistens* was found to be optimally active at pH 10 and showed stability in the range of pH 6–10, therefore, it has a greater potential to be used in the detergent industry. Hasan and others reported 100% stability of lipase produced by *Bacillus* sp., FH5, at pH 10. This enzyme showed promising results when used in combination with different commercially available conventional detergents. *A. radioresistens* produced lipase having optimum activity at pH 10 and stability in the range of pH 6–10 and showed a great potential to be used in the detergent industry. Extracelluar lipase produced by *Micrococcus* sp., ML-1 was commonly used in combination with various commonly used detergents, to enhance the removal of oily stains from various types of fabrics. Soyabean oil-dye stain, followed by dye stain in mustard oil, groundnut oil and coconut oil were removed quite easily. The detergents, Surf, Nirma and Wheel, differed in the oil-dye removes potential stains from cotton, terrrycot and polyester. The ability of detergents to wash, remove fatty food stains and sebum from fabrics, is enhanced by addition of lipases.

SYSCO makes cleaning products for food service industry. Sysco® Action Suds with enzymes removes tough stains and makes the fabric white and gives brightness. This enzyme product has low-alkalinity and works with optical brighteners, soil re-deposition agents and colour-safe bleach to give longer life to fabric. Sysco® enzyme grease and waste digester is a natural, multi-strain bacterial enzyme formulation, effective in both hard and soft water, unclogs drain lines and digests any accumulated grease and waste.

Olin Corporation (Cheshire, UK) invented a detergent composition comprising the microbial lipase SD2, dode-cylbenzene sulphonate and gelatin. Novo-zymes has also marketed lipases (Lipex® and Lipolase ®), used in detergent industry. US Agricultural Research Service scientists obtained an enzyme from a bacterium isolated from shipworm glands that digests proteins in common laundry stains.

Commercial detergent formulations with high-temperature optima have been produced from *P. mendocina* (Lumafast) and *Pseudomonas glumae* are used. In 1995, Lumafast and Lipomax, lipases from *P. mendocina* and *Pseudomonas alcaligenes*, respectively, were produced by Genencor International, AU-KBC Research Centre, Life Sciences, Anna University, Chennai, India. Gerritse and others reported an alkaline lipase from *P. alcaligenes* M-1, capable of removing fatty stains when used in a washing machine.

European Patent Office (EPO) Solvay Enzyme Products, Inc. 1992-01-29/ 1990-07-25, is a nonionic and/or anionic detergent formulation that contains lipase from *Pseudo-monas plantarii*.

Composition of a U.S. Patent 4707291, a detergent, comprise a mixture of an anionic and a nonionic detergent-active compound in combination with a lipase similar to the enzyme produced by *Pseudomonas fluorescens* IAM 1057, *P. gladioli* or *Chromobacterium viscosum* (Detergent formulations containing alkaline lipase, 1992). Lion has released an improved version of its powder type laundry detergent 'Top' with two different acting enzyme cleansing agents.

18.5 Fungal lipases in detergents

During 1980–1990, many different enzymes containing detergents were developed for softening of clothes. In 1988, Novo Nordisk developed a lipase produced by the fungus *Humicola*. Commercial detergent formulations with high-temperature optima have been produced from *Humicola lanuginosa* (Lipolase). As the quantity of this enzyme was less than required for its commercial application, therefore, the yield of enzyme was increased by cloning the gene coding for this lipase in *Aspergillus oryzae*. This fungus produced increased quantity of the enzyme that can be used commercially in detergents. The first commercial lipase, Lipolase, was introduced by Novo Nordisk in 1994. This enzyme was produced from *Trichoderma lanuginosus* and was expressed in *A. oryzae*. They also produced a lipase containing detergent 'LipoPrime®'.

Laundering with lipase containing detergents is generally carried out in alkaline conditions, therefore, alkaliphilic lipases are preferred for example, lipase obtained from the *A. oryzae*. A presoak formulation was developed using an alkaline lipase produced by *Trichosporon asahii* MSR 54, used for removing oil stains at ambient temperature.

ABS fungal lipase is used in a many sanitary and waste treatment liquid formulations. This product exhibits activities in a wide range of pH. Enzymatic methods are preferred to chemical methods in many sanitary applications such as grease trap applications, other applications include detergents, pre-spotters and industrial cleaning compounds.

18.6 Other lipase containing detergents

SEBrite-L detergent containing lipase helps in removing cosmetics, sebum and fatty food stains from garments, salad oil, animal fat and butter, suitable to use at lower washing temperatures and neutral pH. Enzyme-containing detergents are also used in disinfecting and cleaning hard surfaces in food manufacturing plants and buildings, contaminated by bacteria and fungi. SEBalase-M and SEBrite-M detergents with protease and lipase enzymes, respectively, are used to effectively remove proteinaceous debris and fat lubricating oils from medical instruments. Lipases that deal with greasy soil on cloths leave residual 'sick' odours on the cloth. Some detergent compositions are produced which have fragrances and have the ability to counter the bad odours due to lipase in detergent products.

Enzyme containing laundry detergents marketed by Procter and Gamble (P&G) includes, tide, ivory snow, dreft, cheer and era, perform well. The other benefits to consumers are better performance, the abilities to remove fatty stains at low wash temperatures, improve whiteness, soften fabrics, brighten their colours and perform in the presence of chemicals (bleach, builder, surfactant, etc.). P&G searches for micro-organisms/genes which can fulfill consumer need, e.g., an alkaline cold wash enzyme by a limited number of cycles of directed evolution is a major challenge in using mutagenically improved strain of micro-organisms/a gene. Thousands of mutants have to be tested to find the best strain.

Immobilised lipase (Patent # 6265191, issued 07/24/2001) is used on surfaces for easy removal of oil and to change the wettability of fabric surfaces. Lipase in detergents forms a fabric-lipase complex on the surface of cloth and removes oil stain and makes them resist denaturation in the presence of surfactants and deactivation by heat. This complex resists the removal of enzyme from the surface of fabric during washing and it remains active stains after drying of fabric at higher temperatures, during storage and wear of fabric. Lipase impedes the oil re-deposition during laundering of fabric. Oil hydrolysis by-products are removed during washing of fabric at a basic pH or in the presence of a surfactant.

Two laundry products (a liquid and a solid enzyme) by Gateway ProClean Inc. USA, breaks down soil into simpler forms which can then be easily removed. Lipase in these products remove fat-containing stains of frying fats, salad oils, butter, fat-based sauces, soups, human sebum or certain cosmetics. During manufacturing beverages such as wine, fruit juice, beer and milk are commonly filtered through ultra- and nano-filters. These filters are fouled with protein, cellulose, starches, fats and pectin residues and are cleaned and disinfected with caustic and chlorine. These can be effectively cleaned using enzyme-containing detergents like SEB-Prolase P or SEBalase-PB (protease), CelluSEB-T (carbohydrase), SEBamyl-B (alpha-amylase), SEBrite-L (lipase) and ExtractSEB-R (pectinase).

Novozyme launched a low-temperature detergent enzyme suitable that can be used in the detergents usually used for hand-washing in developing countries like China and India. Another product, Polarzyme, can clean clothes in cold water and also it has attractive price.

Enzymes and detergents are used to clean medical instruments since the last two decades. Multiple enzyme/detergent instrument cleaners can also remove the microbial biofilms (Enzymatic instrument cleaners in health care).

Another common application of enzymatic detergents is cleaning of gastrointestinal endoscopes before disinfection. Misuses of enzyme detergents include failure to dilute the detergent, over dilution, use of expired detergent, in-adequate exposure time and failure to rinse the detergent properly from the instrument.

ISI enzymatic detergents contain the enzymes that can remove the stubborn organic materials. A pre-cleaning enzymatic foam developed by ISI, Hemaway enzymatic foam, is ideal for use in the immediate cleaning of surgical instruments. Another product by ISI, Hemaway Tri-Max contain three enzymes for enhanced cleaning and removing tough organic materials from surgical instruments. A unique enzyme formulation, Endozime® AW Plus, can rapidly remove fat, blood, starch, carbohydrates and protein from scopes and surgical instruments, cleans the most difficult instruments (that is, orthopaedic, laproscopic) and is safer to use on the most delicate (that is, ophthalmic, microsurgical) instruments. It is a mixture of enzymes (protease, amylase, lipase and carbohydrase), buffers and nonionic detergents.

Protease and lipase in the lens cleaning system considerably enhances the removal of proteinaceous and lipid material that accumulates on the contact lens. POLYZYM® enzymatic cleaning tablets loosen organic deposits on the lens surface and remove it easily. Clear-Lens (TM) Lipo is also a lipase preparation available for cleaning contact lenses.

Besides increasing detergency, lipase component in detergents also prevents scaling. The cost of production of these enzymes is very high. This limitation may be overcome by employing molecular techniques, thus enhancing the production of these enzymes at high levels and in purified form. Extensive and continued research is going on to develop lipases, which will work under various suitable conditions as fat removers. There is a lot to be done to extend and increase the use of enzymes in detergents globally. Unfortunately, the enzyme detergents are not used much in developing countries which are comparatively more hot and dusty, which requires frequent washing of clothes. Enzymes such as glucose oxidase, lipoxygenase and glycerol oxidase can generate hydrogen peroxide *in situ*. Peroxidases may help enhance the bleaching efficacy of peroxide.

There is still a need of extensive and continuous screening for new micro-organisms that have the ability to produce lipolytic enzymes with enhanced activities at low temperature, have stability at high temperatures, active at alkaline conditions, higher stabilities in the presence of other additives usually present in a detergent. The most important thing among all is to create awareness in the developing world about the 'pros' of enzyme detergents and 'cons' of the chemical detergents they are continuously exposed to.

Developments in enzymology and recombinant DNA technologies, led to the current era of bioengineered enzymes. Many industries of the world are involved in the manufacture and trade of enzymes used in detergents. Presently, three major giants, Novo Nordisk (Novozymes A/S, North America), Genencor International Inc. (California, USA) and DSM N.V. Enzymes (Netherlands), are leading production of industrial enzymes. Novo Nordisk (Novozymes A/S) markets a range of enzymes for various industrial purposes, mostly derived from micro-organisms, such as various strains of alkalophilic *Bacillus* sp., or genetic and/or protein engineered organisms. They also produce enzymes, other than lipases, which are used in detergent industry. Many of these enzymes are produced by fermentation of genetically modified micro-organisms.

19

Enzymes in green chemistry and organic synthesis

19.1 Introduction

Green chemistry is rapidly emerging area of chemical research today which is associated with research and applications of chemical processes which are environmental benign, economical and less polluting. Green chemistry is based on certain principles emphasising on the reduction or elimination, the use and generation of hazardous substances in all steps of chemical synthesis or process. Chemists and scientists are working to minimising the risk for human health and the environment by following all the valuable principles of green chemistry. The green chemistry insists to utilise eco-friendly, non-hazardous, reproducible and efficient solvents and catalysts and chemicals in synthesis of various molecules and in researches. Enzymes are important tools in green chemistry as they act as catalysts which accelerate the chemical reactions, works without using harsh conditions and chemicals and are reusable.

Green chemistry is now a global approach which is contributing to sustainable development through application and extension of its principles. Great efforts are still putting in to design green process starting from and resulting in nonpolluting materials. It is a challenge for the future chemical industry research to design research and to discover safer products and processes. Enzymatic reactions and conversion are seen as a hope for these objectives. The use of enzymes as effective catalysts working under mild conditions results in significant savings in resources such as energy and water for the benefit of both the industry as well as the environment. Enzymatic processes are becoming better financial and ecological alternative to other traditional processes by virtue of being cost effective and more environmental friendly. The success of green chemistry depends on good quality of education and training of a new generation of chemists. So it is necessary at all levels of chemical education to make this new generation familiar with this new 'Green' aspect of chemistry aiming for environment protection.

19.2 Concept of green chemistry

The concept of green chemistry includes a new approach towards the synthesis, processing and application of chemical substances in such a manner which reduce threats for environment and health. This new approach is also called as clean chemistry or environmentally benign chemistry. Green chemistry is the designing of chemical products and processes that reduce or eliminate the use and generation of hazardous substances. For example in green chemistry we can use water as the medium of chemical reactions in the laboratory avoiding pollution and the hazardous effect of the volatile solvent. Green chemistry applies combined approaches of organic chemistry, inorganic chemistry, biochemistry, analytical chemistry and physical chemistry to minimise waste and utilise renewable resources.

19.3 Principles of green chemistry

Green chemistry is a highly effective approach to prevention of environment pollution as for energy conservation as it applies innovative scientific approaches to solve real-world environmental situations. Green chemistry is based on certain principles which are called as 12 principles of green chemistry.

These 12 principles of green chemistry provide a way to implement the strategies of green chemistry action plan which are listed below:

1. Prevention: It is better to prevent waste than to treat or clean up waste after it has been created.
2. Atom economy synthetic methods should be designed to maximise the incorporation of all materials used in the process into the final product.
3. Less hazardous chemical synthesis synthetic methods should be designed, wherever practicable, to use and generate substances that possess little or no toxicity to human health and the environment.
4. Designing safer chemicals and chemical products should be designed to achieve their desired function while minimising their toxicity.
5. Safer solvents and auxiliaries unnecessary use of auxiliary substances (e.g., solvents, separation agents, etc.), should be avoided wherever possible and made innocuous when used.
6. Design for energy efficiency energy requirements of chemical processes should be recognised for their environmental and economic impacts and should be minimised. If possible, synthetic methods should be conducted at ambient temperature and pressure.
7. Use of renewable feedstocks whenever technically and economically practicable, raw material or feedstock should be renewable rather than depleting it.
8. Reduce derivatives unnecessary derivatisation (use of blocking groups, protection or deprotection, temporary modification of physical/chemical processes) should be minimised or avoided if possible, because such steps require additional reagents and can generate waste.
9. Catalysis catalytic reagents (as selective as possible) are superior to stoichiometric reagents.

10. Design for degradation chemical products should be designed so that at the end of their function they break down into innocuous degradation products and do not persist in the environment.
11. Real-time analysis for pollution prevention analytical methodologies need to be further developed to allow for real-time, in-process monitoring and control, prior to the formation of hazardous substances.
12. Inherently safer chemistry for accident prevention substances and the form of a substance used in a chemical process should be chosen so as to minimise the potential of chemical accidents, including releases, explosions and fires.

The greenness of a chemical transformation is generally assessed on the context of its safer application and practice in life. For proper implementation of green chemistry in practice all these principles must be well considered and followed.

19.3.1 Enzymes in green chemistry

Enzymes are biological catalysts existing in living systems. Chemically these are made from proteins and perform very specific biochemical tasks. Enzymes have emerged as preferred tools in green chemistry and increasingly used in industrial processes due to their ability to catalyse reactions with high catalytic efficiency and specificity. Enzymes have found their application in many environment friendly industrial and chemical processes today. Enzymes have applications in several aspects of our life, such as in detergents, leather, paper and pulp, textile, cosmetic, health care and pharmaceutical and food processing industries, for clinical and diagnostics research and also for organic synthesis in non-aqueous media. Enzymes are increasingly used in industrial processes also due to their ability to catalyse reactions with high catalytic efficiency and specificity. Enzymes based reactions are replacing hazardous processes and toxic chemicals/by-products and are powerful and important part of industrial sustainable chemistry, e.g., in production of herbicides and chiral drug intermediates. Today a Google search for 'enzymes in green chemistry' produces 20,00,000 results in 0.37 seconds showing their importance. The world market for enzyme production and application in chemical sector is growing bigger every day and the demand for useful biocatalysts is increasing rapidly.

19.3.2 Benefits of using enzymes in chemical synthesis

Enzymes offer following advantages over other chemical catalysts:

1. Enzymes operate under milder conditions.
2. Enzymes offer high optical purity.
3. Enzymes offer high chemo-selectivity, high region- and stereo-selectivity.

4. Enzymes are alternative for polluting and harsh chemicals.
5. These involve lesser use of hazardous and harsh chemicals as intermediates.
6. Enzymatic reactions are easy to control.
7. Enzymes produce no harmful by-products.
8. Enzymes are biodegradable and reusable.

Some examples of enzymes in chemical synthesis

Enzymatic catalysis is being seen as a promising option in the area of chemical synthesis. Enzyme based processes are today being widely applied for the production of different chemicals.

Some of the examples of usage of enzymes in green chemical processes are given below:

1. Lipases are used for the production of enantiopure alcohols and amides, nitrilases for the production of enantiopure carboxylic acids and acylases for the production of new semi-synthetic penicillins.
2. Acylase are used for synthesis of semi-synthetic penicillin, the use of nitrilase for synthesis of enantiopure carboxylic acids.
3. Enzyme laccase and manganese peroxidise catalyses oxidation of phenols.
4. In oxidative biotransformation reactions mono-oxygenases are used for biotransformation of ketones to esters or lactones, in Baeyer–Villiger oxiation, in biotransformation of terminal alkenes to (R)-epoxides and substituted phenols to hydroxybenzaldehydes.
5. Steroid hydroxylases catalyse the site specific hydroxylation of steroids and many other terpenoid, benzylic or alicyclic compounds.
6. Chloroperoxidase are used in the oxidation of salkylcysteine derivatives to yield sulphoxides.
7. The application of lipases in synthetic biotransformations in non-aqueous media consists a wide range of solvolytic reactions of the carboxyl group, like esterification, transesterification (alcoholysis), perhydrolysis and aminolysis (amide synthesis).
8. Subtilisin, an endoprotease, is used in the enantioselective hydrolysis of *N*-acylamino acid esters into the corresponding (S)-amino acids.
9. Glycosidases hydrolyse glycosidic bonds but they are also widely used as biocatalysts for carbohydrate synthesis *in vitro*, such as in the transglycosylation of lactose with *N*-acetylglucosamine.
10. In reduction of carbonyl groups such as aldehydes and ketones - Alcohol dehydrogenases (ADH) are the preferred catalysts for these reduction reactions with ketoreductases (KRED), carbonyl reductases (CR), which

are frequently used. These are generally termed as dehydrogenase (DH) and are useful in.

(a) Amino acid dehydrogenases (AaDHs) perform the reductive amination of α-ketoacids.

(b) In Reduction of acids tungsten containing aldehyde oxidoreductases (AOR) first reduce the aldehyde, which is followed by an ADH-catalysed further reduction to the alcohol.

(c) For the enantiospecific reduction of conjugated C-C double bonds, the enoate reductases are applied.

19.3.3 Green chemistry and sustainable development

The UN defines sustainable development as 'meeting the needs of present without compromising the ability of future generation.' Green chemistry focuses on how to achieve sustainability through science and technology.

This is possible by:

1. Better understanding and solving the issue of environmental pollution, many approaches and models have been developed for environmental impact assessments.
2. Some of these approaches and models have been successful in predicting impacts for selected chemicals in selected environmental settings.
3. These models have joined air and water quality aspects to point and non-point sources and have been very useful for the development of emission control and compliance strategies.
4. However, some of the approaches and models were aimed primarily at evaluating the quantity of pollutants that could be discharged into the environment with acceptable impact, but failed to focus on pollution prevention.
5. The concept of waste management and strategies such as environmentally conscious manufacturing, eco-efficient production or pollution prevention gained recognition are useful tools to achieve sustainability.

19.3.4 Recent trends in green chemistry

Chemists are practicing green chemistry thinking about reoptimisation time frames, costs, access to established chemicals and also the methods and regulatory issues. The goal is to identify the most environmentally benign conditions for chemical transformations and analyse reactions for four green criteria:

1. The environmental degradation of reagent and degradation products.
2. Solvent compatibility to green solvents and availability from natural feedstock.

3. Atom efficiency of transformation.
4. Toxicity and process safety hazards.

'Green chemistry is just good chemistry, as new science develops, it delivers economic and environmental benefits.'

Some of the current practices in the direction are listed below:

1. Pfizer improved the manufacturing process for sertraline, the active ingredient in Zoloft®. The original three step sequence was reduced to one step, producing chirally pure sertraline in much higher yield and with greater selectively and decreased raw material usage.
2. Employing ethanol as a solvent eliminated the use, distillation and recovery of four more hazardous solvents. The green approach has eliminated nearly 2 million pounds of chemical waste per year, improved safety and material handling, reduced energy and water use and doubled overall product yield.
3. Immediate challenges include reducing chlorinated solvent usage, phasing out toxic and non-economical reagents, modifying wasteful ordering and disposal habits and encouraging use of green conditions.
4. Other effective measures include replacing dichloromethane chromatography with greener solvent systems (like heptanes, ethyl acetate and ethanol) and by replacing normal-phase chromatography with reverse phase medium pressure liquid chromatography, which allows purification of highly polar compounds using aqueous media.
5. Companies like Merck have changed their chemical disposal policy and retaining chemicals with long-term stability rather than discarding all chemicals after three years, thereby saving the cost on disposal.
6. Merck has created a green chemistry e-learning course and a green chemistry tool box. Merck's green chemistry tool box includes a process mass intensity (PMI) and analytical method volume intensity total solvent consumption calculator, solvent selection guide, pathway to greener solvents and a reagent selection guide tool kit for enzymatic and catalytic reactions. This will be beneficial for industries and green chemistry practitioners.
7. Recently 'organic process research and development' has started not to accept papers with chemical solvents like chloroform, benzene and other absolutely non-green solvents.
8. These steps, hopefully, will be beneficial for the green chemistry practice and in turn, for our environment. Enzymes, no doubt will play an important part to achieve the sustainability by performing environmental benign chemical practices.

19.3.5 Future challenges

Sustainable development is now globally accepted by governments, industry and the public as a necessary goal for achieving societal, economic and environmental objectives. For this, green chemistry has a key role to play in maintaining and improving our quality of life, the competitiveness of the chemical industry and the natural environment.

This role for green chemistry is to be recognised by government, the public and the industry. In general chemicals, chemistry and chemists are actually seen by as causes of the problems. So chemists need to design such chemical processes and products which are able to preserve efficacy of the function while reducing toxicity. Chemists design new molecules and new materials. Green chemists have to make sure that the things that we make not only do what they are supposed to do, but they do it safely. In brief it's not only important how chemists make something, it's also important that what they make isn't harmful. In green chemistry making safe, non-toxic products is the major goal. As a part of the search for safer catalysts and reaction and product design and optimisation, enzymes are looked as a future solution for the environmental hazards.

20

Enzymes in clinical applications

20.1 Introduction

Enzymes have found wide and diverse applications at which enzymes increase the rate of reactions which approach to equilibrium. Enzymes play critical role in the metabolic activities of all living organisms whether humans, animals, plants or micro-organisms and are widely applied in microbial technology and their diagnosis processes. Abnormality of the enzyme metabolism system leads to a number of metabolic diseases. It is shown that many diseases associate with many components of the enzyme metabolism systems are now widely applied in clinical examinations as special markers for diseases. An interesting discovery suggesting that new roles of enzymes as a potential link that associates to prevent metabolic disorder.

Enzymes are produced by cellular anabolism, the naturally occurring biological process of making more complex molecules from simpler ones. Source organisms include bacteria, fungi, higher plants and animals. Enzymes may be extracted from a given source organism by a number of different methods. Most of the organisms that produce commercial enzymes are fungi. These organisms are molds *Rhizopus oryzae*, *Aspergillus niger*, *Rhizomucor meihei*, blights such as *Endothia parasitica* and yeasts such as *Saccharomyces* and *Candida* sp.

In cells and organisms most reactions are catalysed by enzymes, which are regenerated during the course of a reaction. These biological catalysts are physiologically important because they speed up the rates of reactions that would otherwise be too slow to support life. Enzymes increase reaction rates, sometimes by as much as one million-fold, but more typically by about one thousand fold.

Catalysts speed up the forward and reverse reactions proportionately so that, although the magnitude of the rate constants of the forward and reverse reactions are increased, the ratio of the rate constants remains the same in the presence or absence of enzyme. Since the equilibrium constant is equal to a ratio of rate constants, it is apparent that enzymes and other catalysts have no effect on the equilibrium constant of the reactions they catalyse.

Enzymes increase reaction rates by decreasing the amount of energy required to form a complex of reactants that is competent to produce reaction products. This complex is known as the activated state or transition state complex for the reaction. Enzymes and other catalysts accelerate reactions by lowering the energy of the transition state. The free energy required to form an activated complex is much lower in the catalysed reaction. The amount of energy required to achieve the transition state is lowered, consequently, at any instant a greater proportion of the molecules in the population can achieve the transition state. The result is that the reaction rate is increased.

While it is clear that enzymes are responsible for the catalysis of almost all biochemical reactions, it is important to also recognise that rarely, if ever, do enzymatic reactions proceed in isolation. The most common scenario is that enzymes catalyse individual steps of multi-step metabolic pathways, as is the case with glycolysis, gluconeogenesis or the synthesis of fatty acids. As a consequence of these lock-step sequences of reactions, any given enzyme is dependent on the activity of preceding reaction steps for its substrate.

In humans, substrate concentration is dependent on food supply and is not usually a physiologically important mechanism for the routine regulation of enzyme activity.

Many enzymes require a coenzyme which functions as group transfer reagents. Many enzymes that catalyse group transfer and other reactions require, in addition to their substrate, a second organic molecule known as a coenzyme, without which they are inactive. Coenzymes expand the repertoire of the catalytic capabilities of an enzyme far beyond those offered by the functional groups alone of the amino acids that constitute the bulk of the enzyme. Coenzymes that are tightly associated with an enzyme through either covalent bonding or non covalent forces are often referred to as prosthetic groups. Coenzymes that are freely diffusible generally serve as continually recycled carriers of hydrogen, flavin adenine dinucleotide (reduced) (FADH), hydride nicotinamide adenine dinucleotide (reduced) (NADH) and nicotinamide adenine dinucleotide phosphate (reduced) (NADPH), or chemical units such as acyl groups (coenzyme A) or methyl groups (folates), shuttling them between their points of generation and consumption.

These latter coenzymes can thus be considered as second substrates. Enzymes that require coenzymes include those which catalyse oxidoreductions, group transfer and isomerisation reactions and reactions that form covalent bonds (IUB classes 1, 2, 5 and 6). Lytic reactions, including the hydrolytic reactions catalysed by the digestive enzymes, do not require coenzymes.

Coenzymes can be classified according to the group whose transfer they facilitate based on the above concept, coenzymes may be classified as follows:

1. Coenzymes involved in transfer of groups other than hydrogen: Biotin, CoA-SH, cobamide (B12) coenzymes, folate coenzymes, pyridoxal phosphate, lipoic acid, sugar phosphates, thiamine pyrophosphate.
2. Coenzymes involved in transfer of hydrogen.

3. Nicotinamide adenine dinucleotide (oxidised) (NAD^+), Nicotinamide adenine dinucleotide phosphate (oxidised) ($NADP^+$).
4. Flavin mononucleotide (FMN), flavin adenine dinucleotide (FAD), lipoic acid, coenzyme Q.

Some enzymes are found exclusively in tissue cells. If they are found in the bloodstream, it indicates damage to that tissue, the extent of cell damage can sometimes be estimated from the magnitude of serum concentration increase above normal levels. The measurement of enzyme concentrations in blood serum is a major diagnostic tool, especially in diseases of the heart, liver, pancreas, prostate and bones.

20.2 Enzyme application in microbial technology

There are many enzymes employed in biotechnological operations. These include:

SI endonuclease: This enzyme is isolated from *Aspergilus cryzae* and it acts exclusively on ssDNA or RNA. It can break supercoiled DNA because it contains ss bubbles. It can also be used to distinguish supercoiled from both non-supercoiled, covalent circles and nicked circular DNA, both of which are resistant to the enzyme.

Restriction endonucleases: Restriction endonucleases are enzymes that cut DNA molecules at specific positions. It recognises 'foreign' (unmodified) DNA at a specific site and degrades it by internal cleavage. Most commonly used are the type II restriction enzymes, which cut within the recognition site. Some examples are EcoRI from *E. coli*, HindIII from *Haemophilus influenzae*, BamHI from *Bacillus amyloliquefaciens*, PstI from *Providencia stuartii*, SmaI from *Serratia marcescens*, Sau3A from *Staphylococcus aureus*, AluI from *Arthrobacter luteus*, TagI from *Thermus aquaticus* and HpaII from *Haemophilus parainfluenza*.

DNA polymerase I: It is primarily known as a 'repair' polymerase, which fills in single-stranded gaps. It is also involved in repair of the gaps formed on the lagging strand during replication. It also possesses both 5^{I} --- 3^{I} and 3^{I} --- 5^{I} exonuclease activity.

Klenow fragment of E. coli DNA polymerase 1: This enzyme is used for sequencing DNA using the Sanger Dideoxy System, filling the 3 recessed termini of restriction enzyme treated DNA and also used for labelling the termini of DNA fragments. The enzyme is also used for second strand cDNA synthesis in the cDNA procedures. 'Klenow fragment' is a proteolytic cleavage of DNA polymerase I. It leaves a fragment that is devoid of 5^{I} ---3^{I} exonuclease activity.

DNA polymerase III: This is the main 'replication' polymerase.

Helicase: It unwinds DNA, for example in conjugal plasmid transfer.

S1 nuclease: This enzyme degrades single-stranded DNA.

T4 DNA polymerase: This enzyme is used in the labelling of DNA fragments for use as hybridisation probes.

RNA polymerase: This enzyme synthesises RNA, using a DNA template.

Primase: This is a special RNA polymerase, which makes a short primer required for DNA synthesis.

Replicase: This is RNA-directed RNA polymerase used in replication of some RNA viruses.

Ribonuclease (RNase): This enzyme degrades RNA molecules.

Rnase H: This is a specific RNAse which cuts RNA---DNA hybrids. It is involved in replication of Col EI-like plasmid.

Exonuclease: This is an enzyme that removes nucleotides from the ends of DNA fragments. A 5^{I}----3^{I} exonuclease removes nucleotides from the 5^{I} end, while a 3^{I} ----5^{I} exonuclease removes nucleotides from the 3Iend. This enzyme recognises the terminal 5^{I}-phosphate of dsDNA for its exonuclease activity. Its primary use is the removal of protruding 5^{I} terminus from dsDNA which is needed for the terminal transferase tailing of DNA.

Exonuclease III: This enzyme is used for generating linear template DNA for the dideoxysequencing technique and generating staggered ends on dsDNA due to its 3^{I}---- 5^{I} exonuclease activity.

Ligases: It seals single-stranded gaps (nicks) in double-stranded DNA. It is also used for the formation of recombinant DNA molecules in gene cloning.

T4 DNA ligase: This enzyme catalysis the formation of a Phosphodiester bond between 3^{I}-OH and 5^{I}- phosphate ends in DNA using DNA molecules with cohesive ends as substrate.

Alkaline phosphatase: In gene cloning, this enzyme is used to remove phosphate groups from the 5^{I} end of DNA molecules and also used as a reporter gene for identification of secretion signals.

Polynucleotide kinase: It transfers a phosphate group from ATP to the 5^{I} OH end of DNA or RNA. T4 Polynucleotide kinase is an enzyme isolated from T4 infected *E. coli* and catalyses the transfer of-phosphate of ATP to a 5^{I}-OH end in DNA or RNA. It is also used for the labelling of 5^{I} termini of DNA for Maxam and Gilbert DNA sequencing and the phosphorylation of DNA lacking 5^{I} P termini.

Reverse transcriptase: This is RNA-directed DNA polymerase. It synthesises DNA (complementary DNA) using mRNA template. For example reverse transcriptase is an enzyme coded for by avian myeloblastosis virus which catalyses the synthesis of cDNA from an RNA template. It can also be used for the labelling of termini of DNA with extended 5I ends.

Topoisomerase: This is a class of enzymes that alters the conformation of DNA, for example by changing the degree of winding or super coiling.

Transposase: This enzyme catalysis the initial steps in transpositions.

Terminal transferase: This adds nucleotides to the 3^{I} end of DNA, without requiring a template strand.

Terminal deoxynucleotidyl transferase: This enzyme is isolated from calf thymus and catalyses the addition of dNTP to the 3^{I} -OH of DNA molecules. One of the primary uses of terminal transferase is the tailing of vectors and cDNA with complementary bases, thus permitting the cloning of the cDNA fragments. It can also be used for labelling of 3^{I} ends of DNA fragments.

20.3 Immobilised enzymes

Since the second half of the last century, numerous efforts have been devoted to the development of insoluble immobilised enzymes for a variety of applications. These applications can clearly benefit from use of the immobilised enzymes rather than the soluble counterparts, for instance as reusable heterogeneous biocatalysts, with the aim of reducing production costs by efficient recycling and control of the process as stable and reusable devices for analytical and medical applications as selective adsorbents for purification of proteins and enzymes as fundamental tools for solid-phase protein chemistry and as effective micro devices for controlled release of protein drugs.

This is an enzyme that is physically confined while it carries out its catalytic function. This may occur naturally, as in the case of particulate enzymes, or it may be produced artificially by chemical or by physical methods.

In the chemical methods, the enzyme is linked covalently to a support. These methods include attachment of the enzyme to a water-insoluble support, incorporation of the enzyme into a growing polymer chain, or cross linking of the enzyme with a multifunctional low molecular weight reagent. In the physical methods, the enzyme is not linked covalently to a support. These methods include adsorption of the enzyme to a water-insoluble matrix, entrapment of the enzyme within either a water-insoluble gel or a microcapsule, or containment of the enzyme within special devices equipped with semi permeable membrane. Expensive enzymes can be recovered and used again. The enzyme can also be used in a variety of configurations of bioreactors that permit continuous operation.

20.4 Enzyme immune system

Enzymes become part of the immune system (alpha-2-macroglobulin), working with it and facilitating its function. The modulation of growth factor binding properties of alpha-2-macroglobulin by enzyme therapy. Ingestion of proteinases was found to trigger the formation of intermediate forms of alpha-2-macroglobulin displaying high affinity of transforming growth factor- beta (TGF-β). Smith and others observed maximum binding of TGF-β 1–2 hr after bolus ingestion and steadily levelled off with time. They concluded that intestinal absorption of proteinase triggers the formation of TGF-β binding species of alpha 2 macroglobulin in blood mediated by this process of high concentrations of TGF-β might be reduced via enhanced clearance of alpha 2-macroglobulin TGF-β complexes. Thus, proteinase therapy may have beneficial effects in treatment of fibrosis and certain cancers accompanied by excessively high TGF-β concentrations. The oral therapy with proteolytic enzymes decreases excessive TGF-beta levels in human blood. In the study, they stated that therapy with oral proteolytic enzymes (OET) with combination drug products containing papain, bromelain, trypsin and chymotrypsin have been shown to be beneficial in clinical settings such as radio therapy induced fibrosis, bleomycin pneumotoxicity and immuno suppression in cancer, all of which are now-a-days knowned to be accompanied by excessive transforming growth factor-beta (TGF-β) production. Finally Smith and others concluded that their results support the concept that OET is beneficial in diseases characterised in part by TGF-β1 overproduction.

20.5 Enzymes as markers for disease

Some enzymes are found only in specific tissues or in a limited number of such tissues. For example, lactase dehydrogenase (LDH) has 2 different forms, called isozymes, in heart and skeletal muscle. Two forms differ slightly in amino acid composition and can be separated on the basis of charge as a result. Since LDH is a tetramer of four subunits, it too can exist in 5 different forms depending on the source of the subunits. An increase of any form of LDH in the blood indicates some kind of tissue damage. A heart attack can usually be diagnosed with certainty if there is an increase of LDH from heart. Also, there are different forms of Creatine Kinase (CK), an enzyme that occurs in the brain, heart and skeletal muscle. Appearance of the brain type can indicate a stroke or a brain tumour, whereas the heart type indicates a heart attack. After a heart attack, CK shows up more rapidly in the blood than LDH. Monitoring the presence of both enzymes extends the possibility of diagnosis, which is useful, since a very mild heart attack might be difficult to diagnose. An elevated level of the isozyme from heart in blood is a definite indication of damage to the heart tissue.

Another useful enzyme assayed is acetyl cholinesterase (AChE), which is important in controlling certain nerve impulses. Many pesticides affect this enzyme, so farm workers are often tested to be sure that they have not received inappropriate exposure to these important agricultural toxins. There are several enzymes that are typically used in the clinical laboratory to diagnose diseases. There are highly specific markers for enzymes active in the pancrease, red blood cells, liver, heart, brain, prostate gland and many of the endocrine glands. Since these enzymes are relatively easy to assay using automated techniques, they are part of the standard blood test veterinary and medical doctors are likely to need in the diagnosis and treatment/management of diseases.

20.6 Enzymes as drug or antibiotics

New antibiotics that are active against resistant bacteria have lived on earth for several billion years. During this time, they encountered in nature a wide range naturally occurring antibiotics or drugs. To survive, bacteria developed antibiotic resistance mechanism. Enzymes as drugs have two important features that distinguish them from all other types of drugs. First, enzymes often bind and act on their targets with great affinity and specificity. Second, enzymes are catalytic and convert multiple target molecules to the desired products. These two features make enzymes specific and potent drugs that can accomplish therapeutic biochemistry in the body that small molecules cannot. These characteristics have resulted in the development of many enzyme drugs for a wide range of disorders.

20.7 Enzymes in the diagnosis of pathology

The measurement of the serum levels of numerous enzymes has been shown to be of diagnostic significance. This is because the presence of these enzymes in the serum indicates that tissue or cellular damage has occurred resulting in the release of intracellular components into the blood. Hence, when a physician indicates that he/she is going to assay for liver enzymes, the purpose is to ascertain the potential for liver cell damage. Commonly assayed enzymes are the aminotransferases: alanine transaminase, ALT (sometimes still referred to as serum glutamate-pyruvate aminotransferase, SGPT) and aspartate aminotransferase, AST (also referred to as serum glutamate-oxaloacetate aminotransferase, SGOT), lactate dehydrogenase, LDH, creatine kinase, CK (also called creatine phosphokinase, CPK), gamma-glutamyl transpeptidase, GGT. Other enzymes are assayed under a variety of different clinical situations. Many enzymes are involved in the clinical diagnoses of various diseases in human and veterinary medicine. These enzymes facilitate or enhance rapid diagnoses of these diseases.

The classification of these enzymes is discussed below:

Alkaline phosphatase: Alkaline phosphatases were the earliest serum enzymes to be recognised to have clinical significance, when in the 1920s, it was discovered that they increase in bone and liver diseases . Since then, they have been the subject of more publications than any other enzyme. Alkaline phosphates are a group of isoforms which hydrolyse many types of phosphate esters, whose natural substrate or substrates are unknown. The term 'alkaline' refers to the optimal alkaline pH of this class of phosphatases *in vitro*. In both humans and animals, the major sources of ALPs are the liver, bone, kidney and placenta. In humans, it is involved in bone and hepatobiliary diseases. ALPs are also of diagnostic importance in animal diseases. Total serum ALP activity has diagnostic value in the hepatic and bone diseases in dogs and cats. It is of little value in hepatic diseases of horses and ruminants because of the broad range of reference values against which the patients' values must be compared. The range of serum ALP value in goats may be 10-fold with no evidence of hepatic damage. Values within the individual are fairly constant for sequential evaluation.

Creatine kinase: Creatine kinase isozymes are the most organspecific serum enzymes in clinical use. They catalyse the reversible phosphorylation of creatine by ATP to form creatine phosphate, the major storage form of high-energy phosphate required by muscle. Creatine kinases are found in many parts of the body like the heart, brain, skeletal muscle and smooth muscle but they have their highest specific activity in the skeletal muscle. In humans, creatine kinase is associated with myocardial infarction and muscle diseases. Increase in creatine kinase in cerebrospinal fluid has been associated with a number of disorders in dogs, cats, cattle and horses. The creatine kinase are such sensitive indicators of muscle damage that, generally, only large increases in serum activity are of clinical significance.

Alanine aminotransferase: It was formerly known as Glutamic Puruvate Transaminase, (GPT). It catalyses the reversible transamination of Lalanine and 2-oxoglutarate to pyruvate and glutamate in the cytoplasm of the cell. ALT can be found in the liver, skeletal muscle and heart. The greatest specific activity of ALT in primates, dogs, cats, rabbits and rats is in the liver. It is a well established, sensitive liver-specific indicator of damage. However, ALT in the tissues of pigs, horses, cattle, sheep or goats is too low to be of diagnostic value. It is used as an indicator of hepatopathy in toxicological studies which use small laboratory rodents as well as dogs.

Aspartate aminotransferase: It was formerly called Glutamic Oxaloacetic Transaminase, (GOT). It catalyses the transamination of L-aspartate and 2-oxoglutarate to oxaloacetate and glutamate. AST is found in skeletal muscle, heart, liver, kidney and erythrocytes and is associated with myocardial, hepatic parenchymal and muscle diseases in humans and animals. The presence of AST in so many tissues make their serum level a good marker of soft tissue but precludes its use as an organspecific enzyme. Red blood cells contain a large amount of AST which leaks into plasma before haemolysis is seen.

Sorbitol dehydrogenase (SDH): It is also called L-iditol dehydrogenase, (IDH). It catalyses the reversible oxidation of D-sorbitol to D-fructose with the cofactor NAD. The plasma activity is low in dog and horse plasma but appreciably greater in cattle, sheep and goat serum. Aside from the testes, it is found in appreciable amounts only in hepatocytes. As a result of this, an increase in plasma SDH is consistent with hepatocyte damage. SDH is liver specific in humans and all species of animals and hepatic injury appears to be the only source of increased SDH activity. Although SDH is liver specific in all species, the already established usage of ALT in dogs and cats has limited SDH as a diagnostic indicator of hepatocellular damage to horses, cattle, sheep and goats.

Lactate dehydrogenases (LDH): It catalyses the reversible oxidation of pyruvate to L(+) lactate with the cofactor NAD. The equilibrium favours lactate formation, but the preferred assay method is in the direction of pyruvate because pyruvate has an inhibitory effect on LDH. Lactate dehydrogenase has isoenzymes. LDH can be found in the heart, liver, erythrocyte, skeletal muscle, platelets and lymph nodes. In humans, it is involved in myocardial infarction, haemolysis and liver disease. LDH isoenzyme profiles were the first isoenzyme profiles used in clinical veterinary medicine in an attempt to detect specific organ damage. The introduction of more highly organspecific procedures has resulted in LDH no longer being in common use in veterinary medicine.

Cholinesterase (ChE): Serum cholinesterase activity is composed of two distinct cholinesterases. The major substrate is acetylcholine, the neurotransmitter found at the myoneural junction. Acetylcholinesterase (AChE, EC 3.1.1.7) found at the myoneural junction is the true ChE and is essential in hydrolysing acetylcholine so that the junction can be re-established and prepared for additional signals. The myoneural junction AChE is also found in red blood cells (RBC), mouse, pig, brain and rat liver. Only a small amount of AChE is found in plasma. The ChE of plasma is a pseudocholine sterase, butyl-cholinesterase (ButChE, EC 3.1.1.8), which hydrolyses butyrylcholine four times faster than acetylcholine and is also located in white matter of the brain, liver, pancrease and intestinal mucosa. Decreases in ButChE have been reported in humans with acute infection, muscular dystrophy, chronic renal disease and pregnancy, as well as insecticide intoxification.

Lipase: Serum pancreatic lipases (EC 3.1.1.3, triacylglycerol lipase) catalyse the hydrolysis of triglycerides preferentially at the 1 and 3 positions, releasing two fatty acids and a 2-monoglyceride. Lipase can be found in the pancrease and hepatobiliary tract and is involved in pancreatitis and hepatobiliary disease.

Amylase: Amylases are calcium-dependent metalloenzymes that randomly catalyse the hydrolysis of complex carbohydrates, e.g., glycogen at the

1-4-linkages. The products of this action are maltose and limit dextrins. The enzyme is a Ca_2^+ metalloenzyme which requires one of a number of activator ions such as Cl-or Br-. Amylase can be found in the salivary glands, pancrease and ovaries and is used as a diagnostic aid for pancreatitis.

Glutamyltransferase: This is a carboxypeptidase which cleaves C-terminal glutamyl groups and transfers them to peptides and other suitable acceptors. It is speculated that GGT is associated with glutathione metabolism. The major sources are the liver and kidney and are involved in hepatobiliary disease and alcoholism. Cholestatic disorders of all species examined result in increased serum GGT activity.

Trypsin: Trypsins are serum proteases which hydrolyse the peptide bonds formed by lysine or arginine with other amino acids. The pancreas as the zymogen trypsinogen, which is converted to tyrosine by intestinal enterokinase or trypsin itself, secretes them.

Glutathione peroxidases: These are metalloenzymes containing four atoms of selenium per molecule of enzyme. They catalyse the oxidation of reduced glutathione by peroxide to form water and oxidised glutathione. Because of the high concentration of selenium in glutathione peroxidases, there is a good direct correlation between the amount of red blood cell GPx activity and the selenium concentration of other organs. Other enzymes with disease diagnosis applications are acid phosphatase (ACP), found in prostate and erythrocytes and are used in diagnosis of prostate carcinoma. Aldolase (ALD), found in skeletal muscle and heart and involved in muscle disease.

Glutamate dehydrogenase (GLDH), found in the liver is used to diagnose hepatic parenchymal disease. Hydroxybutyrate dehydrogenase (HBD), which is the heart form of lactate dehydrogenase is involved in myocardial infarction Just as enzyme assay is used to diagnose diseases in humans and animals, it may also be applied to the investigation of diseases in plants. For example, it has been found that an injury (either mechanical or pathogenic) results in a marked, localised increase in the activity of glucose-6-phosphate dehydrogenase, but not of glucose phosphate isomerase, indicating diversion of glucose break down from glycolysis to the pentose phosphate pathway.

20.8 Enzymes used in immunoassay

Enzymes may also be used as an alternative to radio isotopes as markers in immunoassays have been used for the determination of a variety of proteins and hormones. The role of enzymes in immunoassay used to replace radio-isotopes as markers, since they are not hazardous to health and can be detected by techniques which are more generally available. Any enzyme with a sensitive and convenient assay procedure can be used for this purpose. Two common examples of enzyme immunoassay (EIA) procedures are enzyme-linked immunosorbent assay (ELISA) and enzyme-multiplied immunoassay test (EMIT). ELISA is a highly sensitive assay that can be used to detect either antigen or antibody. Applications of ELISA include diagnostics for non-infectious diseases involving hormones, drugs, serum components, oncofetal proteins, or autoimmune diseases, as well as diagnostics for infectious diseases caused by bacterial, viral, mycotic or parasitic organisms.

The enzymes frequently used in ELISA are Horseradish peroxidase, alkaline phosphatase and β-galactosidase. In EMIT, the activity of malate dehydrogenase is assayed by standard enzyme methodology for the detection of thyroxine by enzyme-labelled immunoassay.

20.9 Enzymes as therapeutic agents

Enzymes have found a few applications as therapeutic agents. Some examples are transfusion of fresh blood or its active components in bleeding disorders, oral administration of digestive enzymes in digestive diseases (e.g., cystic fibrosis), administration of fibrinolytic enzymes (e.g., streptokinase) to recanalise blood vessels occluded by blood clots (thrombi) in thromboembolic disorders (e.g., pulmonary embolism, acute myocardial infarction), treatment of selected disorders of inborn errors of metabolism (e.g., *Gaucher's disease*), and cancer therapy (e.g., L-asparaginase in acute lymphocytic leukemia). For enzymes to be therapeutically useful, they should be derived from human sources to prevent immunological problems. Although enzymes derived from human blood are readily obtainable, enzymes derived from tissues, which would be particularly useful in the treatment of inborn errors of metabolism, are difficult to obtain in adequate quantities. Transport of specific enzymes to target tissues is also a problem, but some recent advances and commercial applications (e.g., propagation of human tissue culture cell lines, isolation and cloning of specific genes) have the potential of overcoming these difficulties. Such techniques have been used in the production of peptide hormones such as somatostatin and insulin, interferon, and tissue plasminogen activator. In the treatment with enzymes or proteins, covalent attachment of an inert polymer polyethylene glycol (PEG) provides many therapeutic benefits.

These include slowing the clearance, diminished immunogeneity, prevention of degradation, and binding to antibody. PEG-enzyme therapy is used in the treatment of immunodeficiency disease caused by adenosine deaminase deficiency and PEG-interferon alfa complex (peginterferon alfa-2a) is used in treatment of chronic hepatitis C infection. Ultimately, however, when a gene is cloned, techniques will be developed to incorporate it into the genome of persons lacking the gene or having a mutated gene.

Section V

Enzymes in nanoparticles

21

Enzymes in biosynthesis of nanoparticles

21.1 Introduction

While a large number of microbial species are capable of producing metal nanoparticles (NPs), mechanism of nanoparticle biosynthesis is very important. The metabolic complexity of viable micro-organisms complicates the analysis and identification of active species in the nucleation and growth of metal NPs. Strategies such as enzymatic oxidation or reduction, sorption on the cell wall and in some cases, subsequent chelating with extracellular peptides or polysaccharides have been developed and used by micro-organisms. Some species can control the membrane transport of heavy metals towards, or their active efflux from, the cell. Metal ion resistance via transport and passive mechanisms leading to extracellular precipitation is more characteristic for prokaryotes.

This chapter discusses several enzymes which play an important role in nanoparticle synthesis by micro-organisms. Variety of techniques for making of metal nanoparticles are chemical recovery using regenerative materials, aerosol technique, electrochemical deposition, photochemical recovery and laser exposure, etc. But as the term of green nanotechnology has emerged that a lot of attention has attracted and includes a wide range of processes that reduce or eliminate toxic substances to restore environment. Green nanotechnology also seek more effective alternatives for energy production (e.g., solar and fuel cells). In green nanotechnology, for the synthesis of nanoparticles micro-organisms are used. It is well known that many micro-organisms, aggregate inorganic material within or outside the cell to form nanoparticles. However, large number of microbial species are capable of producing metal NPs, the mechanism of nanoparticle biosynthesis is very important. Many things about the biochemical and molecular mechanism of these processes remain unknown and should be revealed. In fact, the biochemical mechanisms referred to finding materials like enzymes, which may mediate the biosynthesis mechanism. The studies of the enzyme structure and the genes which code these enzymes may help improve our understanding of how metal nanoparticle synthesis is performed. Improvements in chemical composition, size and shape and dispersity of generated nanoparticles could allow the use of nanobiotechnology in a variety of other applications. Cell walls and cell wall proteins are likely to play an important role in the reduction of metal ions. This chapter discusses the importance of enzymes in several micro-organisms in nanoparticle synthesis process.

21.2 Bioreduction by oxidoreductase enzymes

One mechanism of metal nanoparticles biosynthesis by micro-organisms is bioreduction. In microbial bioreduction processes, myriads of proteins, carbohydrates and biomembranes are involved. Nanoparticles are formed on cell wall surfaces and the first step in bioreduction is the trapping of the metal ions on this surface. This probably occurs due to the electrostatic interaction between the metal ions and positively charged groups in enzymes present at the cell wall. This may be followed by enzymatic reduction of the metal ions, leading to their aggregation and the formation of nanoparticles. The microbial cell reduces metal ions by use of specific reducing enzymes like NADH-dependent reductase or nitrate dependent reductase.

21.2.1 Oxidoreductase in yeast

In the yeast the membrane bound (as well as cytosolic) oxidoreductases and quinones might have played an important role in the process. The oxidoreductases are pH sensitive and work in alternative manner. At a lower value of pH, oxidase gets activated while a higher pH value activates the reductase. Along with this, a number of simple hydroxy/methoxy derivatives of benzoquinones and toluquinones are elaborated by lower fungi (especially *Penicillium* and *Aspergillus* species). Yeast might be treasuring any other such quinine because it belongs to the same class of fungi thereby facilitating the redox reactions due to its tautomerisation. The transformation seems to be negotiated at two distinct levels, at the cell membrane level immediately after addition of the $TiO(OH)_2$ solution which triggers tautomerisation of quinones and low pH sensitive oxidases and makes molecular oxygen available for the transformation. Such a stress generated response had earlier been suggested in case of *Candida glabarata* cell, challenged with cadmium ion in form of elaboration of an enzyme phytochelatin synthase and a protein HMT-1 which effectively aborted the CdS nanocrystals from Cytosol. Once entered into the cytosol, the $TiO(OH)_2$ might have triggered the family of oxygenases harboured in the endoplasmic reticulum (ER), chiefly meant for cellular level detoxification through the process of oxidation/oxygenation. Taking use of the above-mentioned facts was earlier reported synthesis of metallic selenium, cadmium, silver, titanium, as well as antimony oxide.

21.2.2 NADH-dependent reductase

Fusarium oxysporum

Protein assays indicate that an NADH-dependent reductase, is the main responsible factor of biosynthesis processes. This reductase gains electrons from NADH and oxidises it to NAD^+. The enzyme is then oxidised by the simultaneous reduction of metal ions. The enzymatic route of *in vitro* synthesis of silver hydrosol of 10–25 nm using α-NADPH-dependent nitrate reductase (44 kDa) from *F. oxysporum* with capping peptide, phytochelatin was demonstrated recently. The mechanistic aspect was explained by Duran and others that apart from enzymes, quinine derivates of napthoquinones and anthraquinones also act as redox centres in the reduction of silver nanoparticles. A similar finding was also reported in the reduction of gold (III) chloride to metallic gold by α-NADPH-dependent sulphite reductase of molecular mass of 35.6 kDa and phytochelatin. A dimeric hydrogenase enzyme (44.5 kDa and 39.4 kDa) of *F. oxysporum* that showed optimum activity at pH 7.5 and 38°C passively reduced H_2PtCl_6 to platinum nanoparticles was also reported. To date, only very few reports have been documented on optimisation in biological process, a 29-kDa 'gold shape-directing protein (GSP)' present in the extract of green algae. Chlorella vulgaris was used in the bioreduction and in the synthesis of shape-and size-controlled distinctive triangular and hexagonal gold nanoparticles. With increase in the concentration of GSP produced gold plates with lateral sizes up to micrometers. Such mechanistic components should be unraveled for efficient biological processes.

Aspergillus flavus

The fungus, *Aspergillus flavus* also resulted in the accumulation of silver nanoparticles on the surface of its cell wall when incubated with silver nitrate solution. Extracellularly produced nanoparticles were stabilised by the proteins and reducing agents secreted by the fungus. A minimum of four high molecular weight proteins released by the fungal biomass have been found in association with nanoparticles. One of these was strain specific NADH-dependent reductase. However, emission band produced by fluorescence spectra indicate the native form of these proteins present in the solution as well as bound to the surfaces of nanoparticles. Further, the reduction of metal ions and surface binding of the proteins to the nanoparticles did not compromise the tertiary structure of the proteins.

21.2.3 Nitrate/nitrite reductase

Fusarium oxysporum

One other important enzyme that is responsible for this reduction in some micro-organisms is nitrate dependent reductase. In *Fusarium oxysporum,* this enzyme is conjugated with an electron donor (quinine), reduces the metal ion and changes it to elemental form. In the case of rapid extracellular synthesis, because the reduction happens in very few minutes, complex electron shuttle materials may be involved in the biosynthesis process.

Enterobacteriaceae

The culture supernatants of *Enterobacteriaceae* (*Klebsiella pneumonia*, *E. coli* and *Enterobacter cloacae*) also rapidly synthesised silver nanoparticles by reducing Ag^+ to Ag^0. With the addition of piperitone, silver ion reduction was partially inhibited, which showed the involvement of nitroreductase enzymes in the reduction process.

21.2.4 Sulphate and sulphite reductase

Rhodobacter sphaeroides

In the formation of ZnS nanoparticles by *Rhodobacter sphaeroides* (Fig. 21.1) a series of reductase have serious roles: First, soluble sulphate enters into immobilised beads via diffusion and later is carried to the interior membrane of *R. sphaeroides* cell facilitated by sulphate permease. Then, the sulphate is reduced to sulphite by ATP sulphurylase and phosphoadenosine phosphosulphate reductase and next sulphite is reduced to sulphide by sulphite reductase.

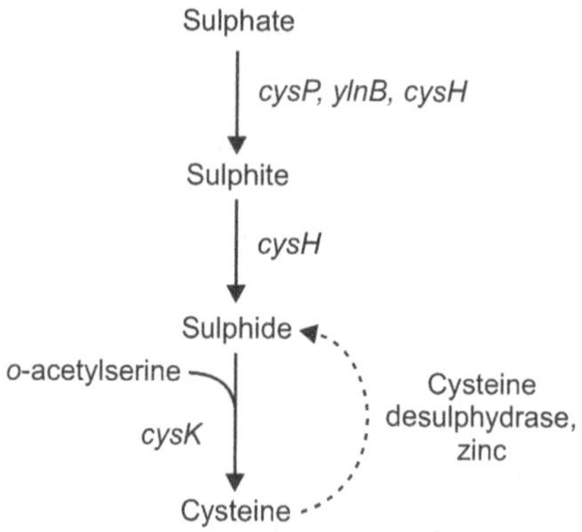

Figure 21.1: Sulphate assimilation action of *Rhodobacter sphaeroides*. The enzymes present in *Rhodobacter sphaeroides* are indicated by the corresponding genes: *cysP*, sulphate permease, *ylnB*, ATP-sulphurylase, *cysH*, phosphoadenosine phosphosulphate reductase, *cysH*, sulphite reductase, *cysK*, *o*-acetylserine synthase.

The sulphide reacts with *o*-acetyl serine to synthesis cysteine via, *o*-acetyl serine thiolyase and then cysteine produces S^{2-} by a cysteine desulphydrase in presence of zinc. After this process, S^{2-} reacts with the soluble zinc salt and the ZnS nanoparticles are synthesised. Finally, ZnS nanoparticles are discharged from immobilised cells to the solution. In the synthetic process, the particle size is controlled by the culture time of the *R.sphaeroides* and simultaneously the immobilised beads act on separating the ZnS nanoparticles from the rhodobacter sphaeroides.

Cyanobacteria

The mechanisms of gold bioaccumulation by cyanobacteria (Plectonema boryanum UTEX 485) from gold (III) - chloride solutions have documented that interaction of cyanobacteria with aqueous gold (III) - chloride initially promoted the precipitation of nanoparticles of amorphous gold (I)-suphide at the cell walls and finally deposited metallic gold in the form of octahedral (III) platelets near cell surfaces and in solutions. Adding further to the mechanism, a sulphate-reducing bacterial enrichment was used to destabilise gold(I) thiosulphate complex to elemental gold and proposed that this could occur by three possible mechanisms involving iron sulphide, localised reducing conditions and metabolism.

The interaction of *P. boryanum* UTEX485 with $Au(S_2O_3)_2^{-3}$ promoted the accumulation in membrane vesicles with 10–25 nm in size in cubic morphology, clustering inside the cell with ≤10 nm and precipitated in solution with <10–25 nm along with admixed AuS nanoparticles of ≤10 nm. But in the presence of $AuCl_4^-$ precipitation was resulted in octahedral gold platelets of 1–10 μm in solution and ≤10 nm inside the bacterial cells.

21.3 Hydrolases in fungi

Proteins have been implicated in nanoparticle formation in a number of different studies found that during the formation of zirconia nanoparticles, the fungus secreted proteins capable of extracellular hydrolysing compounds with zirconium ions and this was confirmed in subsequent studies with silica and titania. It was found that their fungus was also capable of hydrolysing metal halide precursors under acidic conditions. The studies indicated that the proteins involved in the reduction of metal nanoparticles were cationic proteins with molecular weights of around 21–24 kDa also suggested that a cationic protein of around 55 kDa found in extracellular extracts of *Verticillium* sp., might be responsible for the hydrolysis of $[Fe(CN)_6]_3$ and $[Fe(CN)_6]_4$ found proteins bound to the nanoparticle surface and the presence of sodium (S) atoms around the silver nanoparticles was taken to suggest an association between nanoparticles and fungal proteins.

21.3.1 Cysteine desulphydrase in *Rhodopseudomonas palustris*

This enzyme belongs to the family of lyases, specifically the class of carbon-sulphur lyases. The systematic name of this enzyme class is D-cysteine sulphide-lyase. This enzyme participates in cysteine metabolism. A simple route for the synthesis of cadmium sulphide nanoparticles by photosynthetic bacteria *Rhodopseudomonas palustris* has been demonstrated. The purified solution yielded the maximum absorbance peak at 425 nm due to CdS particles in the quantum size regime. Transmission electron microscopic analysis of the samples showed a uniform distribution of nanoparticles, having an average size of 8.01±0.25 nm and its corresponding electro diffraction pattern confirmed the face-centered cubic (fcc) crystalline structure of cadmium sulphide. Furthermore, it was observed that the cysteine desulphydrase producing S^{2-} in the *R. palustris* was located in cytoplasm and the content of cysteine desulphydrase depending on the growth phase of cells was responsible for the formation of CdS nanocrystal, while protein secreted by the *R. palustris* stabilised the cadmium sulphide nanoparticles. In addition, *R. palustris* was able to efficiently transport CdS nanoparticles out of the cell.

21.4 Glutathione

The major molecules that contribute to the detoxification mechanisms in yeast cells are glutathione (GSH) and two groups of metal-binding ligands: metallothioneins and phytochelatins, GSH with the structure *c*-Glu-Cys-Gly is an important tripeptide involved in various metabolic processes in bacteria, yeasts, plants and animals. The unique redox and nucleophilic properties classify this compound as a detoxificator, actively taking part in the bioreduction and defense against free radicals and xenobiotics. Metallothioneins are low molecular-weight and it classified according to the arrangement of these residues. Although the model has been created on the basis of detoxification of Cu^{2+} ions, it is also accurate for other metals in yeast.

There is evidence that metallothioneins play a similar role in some plants and yeast species, e.g., in *S. cerevisiae* and *C. glabrata*. GSH is also a structural unit in phytochelatin molecules, which is one of its major functions. Although, initially, they were described as cadmium binding peptides, phytochelatin formation is induced by a large number of elements such as Cd^{2+}, Pb^{2+}, Zn^{2+}, Sb^{3+}, Ag^{+}, Ni^{2+}, Hg^{2+}, $HAsO_4^{2-}$, Cu^{2+}, Sn^{2+}, SeO_3^{2-}, Au^{+}, Bi^{3+}, Te^{4+} and W^{6+}, when supplemented to the medium. Phytochelatins have the general structure (*c*-Glu-Cys) *n*-Gly, where $n = 2$–11 and a multitude of structural variants has been described in the scientific literature. The enzyme phytochelatin synthase or *c*-Glu-Cys dipeptidyl transpeptidase (EC 2.3.2.15) catalyses the reaction of transpeptidation of *c*-Glu-Cys dipeptide from a GSH molecule to a second molecule of GSH, resulting in phytochelatin PC2, or to a phytochelatin molecule, resulting in an $n + 1$ oligomer. Phytochelatin synthesis begins within minutes after exposing yeast cells to cadmium ions and is regulated by enzyme activation in the presence of metal ions.

The best activators are cadmium ions, followed by ions of Ag, Bi, Pb, Zn, Cu, Hg and Au. In yeasts, Cd-phytochelatin complexes are formed in the cytosol but accumulate in vacuoles. Detailed studies of the fission yeast *S. pombe* revealed that nearly the whole cadmium and phytochelatin amounts are located in vacuoles. Compared to metallothioneins, phytochelatins feature many advantages, derived from their unique structure and especially the repeated *c*-Glu-Cys units. For example, they have better metal-binding capacity. In addition phytochelatins can incorporate large amounts of inorganic sulphur, resulting in increased capacity of these peptides to bind cadmium. For crystal lattice of CdS consists of 85 CdS pairs covered by 30 (*c*-Glu-Cys)*n*-Gly peptides, with $n = 3$–5. In the presence of heavy metal stress, yeast cells increase cellular pools of glutathione and glutathione-like compounds called phytochelatins. The resulting metal thiolate complex formation neutralises the toxicity of heavy metal ions and traps them inside the cell. Sulphide anions are readily incorporated into these cadmium- glutathione complexes, resulting in the formation of nanocrystals.

Thus, today, nano metal particles have drawn the attention of scientists because of their extensive application to new technologies in chemistry, electronics, medicine and biotechnology. Beside many physical and chemical methods which have been developed for preparing metal nanoparticles, nanobiotechnology also serves as an important method in the development of clean, nontoxic and environmentally friendly procedures for the synthesis and assembly of metal nanoparticles. This new biotechnological method has important advantages in comparison to conventional methods. For instance, it is an easier and cheaper procedure. To be utilised in different scientific fields, biological synthesis requires an understanding of the biochemical and molecular mechanisms of the reaction for obtaining better chemical composition, shape, size and monodispersity.

As already discussed in nanoparticle biosynthesis, many enzyme such as redutases, syntases and hydrolases, etc., are important and describe specific genes and characteristisation of enzymes involved in the biosynthesis of nanoparticles is also required. Therefore, a complete knowledge of the molecular mechanisms involved in the microbial synthesis of nanoparticles is neseccery to control the size, shape and crystallinity of nanoparticles. In regard with nanoparticle biosynthesis considerable advantages, with improvement of these methods they could be the leading large scale production method for nanoparticles in future.

22

Nanoparticles in enzyme immobilisation

22.1 Introduction

Macro, micro and nanosized chitosan particles are suitable as carriers for enzyme immobilisation. Chitosan nanoparticles due to their highest specific surface area are much proper for immobilisation of higher amount of enzymes. As compared to chitosan macro and microparticles, higher activity values of immobilised enzyme onto chitosan nanoparticles is explained again by better distribution of the enzyme onto the larger surface area of nanoparticles. Higher magnetic property of chitosan nanoparticles reduces self aggregation of immobilised enzymes onto them and increases the stability of immobilised enzymes. Further studies will be needed to explore the kinetic of immobilised enzymes onto chitosan nanoparticles. The immobilisation of enzymes onto nanoparticles and the subsequent attachment of the nanoparticles onto an electrode is also an attractive alternative in biosensor researches and developments, especially in the case of magnetic nanoparticles which can be removed from the electrode by the action of magnetic fields.

Enzymes are proteins that catalyse chemical reactions. Unlike more traditional organic and inorganic catalysts, enzymes are large and fragile molecules. Instability and sensitivity to process conditions are disadvantages of using soluble enzymes. Therefore, application of solid phase biocatalysts has become more and more important during the last decades. Immobilised enzymes have a great importance in industrial bioprocesses especially in food, nutritional and pharmaceutical technologies. There are several reasons for using an enzyme in an immobilised form. In addition to more convenient handling of the enzyme, it provides for its facile separation from the product. It helps to prevent the contamination of the substrate with enzyme/protein or other compounds, which decreases purification costs. Immobilisation also facilitates the efficient recovery and reuse of costly enzymes, with longer half lives and less degradation, in successive batches, or the process can eventually be carried out in a continuously operating reactor.

There are several methods used to immobilise enzymes and three of the most common methods are physical adsorption, entrapment (encapsulation) and cross-linking or covalently binding to a support. For practical purposes, carrier beads with size falling into millimeter range are mainly used. However, more and more results are reported on immobilisation of enzymes onto microparticles possessing high specific surface area and numerous active sites available for the enzyme molecules to be fixed. Moreover, because of the smaller size of the support particles, the internal diffusion hindrance diminishes. Hindrance can be described as a function of solute size and geometric properties of the porous network. Diffusion of macro-molecules in porous structures is hindered in comparison to diffusion in free solution. Various supports have been used for enzymes immobilisation such as synthetic organic polymers, biopolymers, hydrogels, smart polymers and inorganic supports. A variety of biopolymers, mainly water insoluble polysaccharides such as cellulose, starch, agarose and chitosan and proteins such as gelatin and albumin have been widely used as supports for immobilising enzymes.

Chitosan with some primary amino groups is derived by deacetylation of chitin. Its pK_a is about 6.3. At lower pH solutions ($<pK_a$), most of the amino groups are protonated, making chitosan a water soluble polyelectrolyte. When the pH is higher than pK_a, the amino groups are deprotonated and chitosan becomes insoluble. Because of its excellent film forming ability, biocompatibility, nontoxicity, high mechanical strength, cheapness and a susceptibility to chemical modifications, chitosan has been extensively used for immobilisation of enzymes. In recent years, nanotechnology has shown a significant attraction to the preparation of immobilised enzymes. Nanomaterials due to their small size and large surface area to volume ratio have special characteristics, which make them favourable for enzyme immobilisation.

22.2 Immobilised enzyme supports

Immobilisation is often the key to optimising the operational performance of an enzyme in industrial processes, particularly for use in non aqueous media. The immobilisation of enzymes has been a growing field of research, because it has allowed enzymes to be easily reused multiple times for the same reaction with longer half lives and less degradation and has provided a straight forward method of controlling reaction rate as well as reaction start and stop time. The properties of immobilised enzymes are governed by the properties of both the enzymes and the support materials. The interaction between the two lends an immobilised enzyme specific physico-chemical and kinetic properties that may be decisive for its practical application and thus, a support judiciously chosen can significantly enhance the operational performance of the immobilised system.

Suitable materials used as a carrier should have chemical, physical and biological stability during processing, as well as in the reaction conditions, sufficient mechanical strength, specially for its utilisation in reactors and industries, should be nontoxic both for the immobilised enzyme/bioparticle, as well for the product, also should have adequate function groups for binding biocatalyst and high loading capacity. Other criteria, such as structural characteristics (porosity, swelling, compression, material and mean particle behaviour), as well as possibility for microbial growth, biodegradability, high affinity to proteins, solubility, regenerability and ease of preparation in different geometrical configurations that provide the system with permeability and surface area are more application specific. Chemical structure of carrier material determines interaction with enzymes. If the support material is highly porous, pore size and pore size distribution will play an important role in determining the immobilised enzyme properties. A small pore size can cause diffusion limitation resulting in structural rearrangement of the enzymes and subsequent inactivity. However, for very large pore sizes, enzymes can cluster together and thus lose activity. Enzymes immobilisation onto carriers has been extensively studied and applied in many fields, such as biocatalysts medical devices, drug delivery systems and biosensor. Several carriers such as synthetic organic polymers (e.g., Eupergit C and polyurethane), biopolymers (e.g., alginate), hydrogels (e.g., polyvinyl alcohol), smart polymers (e.g., poly-*N*-isopropylacrylamide) and inorganic supports (e.g., alumina, silica and zeolites) have been used in immobilisation of enzymes. Biopolymers are one of the most applied supports for immobilising enzymes, includes two types of mainly water insoluble proteins such as gelatin and albumin and polysaccharides such as starch, alginate and chitosan.

22.3 Enzyme immobilisation techniques

Immobilisation often stabilises structure of the enzymes, thereby allowing their applications even under harsh environmental conditions of pH, temperature and organic solvents and thus enable their uses at high temperatures in non aqueous enzymology and in the fabrication of biosensor probes. Different methods for the immobilisation of enzymes have been critically reviewed.

These methods are divided into two main categories namely, reversible and irreversible immobilisation. Adsorption, disulphide bonding and chelation or metal binding are the reversible methods for enzyme immobilisation.

Adsorption: Adsorption is the elementary and probably the simplest method of immobilisation. This method is based on weak forces (e.g., van der Waals), however still enabling an efficient binding process. A wide range of both organic and inorganic materials can be used as a support in this method.

Disulphide: Disulphide binding may be seen as a variation of covalent bonding, as there are stable covalent bonds formed between activated support and free thiol group (e.g., on cysteine) in the biocatalyst.

Chelation: Chelation is based on the ability of the side chains of certain amino acids (e.g., histidine, tryptophan, tyrosine, cysteine and phenylalanine) to substitute weakly bonded ligands in the metal ions that have been immobilised by chelating group covalently bound to a solid support. Schematics of reversible enzyme immobilisation methods are shown in Fig. 22.1.

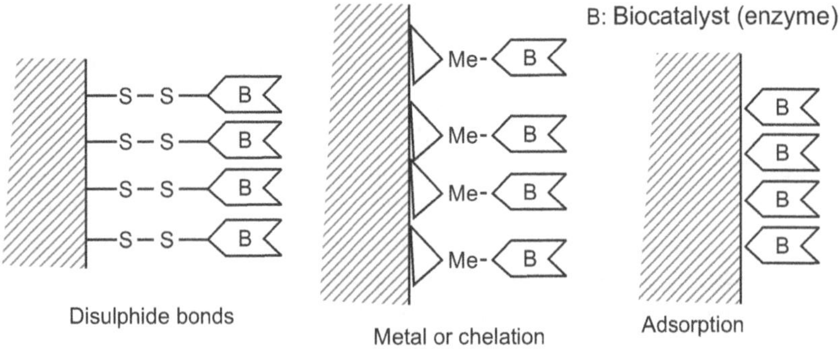

Figure 22.1: Reversible methods of enzyme immobilisation.

Covalent bonding, entrapment, encapsulation and cross-linking, fall into the irreversible enzyme immobilisation methods. Covalent bonding is one of the most widely used methods for enzymes immobilisation. Popularity of this approach is mainly connected with the stability of the bonds formed between the enzyme and the support, which prevents enzyme release into the environment. Entrapment is based on incorporating enzymes into the lattices of a semi permeable gel or enclosing the enzymes in a semi permeable polymer membrane. Encapsulation is similar to entrapment. In this process, enzymes are restricted by the membrane walls (usually in a form of a capsule).

Cross-linking method is widely used to immobilise enzymes and is based on intermolecular cross-linking of enzymes by bi-functional or multi functional reagents. In fact, this method does not require a support and this is the advantage of this method. Immobilisation of an enzyme on a carrier often leads to the loss of more than 50% native activity, specially at high enzyme loadings. The design of carrier bound immobilised enzymes also relies largely on laborious and time consuming trial and error experiments, because of the lack of guidelines that link the nature of a selected carrier to the performance expected for a given application. Schematics of irreversible enzyme immobilisation techniques are shown in Fig. 22.2.

22.4 Chitosan

Chitin is a copolymer of *N*-acetyl-β-glucosamine and D-glucosamine units linked with β-(1-4) glycosidic bond, where *N*-acetyl-d-glucosamine units are predominant in the polymeric chain. The deacetylated form of chitin refers to chitosan. Basically, the process consists of deproteinisation of the raw shell

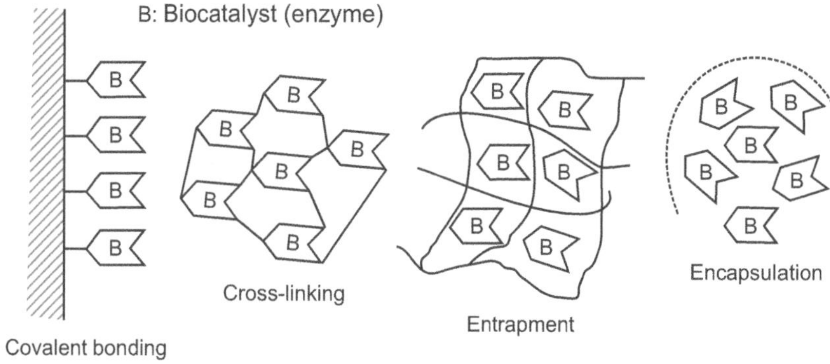

Figure 22.2: Irreversible methods of enzyme immobilisation.

material with a dilute NaOH solution and decalcification with a dilute HCl solution. To result in chitosan, the obtained chitin is subjected to *N*-deacetylation by treatment with a 40–45% NaOH solution, followed by purification procedures. Thus, production and utilisation of chitosan constitutes an economically attractive means of crustacean shell wastes disposal sought worldwide.

These renewable resources are found in many naturally occurring organisms (e.g., fungi and yeast) and is the principal component in the exoskeleton of sea crustaceans such as crabs, shrimps, lobsters and krills. Chitosan represents long chain polymers having molecular mass up to several million Daltons. Chitosan is relatively reactive natural biopolymer due to presence of reactive amino and hydroxyl functional groups in its structure and can be produced in various forms such as powder, gels and films, sponges, intragastric floating tablets and especially spherical micro and nanoparticles. If degree of deacetylation and molecular weight of chitosan can be controlled, then it would be a material of choice for developing micro and nanoparticles. Chitosan has ability to control the release of active agents and avoid the use of hazardous organic solvents while fabricating particles since it is soluble in aqueous acidic solution. It is a linear polyamine containing a number of free amine groups that are readily available for cross-linking and its cationic nature allows for ionic cross-linking with multivalent anions. Chitosan has muco-adhesive character, which increases residual time at the site of absorption. Chemical structures of chitin and chitosan are shown in Fig. 22.3.

22.5 Applications of chitosan

Chitosan due to its unique physico-chemical and biological properties is an attractive material for use in various applications. These properties include: biodegradability, lack of toxicity, antibacterial and antifungal effects, wound healing acceleration and immune system stimulation. Because of excellent

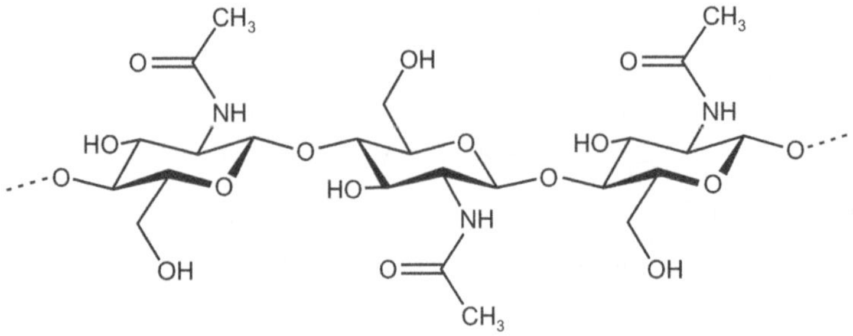

(a)

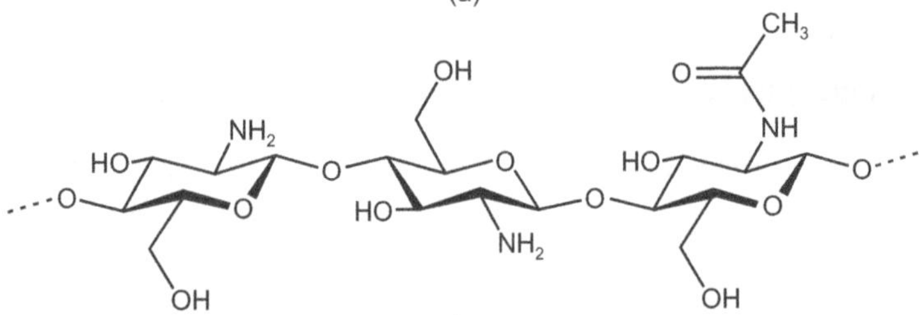

(b)

Figure 22.3: Chemical structure of (a) chitin and (b) chitosan.

biological and physico-chemical properties of chitosan, it has the ability to bind to particular materials including cholesterols, fats, proteins, metal ions and even tumour cells. This allows chitosan to be used as a chelating agent in various applications.

Over the last decade, a number of potential products based on chitosan have been increasingly produced and applied in several areas such as: wastewater treatment, due to its high adhesive and insolubility properties (e.g., removal of heavy metal ions and membrane purification processes), food industries (e.g., anticholesterol and fat binding, preservative, packaging material and animal feed additive), agriculture (e.g., seed and fertiliser coating, controlled agrochemical release), pulp and paper industries, because of its wet strength to paper (e.g., surface treatment, photographic paper) and cosmetics and toiletries due to its fungicidal and fungistatic properties (e.g., moisturiser, body creams, bath lotion). Fibres made of chitosan are useful as absorbable sutures and wound dressing materials. Chitosan has also important applications in photography due to its resistance to abrasion, optical characteristics and film forming ability. Due to its unique molecular structure, chitosan has an extremely high affinity for many classes of dyes, including disperse, direct, reactive, acid, vat, sulphur and naphthol dyes.

Chitosan ability to development of drug delivery systems is one of the most well known of its characteristics. Different types of chitosan made drug carriers have been conceived for various administration routes, such as oral, nasal, transdermal, parenteral, vaginal, cervical and rectal due to its biocompatibility, biodegradability and ecological safety. Tissue engineering is an important therapeutic strategy for present and future medicine. Recently, functional biomaterial researches have been directed towards the development of improved scaffolds for regenerative medicine. Chitosan and its derivatives are promising candidates as a supporting material for tissue engineering applications owing to their porous structure, gel forming properties, ease of chemical modification, high affinity to *in vivo* macro-molecules and so on. Moreover, it is one of the most promising immobilisation matrices with excellent characteristics as a support matrix for enzyme immobilisation.

22.6 Immobolised enzymes onto chitosan

Chitosan has shown favourable biocompatibility characteristics, as well as the ability to increase membrane permeability. Moreover, it is one of the most promising immobilisation matrices due to an excellent membrane forming ability, good adhesion, low cost, low immunogenicity and nontoxicity, high mechanical strength and hydrophilicity as well as the improvement of stability. These properties have prompted extensive applications of chitosan as a matrix for enzyme immobilisation. In fact, polycationic nature of chitosan and presence of large primary amine groups in its chemical structure (Fig. 22.3b) make it potentially useful in immobilisation of various enzymes. Results of several studies have been shown that after selection of suitable enzyme immobilisation method, chitosan reactive amino and hydroxyl groups offer a good enzyme coupling efficiency.

Chitosan has been extensively used in immobilisation of various enzymes groups because of its many characteristics like improved resistance to chemical degradation, avoiding disturbance of metal ions to enzyme and antibacterial property. Simsek-Ege and Vaillant used chitosan to immobilise carbonic anhydrase and pectin lyase, respectively, which both of them are classified in lyase enzymes group.

Applications of chitosan in immobilisation of oxidoreductase, hydrolase and transferase enzymes are summarised in Tables 22.1 to 22.3, respectively.

Table 22.1: Some of immobilised oxidoreductases onto chitosan.

Enzyme	*Support*	*Type of immobilisation*
Alanine dehydrogenase	Chitosan beads	Covalent binding with G
Alcohol dehydrogenase	Chitosan beads	Covalent binding with G
Alcohol oxidase	Chitosan beads	Covalent binding with G
Catalase	F3GA attached-chitosan beads	Adsorption on support
Galactose oxidase	Chitosan membrane	Covalent binding with G

(Cont'd...)

Enzyme	*Support*	*Type of immobilisation*
Glucose oxidase	Chitosan capsules	Micro encapsulation
Glutamate dehydrogenase	Chitosan membrane	Covalent binding with G
Glutamate oxidase	Chitosan membrane	Covalent binding with G
Horseradish peroxidase	Chitosan membrane	Covalent binding with G
Laccase	Magnetic chitosan microspheres	Adsorption followed by cross-linking with G
Lactate oxidase	Chitosan-enzyme beads	Gel inclusion
Leucine dehydrogenase	Chitosan capsules	Micro encapsulation
Octopine dehydrogenase	Chitosan beads	Covalent binding with G
Oxalate oxidase	Chitosan powder	Adsorption followed by cross-linking with G
Putrescine oxidase	Chitosan beads	Covalent binding with G
Sulphite oxidase	Chitosan-PHEMA membrane	Adsorption on support
Uricase	Chitosan membrane	Covalent binding with agents other than G
Xanthine oxidase	Chitosan beads	Covalent binding with G

G: Glutaraldehyde

Table 22.2: Some of immobilised hydrolases onto chitosan.

Enzyme	*Support*	*Type of immobilisation*
Acid phosphatase	Chitosan beads	Adsorption on support
Alkaline phosphatase	Chitosan beads	Covalent binding with G
Alkaline protease	Chitosan powder	Covalent binding with G, adsorption on support
Aminoacylase	Chitosan-coated alginate beads	Gel inclusion
α-amylase	Chitosan beads	Covalent binding with G
β-amylase	Chitosan beads	Adsorption on support
α-l-arabino-furanosidase	Chitosan powder	Adsorption, cross-linking and covalent binding with G
Bromelain	Chitosan beads	Covalent binding with agents other than G
Candida ntarctica lipase	Chitosan-based hydrogels	Covalent binding with G
Cellulase	Chitosan beads	Adsorption
Chitinases	Chitosan beads cross-linking with G	Adsorption followed by
α-Chymotrypsin	Chitosan beads cross-linking with G	Adsorption followed by
Creatininase	Chitosan-g-polyaniline with iron oxide	Covalent binding with G, adsorption on support

(Cont'd...)

Enzyme	*Support*	*Type of immobilisation*
Creatinine deaminase	Chitosan membrane	Adsorption on support
Endo-1,4-β-Xylanase	Chitosan powder	Adsorption on support
Ficin	Chitosan beads	Covalent binding with agents other than G
β-galactosidase	Chitosan beads	Covalent binding with G
α-glucosidase	Chitosan beads	Covalent binding with G
β-glucosidase	Chitosan powder	Covalent binding with G
β-glycosidase and precipitate	Chitosan powder	Adsorption followed by cross-linking with G
Invertase	Chitosan powder and solution	Adsorption on support
Lipase	Chitosan flakes and beads	Adsorption on support
Neutral proteinase	Chitosan precipitate	Adsorption followed by cross-linking with G
5-Nucleotidase	Chitosan beads cross-linking with G	Adsorption followed by
Papain	Chitosan beads cross-linking with G	Adsorption followed by
Pectinase	Chitosan beads	Adsorption on support
Pepsin	Chitosan beads cross-linking with G	Adsorption followed by
Phospholipase A2	Chitosan beads	Covalent binding with agents other than G
Proteases	Chitosan beads	Covalent binding with G
Urease	Chitosan-triphosphate beads	Covalent binding with G
β-xylolidase	Chitosan powder and beads covalent binding with G	Adsorption on support,

G: Glutaraldehyde

Table 22.3: Some of immobilised transferases onto chitosan.

Enzyme	*Support*	*Type of immobilisation*
Cyclodextrin glycosyltransferase	Chitosan powder	Adsorption on support
Limonoid glucosyltransferase	Chitosan powder	Covalent binding with G
Nucleoside phosphorylase	Chitosan beads	Covalent binding with G
Transglutaminase	Chitosan beads	Covalent binding with G
W-transaminase	Chitosan beads	Covalent binding with G

G: Glutaraldehyde

22.7 Chitosan nanoparticle

In recent decades, nanostructured materials have attracted much attention because of their unique properties and interesting applications. Nanotechnology is the ability to work on a scale of about 1–100 nm in order to understand, create, characterise and use material structures, devices and systems with new properties derived from their nanostructures.

Reducing the particle size of materials is an effective method for improving their properties. Nanoparticles have proportionally larger surface area and consequently more surface atoms than their microscale counterpart, which in turn affects their physico-chemical, optical, catalytic and other reactive properties.

In polymer composites conjugated with nanoparticles, a uniform dispersion of nanoparticles leads to a very large matrix/filler interfacial area, which changes the molecular mobility, the relaxation behaviour and the consequent thermal and mechanical properties of the materials. In contrast to other biopolymers, chitosan is a hydrophilic polymer with positive charge that comes from weak basic groups, which gives it special characteristics from the technological point of view. Recently, chitosan nanoparticles have attracted great attention in several fields due to their unique physico-chemical and biological properties.

22.8 Chitosan nanoparticles preparation methods

Several methods have been used to prepare chitosan nanoparticulate. The selection of any of these methods depends on shape and particle size requirements. Emulsion cross-linking, emulsion-droplet coalescence, coacervation/precipitation, ionotropic gelation, reverse micelles, template polymerisation and molecular self assembly are the main preparation methods of chitosan nanoparticles.

In emulsion cross-linking method, the reactive functional amine group of chitosan crosses link with aldehyde groups of the cross-linking agent. In fact, chitosan solution is emulsified in oil phase (water-in-oil emulsion) and the aqueous droplets are stabilised using a suitable surfactant. The stable emulsion is then reacted with an appropriate cross-linking agent such as glutaraldehyde to stabilise the chitosan droplets. The nanoparticles are then washed and dried.

Ionotropic gelation method is commonly used to prepare chitosan nanoparticles. This method is based on the electrostatic interactions between the chitosan amine group and a polyanion such as tripolyphosphate. In this method chitosan is dissolved in water or in weak acidic medium. This solution is then added drop wise under constant stirring to the solutions containing other counter ions.

Due to the complexation between oppositely charged species, chitosan undergo ionic gelation and precipitate to form spherical nanoparticles. Much researches have been focused on the preparation of chitosan nanoparticles using inotropic gelation method.

In coacervation/precipitation method, the chitosan solution is spraying into sodium hydroxide, NaOH methanol or ethanediamine alkaline solutions using compressed air, which in turn originates coacervated chitosan droplets in the form of nanoparticles. Emulsion-droplet coalescence method involves both emulsion cross-linking and precipitation. A stable emulsion containing the aqueous chitosan solution in oil and a second emulsion, containing a NaOH solution, is produced. By mixing the both emulsions under high speed stir, droplets of each emulsion collide at random, coalesce and finally precipitate as small size particles.

In the reverse micelles method, a surfactant is dissolved in organic solvent to prepare reverse micelles. The aqueous phase containing the chitosan is added to this emulsion with constant vortexing and the nanoparticles forms in the core of the reverse micelles.

In template polymerisation, chitosan is first dissolved in an acrylic monomer solution under magnetic stirring. Due to the electrostatic interaction, the negatively charged acrylic monomers align along the chitosan molecules. After complete dissolution of chitosan, the polymerisation is started by adding the initiator ($K_2S_2O_8$) under stirring at 70°C. The complete polymerisation leads to the appearance of an opalescent solution, indicating the nanoparticles formation.

Molecular self assembly is based on cationic and hydrophobic properties of chitosan. This method is characterised by diffusion followed by specific association of molecules through non-covalent interactions, including electrostatic and/or hydrophobic associations.

22.9 Applications of chitosan nanoparticles

Chitosan nanoparticles are natural materials with excellent physico-chemical, antimicrobial and biological properties, which make them a superior environmentally friendly material and they possess bioactivity that does not harm humans. Due to these unique properties, chitosan nanoparticles are being used in a vast array of widely different products and applications, ranging from pharmaceutical, tissue engineering and food packaging to biosensing, enzymes immobolisation and wastewater treatment.

The potential use of chitosan nanoparticles as carriers has led to the development of many different colloidal delivery vehicles. The main advantages of this kind of systems lie in their capacity to cross biological barriers, to protect macro-molecules, such as peptides, proteins, oligonucleotides and genes from degradation in biological media and to deliver drugs or macro-molecules to a target site with following controlled release. Chitosan nanoparticles have frequently used as a controlled release drug carrier for gene transfer in artificial organs and for immune prophylaxis. In addition, chitosan nanoparticles have been used to improve the strength and washability of textiles and to confer antibacterial effects. Several researches have been demonstrated that chitosan nanoparticles are able to improve drug bioavailability, modify pharmaco-kinetics and protect the encapsulated drugs.

In tissue engineering, chitosan nanoparticles improve transmucosal permeability enhancing transport through the paracellular pathway due to good bio and muco-adhesive properties of the nanoparticles and to an induced structural reorganisation of tight junction associated proteins.

During the past decade, there was an increasing interest to develop and use bio-based active films which are characterised by antimicrobial and antifungal activities in order to improve food preservation. Chitosan based films have attracted serious attention in food preservation and packaging technology. This is mainly due to the excellent film forming and gas barrier properties of chitosan and its high antimicrobial activity against pathogenic and spoilage micro-organisms, including fungi and bacteria.

Addition of chitosan nanoparticles into the coating and film formulations, improve film tensile properties and their permeability towards water vapour and simple gases (e.g., oxygen and carbon dioxide). Electrochemical biosensor has been considered as the best choice for the *in situ* monitoring of active compounds (e.g., phenolic) by virtue of its high sensitivity, simple instrumentation, low production cost and promising response speed.

Excellent membrane forming ability of chitosan nanoparticles and their small response time and high sensitivity and stability (due to their high surface to volume ratio), low cost and hydrophilicity making them suitable for biosensor applications that are mostly concerned with working of enzymes for detection mechanisms.

Arsenic, molybdenum, lead and copper widely used in industries and are released through industrial wastewater. Sorption processes are found to be capable of adsorbing large number of metal ions from contaminated wastewater. In many studies such as adsorption of metals (e.g., copper and zinc), polymer supported nanoparticles has been prepared and used for selective removal of metal compounds and target metal contaminants. Chitosan due to its high hydrophilicity and presence of a large number of hydroxyl and amino groups with high activity as adsorption sites is found to be more efficient for the removal of metals such as uranium, copper, vanadium and molybdenum. Now-a-days, chitosan nanoparticles have been used effectively in wastewater treatment to remove toxic metal ions such as arsenic. Nano carriers can be effectively controlled by the application of nanotechnology. The catalytic efficiency and stability property of immobilised enzymes can be greatly improved. Furthermore, immobilisation rate of enzyme can be improved using chitosan nanoparticles.

22.10 Chitosan nanoparticles as enzyme immobilisation support

The results of immobilisation, including the performance of immobilised enzymes, strongly depend on the properties of supports, which are usually referred to as material types, compositions and structures. Chitosan is known as an ideal support material for enzyme immobilisation because of its many characteristics like improved resistance to chemical degradation and avoiding disturbance of metal ions to enzyme.

Enzymes have been immobilised onto chitosan supports with different particle size. For practical purposes, carrier beads with size falling into millimeter range are mainly used.

Under the scale of nano, nanomaterials have characteristics, such as magnetism and large surface area to volume ratio. These characteristics are in favour of immobilisation of the enzymes. Recent studies indicated that the performance of enzyme immobilisation onto chitosan nanoparticles is higher than that of the biocatalyst immobilised onto chitosan microparticles. The result can be related to better distribution of the enzyme onto the support due to high specific surface area and numerous active functional groups available for fixing the enzyme molecules.

Activity of immobilised enzymes strongly affects by size and size distribution of the support particles. Specific surface available to bind the applied enzyme and the contacting area with the substrate during a reaction mainly depends on particle size. Therefore, a lot of efforts were made to minimise the particle size and maximise the specific surface area. In addition, the physical characteristics of nanoparticles such as enhanced diffusion and particle mobility can impact inherent catalytic activity of attached enzymes.

Biro and others indicated that the activity of β-galactosidase immobilised onto chitosan nanoparticles was higher than that of the biocatalyst immobilised onto chitosan micro and macroparticles. Research studies have also indicated that the stability of immobilised enzymes onto chitosan nanoparticles drastically affects by method of chitosan nanoparticles preparation and type of surfactant.

Immobolisation of enzymes onto biopolymers nanoparticles has shown some benefits like improving their stability to pH and temperature, resistance to proteases and other denaturing compounds, as well as an adequate environment for their repeated use or controlled release. In the last decades, immobilised enzymes onto nanoparticles have been also considered for biosensors and food packaging applications.

Applications of chitosan nanoparticles in immobilisation of some enzymes are summarised in Table 22.4.

Table 22.4: Some of immobilised enzymes onto chitosan nanoparticles.

Enzyme	*Support*	*Preparation method*
Alcohol dehydrogenase	Fe_3O_4–chitosan nanoparticles	Covalent binding with glutaraldehyde
β-d-galactosidase	Fe_3O_4–chitosan nanoparticles	Covalent binding with glutaraldehyde
β-galactosidase	Chitosan nanoparticles	Precipitation, emulsion cross-linking, ionic gelation
Glucoamylase	Fe_3O_4–chitosan nanoparticles	Ionic adsorption
Glucose oxidase	Chitosan nanoparticles	Covalent binding
Laccase	Chitosan nanoparticles	Reversed phase suspension
l-Laccase	Fe_3O_4–chitosan nanoparticles	Ionic adsorption and covalent binding
Lipase	Fe_3O_4–chitosan nanoparticles	Cross-linking with trypolyphosphate
Neutral proteinase	Chitosan nanoparticles	Ionic gelation
Pullulanase	Fe_3O_4–chitosan nanoparticles	Photochemistry in aqueous suspension
Trypsin	Linolenic acid-modified chitosan nanoparticles	Covalent binding with glutaraldehyde

Liu and others studied trypsin immobilised on linolenic acid-modified chitosan nanoparticles using glutaraldehyde as cross-linker. Their results indicated that the kinetic constant value (K_m) of trypsin immobilised on nanoparticles (71.9 mg mL^{-1}) was higher than that of pure trypsin (50.2 mg mL^{-1}). Tang and others used Chitosan nanoparticles as support to immobilise and protect activity of neutral proteinase. Their results indicated that chitosan nanoparticles could improve 13.17% of neutral proteinase activity than that of free neutral proteinase. Nakorn used glucose oxidase immobilised onto chitosan nanoparticles in biosensor for glucose determination.

The coated electrode with immobilised glucose oxidase exhibits a rapid and sensitive current response for the changes of glucose concentration in the prepared solutions and indicates the excellent electrocatalytic behaviour of the electrode. Li and others immobilised *Saccharomyces Cerevisiae* Alcohol Dehydrogenase (SCAD) to magnetic Fe_3O_4-chitosan nanoparticles. For reduction of phenylglyoxylic acid by immobilised SCAD, the kinetic analysis data indicated that the immobilised SCAD retained 48.77% activity of its original activity. Furthermore, the immobilised SCAD enhanced thermal stability and good durability in the repeated use after recovered by magnetic separations. Fang and others prepared magnetic chitosan nanoparticles and then immobilised laccase onto them. The immobilised laccase exhibited an appreciable catalytic capability (480 units · g^{-1} support) and had good storage stability and operation stability. The K_m of immobilised and free laccase for ABTS [2,2′-azino-*bis* (3-ethyl benzthiazoline-6-sulphunate)] were 140.6 and 31.1 μM in phosphate buffer (0.1 M, pH 3.0) at 37°C, respectively. Pan and others developed a novel and efficient immobilisation of β-d-galactosidase by using magnetic Fe_3O_4–chitosan nanoparticles. The immobilised β-d-galactosidase showed the same or even higher activity in wider ranges of temperature and pH than that of its free form. In addition, the immobilised enzyme could be stored for a long time with little activity loss. Furthermore, the immobilised enzyme retained 92% of its initial activity after successively utilisation for 15 cycles. Wu and others prepared magnetic Fe_3O_4–chitosan nanoparticles and immobilised lipase onto them.

The immobilisation of lipase onto the nanoparticles showed good loading ability and little loss of enzyme activity and the stability of the catalyst was very good. In fact, it only lost 12% of enzyme activity after five batches. Zhang and others prepared magnetic chitosan beads (with mean particle size of 50 ± 3 nm) and immobilised pullulanase on them. Results indicated that the kinetic constant value (K_m) of immobilised pullulanase was three times higher than that of free pullulanase. However, the thermal and operational stabilities of immobilised pullulanase were improved greatly. Kalkan and others immobilised laccase onto chitosan-coated magnetic nanoparticles by adsorption and covalent binding after activating the hydroxyl groups of chitosan with carbodiimide or cyanuric chloride. The results indicated that the immobilised enzyme retained more than 71% of its initial activity at the end of 30 batch uses. Wang and others immobilised glucoamylase onto Fe_3O_4–chitosan nanoparticles. The results from characterisation and determination remarkably indicated that the immobilised glucoamylase obtained presents excellent storage stability, pH stability, reusability, magnetic response and regeneration of supports.

22.11 Challenges in applications of nanoparticles to immobilise enzymes

Application of nanoparticles in enzyme immobilisation has been associated with two main disadvantages. The first is lower storage stability of immobilised enzyme onto nanoparticles as compared to that of the immobilised enzyme onto microparticles. This can be explained by the fact that nanoparticles are aggregated during the storage. In fact, if the nanoparticles are not stabilised, they try minimising their surface energy by clustering together. Biro and others indicated that the stability of β-galactosidase immobilised onto chitosan nanoparticles was lower than that of the biocatalyst immobilised onto chitosan micro and macroparticles. The second disadvantage of using nanoparticles to immobilise enzyme is difficult separation of them from the reaction mixture at the end of the biochemical process, due to their small size. In this case, suitable methods are necessary to apply to facilitate the separation of the catalyst particles. Therefore, the general use of nanosized catalysts in industrial biotechnology needs considerable research efforts yet.

To overcome these issues, nanosized magnetic particles have received increasing attention because of their larger specific surface area for the enzymes immobilisation, their super paramagnetic nature for the reduction of self aggregation and easy separability from the reaction mixture by the application of a magnetic field. Magnetite (Fe_3O_4), silica, Zinc Oxide (ZnO), cellulose and chitosan are some of magnetic materials and their nanoparticles have been used to immobilise of various enzymes.

Chitosan can be used as a material for magnetic carriers, since it has a variety of functional groups which can be tailored to specific application. Additionally, glutaraldehyde cross-linking is shown to be the simple and efficient method to immobilise enzymes. Glutaraldehyde can react with several functional groups of proteins and supports, such as amino groups. Each of glutaraldehydes is expected to form Schiff bases upon nucleophilic attack by the primary amino groups in enzymes and chitosan and due to the linkage formed by the Schiff base reaction, the enzyme stability is improved. Several methods have been developed to synthesise magnetic chitosan nanoparticles, such as micro-emulsion polymerisation, emulsion polymerisation and *in situ* polymerisation.

As clearly observed in Table 22.4, magnetic Fe_3O_4–chitosan nanoparticles have been used as support to immobilise the most of enzymes. In fact, magnetic Fe_3O_4 nanoparticles tend to aggregate in liquid media due to the strong magnetic dipole–dipole attractions between particles. Thus, chitosan with specific functional groups have been used as stabiliser to modify and increase their stability. In addition to, coating magnetic Fe_3O_4 nanoparticles with chitosan can protect the nanoparticles against corrosion an also offers flexibility, favourable functional groups and features (e.g., their ability to fold into a globular state or structure) for various applications.

Magnetic nano and micron sized particles with specific modifications are widely used in biomedical applications such as diagnostics, magnetic separation and purification of biomolecules. In fact, magnetic nanoparticles coated with biopolymers and immobilised with ligand were found to have promising characteristics for the application of these adsorbents in bioseparation processes, particularly in antibody, nucleic acids and enzymes purification.

Section VI

Risk assessment and safety regulation of enzymes

23

Risk assessment guidance for enzyme containing products

23.1 Introduction

Enzymes are proteins produced by all living organisms. They act as catalysts to increase the rate of chemical reactions. They are generally named after the reactions they catalyse. Amylases catalyse reactions with carbohydrates, cellulases react with cellulose, lipases with lipids (e.g., fats and oils) and proteases with proteins. Enzymes used in finished products usually 'break down' certain materials into their basic components. For instance, protease in laundry detergents breaks peptide bonds in proteins that make up food stains. Other ingredients in the product are able to remove the break down products formed from the action of enzymes easier than if the enzymes were not in the product. In a totally different application, protease enzymes might be formulated in skin care products to facilitate the removal of dry skin flakes.

Enzymes speed reactions that usually occur at a much slower rate. The remarkable property of enzymes is that they complete reactions without being destroyed, allowing a single enzyme molecule to catalyse many individual reactions. Therefore, they can be used at relatively low levels in consumer products and still contribute to product performance.

The first step in an enzymatic reaction is substrate binding to the enzyme at its active site. In this example, a protein binds to the reactive site of a protease. This portion of the enzyme is shaped specifically to allow entrance of only certain substrates with the corresponding shape, much like a key fits into a lock. If the substrate does not fit the shape of the active site, it cannot attach itself to the enzyme and no reaction occurs. The enzyme breaks down the substrate bound to the active site into smaller parts while the enzyme itself is unchanged. Once the enzyme breaks the substrate, the resulting products are released from the enzyme. Thus the enzyme reaction process can take place over and over as the enzyme repeatedly binds to more substrate molecules. Some other types of enzymes can bind two substrates at a time and catalyse a reaction to link the substrates together.

23.2 Risk assessment

Risk assessment is the process of identifying the hazard profile of a given material and gauging the likelihood of adverse effects occurring during handling or use. Risk assessment is important for enzyme-containing consumer products since it helps ensure the continued safety of currently marketed products and is a basis for determining the safety of new applications under commercial development.

23.2.1 Steps of risk assessment

Risk assessment can be divided into four areas: hazard identification, dose-response assessment, exposure assessment and risk characterisation. The risk assessment process for enzymes follows this general approach, but benchmark doses to define effect and no-effect thresholds are used instead of classic dose-response curves. Figure 23.1 outlines the risk assessment process used for enzyme-containing products.

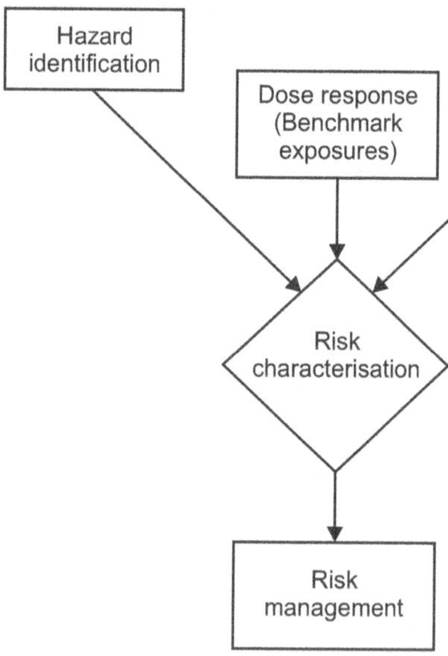

Figure 23.1: Risk assessment paradigm.

The generation and collection of data on the inherent toxicity of a chemical is needed to assess and define the hazard. Hazard information can be generated from clinical studies, animal tests, *in vitro* tests, structure-activity models, etc. In the dose-response assessment, the relationship between the magnitude of exposure (i.e., the dose) and the response is investigated.

In exposure assessment, routes and magnitude of the exposure under various product use, forese-eable misuse and accident conditions in the population are characterised. High levels of uncertainty in the assessment may require measuring exposure under simulated use conditions or during actual consumer use to learn about potential enzyme exposure in a new product. The data gathered in the first three areas can then be used in risk characterisation to make a determination of the likelihood of humans experiencing adverse effects from using a product.

23.2.2 Hazard identification

Hazard identification is the characterisation of the fundamental physical, chemical and biological effects of a material. It is generally conducted without regard for dose-response. It is meant to define the potential consequences of exposure to a material and identify target organs and systems. The health hazards of commercially important enzymes have been well characterised through toxicological, epidemiological and case studies. The toxicology of enzymes is unremarkable. Acute and sub-chronic toxicity is not of concern for enzymes. To some extent, acute toxicology data generated on one enzyme may be applied to the evaluation of other enzymes in the same class with similar activity. The most significant hazard identified to date is an adverse immune response defined as Type 1 hypersensitivity. In addition, some enzymes may produce other undesirable effects, such as skin or eye irritation, inhalation of high concentrations of some enzymes may cause irritation of the upper respiratory tract due to the proteolytic activity. Type 1 hypersensitivity is defined as an adverse immune response mediated by allergen-specific immunoglobulin E (IgE) antibody formation in response to exposure to a foreign protein, or allergen. The symptoms of this response are typically seen in allergic reactions to pets, dust mites, pollens, etc. Symptoms of Type 1 hypersensitivity can occur immediately or up to several hours after exposure to the allergen. Many enzymes are foreign proteins to the human immune system and can act as allergens. Data on health hazards associated with enzymes can be obtained from a variety of reports (e.g., toxicology, epidemiology and case reports) and a variety of sources, including enzyme suppliers, published reports (e.g., journal articles, textbooks) and unpublished studies. For enzymes used in industrial applications and in food applications the health hazards have been well characterised and are summarised below.

Potential routes of exposure via normal use, forese-eable misuse, or accidental contact with end use products are: (i) inhalation, (ii) skin, (iii) eye and (iv) oral. The potential hazards from these exposures are discussed in the following subsections.

Effects from inhalation exposure

Type 1 hypersensitivity: Inhalation of enzymes and other high molecular weight proteins has been shown to cause the development of allergen-specific IgE antibodies and elicit symptoms of Type 1 hypersensitivity in certain individuals. It is important to note that the ability of an enzyme to cause this sort of reaction is not dependent on enzymatic activity but on the protein structure. There are two main stages in the development of Type 1 hypersensitivity, as shown in Figs 23.2a,b. The first stage, called induction, occurs during the initial exposure of an individual to an allergen. The induction stage is defined as the development of allergen-specific IgE antibodies that specifically bind to the allergen. Development of allergen specific IgE antibodies may result from a single high exposure or from repeated exposure over a period of weeks, months, or even years. Further, exposure levels causing IgE development in one individual may not cause it in another due to variability in the immune response that exists within the human population.

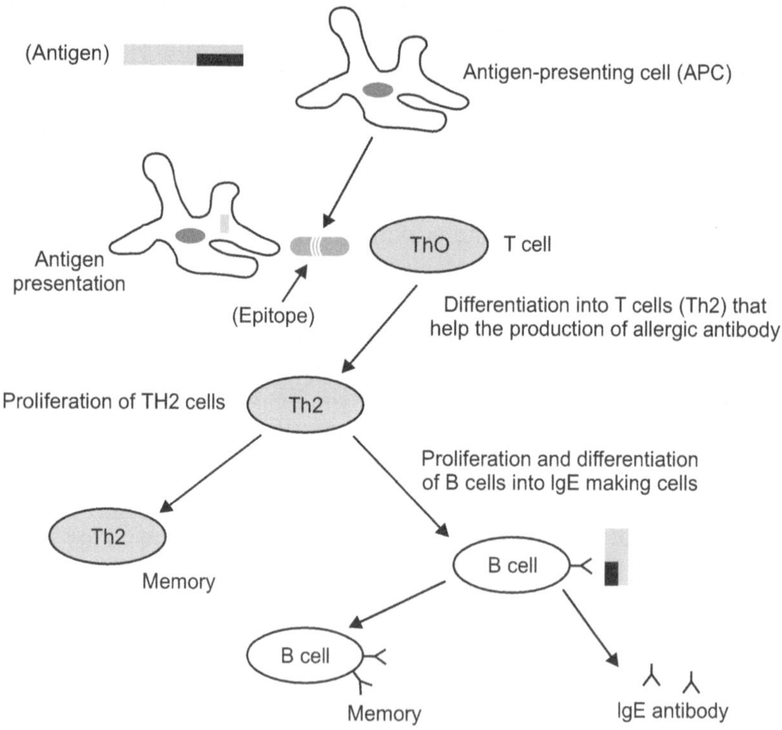

Figure 23.2a: Induction of allergic antibody to protein antigen.

In order to generate allergen-specific IgE antibodies, the enzyme is either inhaled or enters the body through other mucous membranes. The immune system recognises the enzyme as a foreign protein, or antigen, triggering a series of events that lead to the initial production of antigen-specific antibodies. Antigen-presenting cells in the airways engulf the protein and process it into smaller peptide fragments that are associated with peptide-presenting molecules. These molecules make their way to the surface of the antigen-presenting cell where the peptides can be seen by other white blood cells (naïve T cells or ThO cells and B cells). Peptides that are recognised by T cells and B cells are called epitopes. Epitope recognition by T cells and B cells leads to a series of events that result in the development of and secretion of antigen-specific IgE antibodies. IgE antibodies will bind to the surface of cells involved in the inflammatory process such as mast cells found in tissue. At this point a person is considered to be sensitised, they have developed IgE antibodies to a specific enzyme. In addition, a subset of antigen-specific T cells and B cells remain in the secondary lymphoid tissues as memory cells. Re-exposure to the antigen leads to stimulation of memory cells with an accelerated immune response. During the induction stage (Fig. 23.2a), the exposed person shows no clinical symptoms. Thus, the induction of allergen-specific IgE antibodies is not a clinical outcome or disease, it is an indication of exposure. Subsequent exposure to the antigen (in this case, enzyme) may lead to clinical symptoms in some sensitised people (Fig. 23.2b). Not all sensitised people will develop symptoms of allergy when re-exposed to the enzyme.

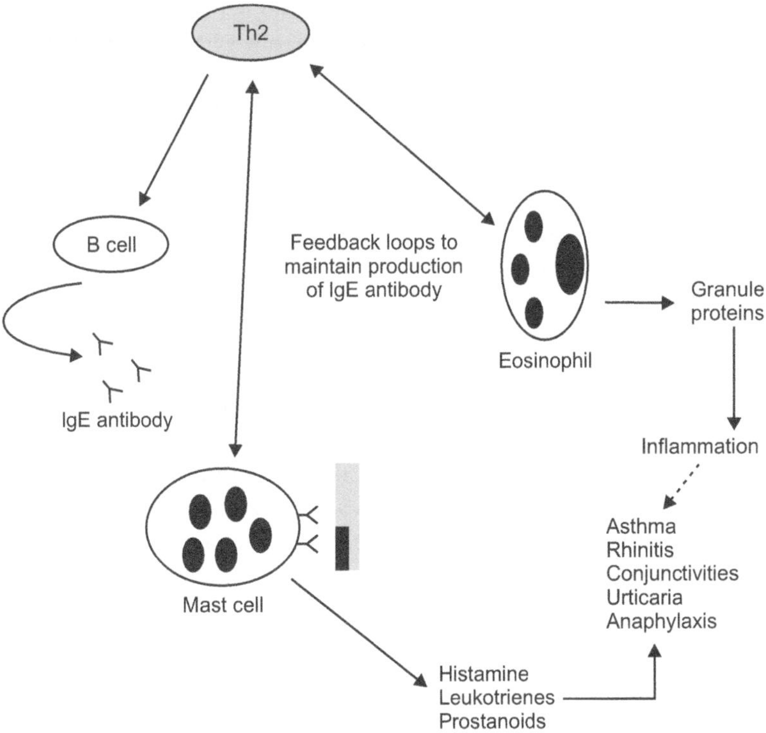

Figure 23.2b: Elicitation of an allergic response upon re-exposure to the protein antigen.

The symptomatic stage is called elicitation, which is the disease state of Type 1 hypersensitivity. The IgE antibodies specific for the enzyme (produced during the induction stage) are localised in tissue on mast cells throughout the body, including the lung. Mast cells in the tissues have granules that contain chemical compounds that mediate the inflammatory events that lead to the symptoms of allergy. Once the enzyme allergen binds IgE molecules on the mast cell, the cell is activated to release pro-inflammatory chemical mediators (including histamine) that cause the allergic symptoms. These mediators can have a direct effect on tissue as well as attract inflammatory cells (e.g., eosinophils) to the area. Symptoms of allergy range from rhinitis (watery eyes, a runny nose and a scratchy throat) to urticaria (hives), to asthma, or in extreme cases, to anaphylaxis. For a population of people exposed to allergens, certain predisposing factors may make some people more susceptible to developing IgE antibodies and possibly clinical symptoms.

For example, the genetic predisposition to make IgE antibodies to common proteins found in the environment (e.g., pollen) can make some individuals more likely to make IgE antibodies to an enzyme they encounter in their workplace. Smoking is another risk factor for making people more susceptible to making IgE antibodies to a protein. However, it is very difficult to predict if a particular individual will develop IgE antibodies and possibly clinical symptoms. Indeed, individuals who are found to have allergen-specific antibodies using diagnostic methods, such as the skin prick test, may never develop allergic symptoms. This is true for enzymes as well as for all other allergens.

Irritation: Inhalation of high concentrations of some enzymes (e.g., proteases) may cause irritation of the respiratory tract due to the proteolytic activity of the enzyme.

Effects from skin exposure

Irritation: Proteases can be irritating to the skin of animals and humans when applied at high concentrations, mainly due to the proteolytic action on the skin. Other enzymes have been found generally to be non-irritating to the skin of humans or animals. Understanding the irritation profile of enzymes (particularly proteases) can be applied to assessing the irritation potential of semi-purified mixtures that may contain contaminant proteases. Furthermore, the encapsulated or formulated enzyme preparation may also contain other ingredients that may cause skin irritation. These should be included as part of the overall assessment of the product.

Immunologic contact urticaria (Type 1 hypersensitivity): There are two types of urticaria, immunological and non-immunological. The majority of contact urticaria is not mediated by the immune system. In very rare cases, enzymes may elicit symptoms of Type 1 allergy on the skin, called immunologic contact urticaria. The major symptom is the classic 'hives' reaction of redness, swelling and itching. This rare condition can be elicited in people who had already made allergen-specific antibodies to the enzyme, most likely in response to inhalation exposure.

Allergic contact dermatitis (Type 4 hypersensitivity): Allergic contact dermatitis is caused by specific T cell responses to a material that comes in contact with the skin. Symptoms start to show 24 to 48 hr after skin contact. The inability of enzymes to induce allergic contact dermatitis (ACD) has been confirmed in numerous studies in which volunteers were tested by one of a variety of standard tests used to assess skin sensitisation (e.g., human maximisation test, modified draize test, human repeat insult patch test). Most of these tests were conducted with proteases, but other enzymes (including amylases, cellulases and lipases) have also been investigated. In addition, many similar tests conducted on detergents containing enzymes have shown that the presence of enzymes in detergents does not result in contact skin sensitisation.

Effects from eye exposure

Irritation: Proteases have been shown to be irritating to the eye when applied in high concentrations for much the same reason that these enzymes are irritating to the skin. Other classes of enzymes are either less or not at all irritating to the eye as compared to proteases. Information on the eye irritation potential of ingredients used in enzyme preparations and product matrices should be evaluated.

Type 1 hypersensitivity: Although development of allergen-specific antibodies through contact with the eyes is believed to be rare, the potential still exists, as illustrated by reports of allergy to enzymes used in contact lens cleaning solutions. Papain-specific IgE antibodies were found in people with symptoms of rhino conjunctivitis.

Effects from oral exposure

The oral toxicity and allergic potential of the ingestion of enzymes have been evaluated and reviewed. Enzymes, at the levels that have been traditionally found in food, are inherently nontoxic via, the oral route and development of allergen-specific antibodies by ingestion of enzyme has not been documented. Guidelines and principles for assessing the safety of traditional food enzymes and food enzymes derived by modern biotechnology have been developed and should be referred to when evaluating an enzyme for food use.

Thus, hazard identification is important because it helps determine the endpoints to focus on in risk characterisation. For all enzymes, the primary hazard is Type 1 hypersensitivity due to exposure via inhalation. For proteases, skin and eye irritation is also a hazard.

23.3 Dose-response assessment

In this step of the risk assessment process, the relationship between the level of exposure and the specific biological effect is characterised. The dose-response assessment consists of determining the amount of exposure to relevant tissues (i.e., the delivered dose) and the corresponding biological effect. The delivered dose will be a function of the level, duration, pattern and route of exposure. This process is not trivial, since the dose-response relationship for enzymes is not clearly defined for Type I hypersensitivity. For enzymes, there are limitations in our knowledge on safe levels of exposure and the role of peak exposures in the development of IgE antibodies and elicitation of symptoms. Therefore, benchmark values rather than more traditional dose-response measures are generally used to support decisions in enzyme risk assessments. Such benchmark values are based on studies in which measured or estimated exposure levels are associated with a demonstrated effect or the lack of an effect in the people exposed. Other end points, including eye and skin irritation and oral toxicity, do have dose-response relationships defined.

23.3.1 Dose-response estimations and benchmarks

Dose-response estimation is inherently a statistical process. Ideally, a mathematical model is developed that relates patterns of exposure to the likelihood of biological or adverse effects. In the absence of that level of predictability, exposures are estimated or measured empirically to establish uses that avoid biological effects and adverse effects. These values are known as benchmarks. A benchmark is a value derived from a study or studies in which a specific biological effect (e.g., production or no production of allergen-specific antibodies) is associated with an exposure level. In the risk assessment process, the exposure level estimated for a use application is compared to benchmark values to judge risk.

For enzymes, there are very little data on dose-response, so a benchmark approach can be used to assess risk. There are some data on dose-response to enzymes related to Type 1 hypersensitivity, but not enough to develop predictive models. In these models, values of estimated or measured exposures are compared to the highest exposure level previously shown not to induce the generation of allergen-specific antibodies (the 'no observed effect concentration,' or NOEC), or to the lowest exposure level previously shown to induce the generation of allergen-specific antibodies (the 'lowest observed effect concentration,' or LOEC). The threshold for inducing the generation of IgE antibodies presumably lies between these two levels. Such comparisons require a consideration of the uncertainty in estimated exposures, as well as uncertainty in the NOEC or LOEC. The existence of a threshold for allergen-specific antibody production to enzymes must be considered a reasonable assumption, as similar thresholds are generally assumed for most biological effects. From occupational data, a decrease in exposure to enzymes led to a sharp decline in the incidence of allergic symptoms among workers to the point where the symptoms were eliminated. In addition, the rate at which workers developed IgE antibodies to enzymes also declined with a decline in exposure. These studies demonstrated a dose-response relationship for antibody production and elicitation of symptoms and support the existence of thresholds for both events. The occupational data also point to peak exposure levels as playing an important role in antibody production and symptoms. It is reasonable to assume that such thresholds and dose-response relationships exist for consumer exposures.

Another issue with building dose-response models for enzymes (antibody production and/or symptoms) is a poor understanding of the delivered dose. When people are exposed to enzymes in air (e.g., in a workplace or through product use), the 'delivered dose' is not readily determined. Measurements of concentrations of enzyme in air can be used as a surrogate for more relevant, but less observable, variables like the 'inhaled dose' or 'concentration in bronchial fluids.' The amount of antigen that gets to antigen-presenting cells in the airways and is presented to T cells and B cells affects the immune response. The delivered dose of enzyme is dependent on a number of factors such as its concentration in the air, the rate of respiration, the particle or droplet size and the duration of exposure. Consideration of all the variables involved in determining 'delivered dose' can help in estimating exposure.

A third issue in building a dose-response model for enzyme allergy is the limitations in air monitoring technology. When air monitoring is used to determine delivered dose, the time of exposure is a direct controlling factor. Ideally, the entire exposure duration should be recorded. However, limitations in air monitoring technology preclude minute-to-minute monitoring of air concentrations of proteins. Instead, time-weighted averages along with activity patterns of exposed individuals are measured.

Currently, there are no universally accepted models applicable to humans that permit the determination of the dose-response relationship or relative potency of enzymes for causing production of allergen specific antibodies or symptoms of allergy via the inhalation route. Although the generation of IgE antibodies to enzyme, as measured by skin prick testing, is often used as a component of the benchmark value, it is not a disease state. Therefore, the generation of IgE antibodies is a conservative endpoint on which to establish a benchmark. Consequently, the alternatives are to:

1. Compare values of exposure for the new enzyme uses to benchmarks obtained from occupational and consumer studies. Ideally, the benchmark used should have well-characterised exposure and clinical endpoints as

well as a duration and frequency of exposure that are comparable to the new enzyme use in order to be relevant to the new exposure value being evaluated. The majority of data on consumer experience with enzyme-containing products are from the use of laundry detergents.

2. Examine the propensity of an enzyme to give rise to allergen-specific antibodies in animal models and compare the data with that obtained for suitable benchmark enzyme(s).

23.3.2 Animal models and use in benchmark assessments

Animal studies have been conducted to assess the allergen-specific antibody response to various enzymes in order to develop occupational exposure guidelines to control employee exposure and prevent allergic symptoms. These tests assess the relative allergenic potential (potency) of enzymes by comparison of the response to a reference enzyme, usually Alcalase. Although animal models might predict enzyme allergenicity/antigenicity in humans, the degree of correlation between relevant allergy epitopes in humans and experimental animals is not yet fully understood.

Exposure routes in animal studies include inhalation, intradermal (ID), intranasal (IN) and intratracheal (IT). While inhalation is the preferred route because of the similarity to human exposure, inhalation studies are labour-intensive, time-consuming and expensive and they present difficulties in delivering an accurate dose. Therefore, IT, IN and ID methods have been developed as surrogates for inhalation studies. Guinea pigs are usually the species of choice for sensitisation studies, although a mouse model is being developed for IT and IN studies. Antibody responses of guinea pigs showed that a dose inhaled over six hours/day has approximately the same effect as a single IT dose. This observation is important because it helps to answer the common question of, 'How should one deal with differences in temporal exposure patterns in the study population and the population for which risk estimates are required?'.

The animal models can provide insight regarding the relationship of dose, duration and frequency. These models can also be used to predict the effects induced by exposure to multiple enzymes or enzymes in different product matrices. It is known that the presence of proteases can increase the potency of non-protease enzymes when presented concurrently to guinea pigs. Similarly, surfactants in detergent matrices enhance the potency of subtilisin. It may be appropriate to test the particular enzyme in the desired product matrix to account for the potential effects of the product matrix.

Animal studies have been conducted to assess the type and extent of reaction of the eyes to enzymes applied neat or to products containing enzymes. Other classes of enzymes, such as cellulases or amylases, are less irritating or non-irritating to the eye, probably because of the lack of appropriate substrate in the eye or lack of access to substrate. In addition, a number of animal studies have also investigated the dose-response relationship of enzymes and skin irritancy. In general, most classes of enzymes other than proteases do not irritate the skin.

23.3.3 Clinical studies for obtaining benchmark data

The three main types of clinical studies used to obtain benchmark data are the prospective test, retrospective test and provocative tests. Study designs can be mixed to combine any or all of the test types. All clinical testing should be conducted in accordance with the principles of the Helsinki agreement. The parameter usually measured in clinical studies of Type 1 hypersensitivity is the development of allergen-specific IgE antibodies. Detection of specific IgE antibodies is an indication of exposure and not disease. IgE antibodies can be detected by serological tests such as the radio allergo-sorbent test (RAST) or by the skin prick test (SPT). Methods for these particular tests are found in the general allergy literature. In addition, exposure assessment should be made under conditions of use and misuse of the product. These exposures can be calculated or measured.

A prospective study is conducted with individuals who have never been previously exposed to the enzyme or who have never developed allergen-specific antibody to the enzyme. The prospective test can be one of the best tests to conduct since it evaluates a selected population over time. Important considerations when designing such a study include its length, number of test subjects, type and frequency of measurements undertaken and target population (e.g., include atopics). The size of the study is dependent on the level at which one needs to predict the risk of allergen-specific antibody production occurring in the population. For example, to predict the risk of development of enzyme-specific IgE antibodies occurring at 0.1% rate at the 95% confidence level, a study would need to include 3000 test subjects. It is appropriate to consult various experts including clinicians, statisticians and epidemiologists before conducting clinical tests.

A retrospective study is a 'look back' at effects, or lack thereof, in a population exposed to enzymes. A retrospective study can be used to help position a new use versus an existing use of an enzyme. These studies are limited since they evaluate responses at a single point in time. In addition, retrospective tests tend to involve 'healthy' self-selected individuals. Individuals who have selected themselves out of a population due to adverse effects from exposure will be difficult to recruit into retrospective studies. In any retrospective study there is the potential for biases that may arise if the test subjects are not representative of the general population.

A provocative test is conducted among individuals confirmed to have allergen-specific antibodies to the enzyme being considered for use in the product. For enzymes, this population is found mainly among occupationally exposed individuals. Provocative tests should only be done under very special circumstances. Clinical testing in a population with allergen-specific antibodies to the enzyme must be done with great care since these individuals are at high risk of developing symptoms. Positive responses in the clinical test can have an impact on the ability of an individual to work and use products. The study size becomes limited due to the finite pool of eligible individuals. Therefore, the statistical power of a provocative test may be limited.

Although data from occupational exposure scenarios in facilities producing enzymes or enzyme-containing products may be useful, caution should be exercised when applying it to finished product safety evaluations because of differences in frequency, duration, route of exposure and possible variations in the exposure to specific enzymes in the workplace.

Occupational exposure guidelines for workplaces handling enzymes may not be appropriate benchmarks for consumer exposure to enzymes, they should be used for general guidance only.

23.4 Exposure assessment

Exposure assessment establishes the amount of enzyme the user may be exposed to during intended use, forese-eable misuse and accidents. This value is then compared to the benchmark exposure to make risk decisions. Measuring or even estimating exposure to enzymes is not a trivial matter. However, the determination of these consumer exposure values is important for thorough risk assessment. In the absence of good quality exposure data, conservative worst-case assumptions and uncertainty factors are employed, which may lead to an overestimation of exposure levels and thereby inappropriately limit the amount and type of enzyme that can be used in a consumer product. Therefore, it is important that the exposure assessment be conducted thoroughly to enable the optimum use of enzymes in consumer products.

This section describes methods and approaches used to estimate exposure to enzymes from the use of enzyme-containing products. Since the primary hazard of enzymes is Type 1 allergy from inhalation, this section will focus on those exposures that can lead to inhalation of enzyme aerosols.

Product formulation: The physical and chemical properties of a formulation influence the exposure potential. The potential for aerosolisation of liquid sprays and powders, leading to inhalation and contact with mucosal membranes, should be evaluated during product development. This can be affected by delivery mechanism and viscosity of the product.

Aerosols should be characterised in terms of their droplet or particle size distribution. The size of droplets or particles, along with their density, determines their rate of settling and thus, the concentration of enzymes in the air during and after use. Large droplets or particles have the advantage of settling out of the air quickly. However, droplet and particle size can change during application. For example, liquid droplet size can decrease after impact on a surface during spray application.

Delivery systems: When formulating a product, consideration should be given to how the design of a delivery system can affect user exposure. The product delivery system should minimise the amount of product that can be inhaled or exposed to mucous membranes. Packaging can have a big impact on the extent and route of exposure to the product. A spray delivery system has the highest potential for inhalation exposure and should be designed carefully to minimise the production of inhalable mists. The delivery system should minimise available enzyme by limiting the production of particles small enough to be captured in the inhaled air stream.

Proper using of products: The physical environment in which the product is used also influences the extent of exposure. Several factors should be considered with respect to the physical environment. For example, factors such as room size and ventilation will affect exposure. Use of a product outdoors where there are air currents can lead to a different exposure in the breathing zone of the user as compared to the use of a product in a small room with poor ventilation. The orientation of the consumer relative to the product during use, i.e., breathing zone relative to the source of enzyme aerosols, will influence exposure.

23.5 Risk characterisation

Risk characterisation is the examination of the relationship between human exposure (calculated or measured) and the inherent toxicity of a substance to assess the likely incidence and severity of any effect. This step is important because it integrates information regarding the hazard identification and exposure assessment associated with use and forese-eable misuse of a product. In traditional risk assessment, dose-response analysis is typically used to estimate risk to humans from a material based on its hazard profile and the exposure level at which effects are expected. However, adequate dose-response relationships are not currently available for Type 1 hypersensitivity to enzymes. Instead, the risk characterisation process for enzymes relies on comparing potential exposure to benchmark values causing irritation or development of allergen-specific antibodies.

23.5.1 Risk characterisation process

The components necessary for risk characterisation of enzymes include hazard identification, a dose-response relationship, an estimate of potential exposure, comparison of exposure to benchmarks and an application of general knowledge regarding IgE antibody production and allergy. Type 1 hypersensitivity being the predominant hazard for most classes of enzymes. When evaluating or assessing hazard, differences in allergenic potency of enzymes as well as the potential adjuvant properties of the product matrix should be considered. For example, as product matrix may have adjuvant effects and enhance allergen-specific antibody responses to enzymes, significant changes to a product matrix can affect the interpretation of comparisons with benchmark data. Although, in some cases, it is assumed that the product matrix adjuvant effects and the potency correlation for enzymes are good between animals and humans, there is in general a paucity of reliable information in this area.

General information on the relationship among exposure, development of IgE antibodies and development of allergic symptoms can be applied to the risk characterisation process. For example, it is well recognised from the occupational literature from enzyme processing or product formulation facilities that the intensity of exposure (e.g., magnitude, duration, frequency) is associated with the development of occupational allergy and asthma. As the intensity of exposure drops, the likelihood of having symptoms also drops. In the detergent enzyme industry, more intense exposures have been associated with symptoms, while less intense exposures have been associated with production of allergen-specific antibodies.

23.5.2 Generation of additional data

Since risk characterisation involves evaluation of substantial amounts of information from a variety of sources, there may be some uncertainty in the final assessment. Sometimes, the overall confidence in the database is low and consideration has to be given to obtaining additional data on likely human exposure or toxicity. Judgment is required to decide whether or not refinement of the risk characterisation is warranted and if so, whether or not development of additional data is practical.

It is recommended that care be taken when considering new applications for an existing enzyme that may already have proved acceptable for another use. Data may need to be generated to support the safety of enzyme use in these new applications. For some new applications, a benchmark may not exist. For example, the introduction of a subtilisin enzyme with a history of safe use in laundry applications into prototype beauty bar soap led to production of allergen-specific antibodies among test subjects in a pilot clinical study.

Similarly, introducing a new enzyme, possibly of a different functional class, for a new application should be carefully considered, particularly when the enzyme characteristics are different or when the potency may be altered because of the presence of adjuvant-like materials in the new formulation. If clinical data need to be generated, one should weigh the ethical issues of intentional exposure of human test subjects with the risk of inducing allergen-specific antibodies and possibly eliciting symptoms in this population.

23.6 Risk management

The objectives of the risk management process are to determine the significance of risks to human health, to ensure that the product use is and remains within the acceptable risk levels and to effectively communicate risks, or lack thereof, to appropriate audiences.

23.6.1 Determining the relevancy of a risk assessment

Risk assessment provides useful information to weigh alternatives and analyse tradeoffs in addition to providing a means of organising relevant information in order to estimate the potential impact on human health. In doing so, assessments may convey a level of precision that fails to reflect the shortfalls of the underlying assumptions and the uncertainties that may characterise the risk assessment. The quality and reliability of the risk assessment is dependent on and is only as good as the data used to conduct the assessment. Uncertainties may exist in dose-response relationships, exposure data and estimates from exposure models. Assumptions and estimations need to be stated clearly as they can affect the reliability and quality of the risk assessment. It is important to consider these points when evaluating information from the risk assessment in determining whether or not the risk is considered acceptable.

23.6.2 Acceptable level of risk

The risk assessment process does not define an acceptable level of risk. No numerical level of risk will receive universal acceptance. Further, it is impossible to eliminate all risks associated with a particular activity and this is also true for the use of enzymes in finished products.

Risk management approaches should be based on critical evaluation of the risks associated with the use of the product and the data generated from the quantitative risk assessment process. If screening level assessments based on estimates of exposure and available hazard information are not sufficient to support the safe use of the product, then the collection or generation of additional data as discussed in the risk characterisation section could be considered. If the completion of the risk assessment results in the determination that the risk is unacceptable, then appropriate risk control measures should be carried out to reduce the exposure to within acceptable risk levels. Risk assessments for a given product and usage may not be applicable for another product or application. It is important to understand these differences as well as the effect of other exposure factors, such as frequency and duration of exposure, on the development of allergen-specific antibodies. Inter-individual variability and susceptible sub populations that are predisposed to allergy development are important factors to consider in the risk management process and can further complicate the establishment of an acceptable level of risk.

23.6.3 Risk control

In general terms, the risk control step of the risk management process should strive to reduce the risk by limiting exposure to enzymes from the product. Risk reduction options may include product modification, product use restrictions or a decision not to market the enzyme-containing product.

Modification options may include changing the matrix or delivery of the enzyme product, reducing the enzyme concentration in the product, substituting other ingredients that may be affecting the potency of the enzyme or a combination of these approaches. In the detergent industry, great steps have been taken to minimise risk through product modification. For example, enzymes are now encapsulated to limit consumer and worker exposure. This risk control method was relatively easy to achieve for consumer laundry products and in turn, provided a reduction of risk in the work environment. Alternatively, appropriate labelling to restrict certain end uses may be considered and weighed against the likelihood that consumers will read the label.

As stated previously, the goal of risk control is to decide, based on an acceptable risk level, whether product modification or restrictions on its use is necessary. If product modification or restriction on use are not alternatives and there is a likelihood of an adverse experience or event, not selling the product is also an option to be considered.

23.6.4 Risk communication

An integral part of the risk management process is to effectively communicate the potential risks to appropriate audiences. There are two important audiences to target in designing a risk communication programme: (i) users of the company's products and (ii) other stake holders, such as the general public and public interest groups.

Product users

Product labels are the primary means of communication with consumers. For enzyme-containing products, as with all consumer products, many countries require that the label include appropriate warning statements. In the U.S., the regulations of the Consumer Product Safety Commission apply, in Canada, those of the Consumer Chemicals and Containers Regulations apply and in Europe, those of the Dangerous Preparations Directive usually apply. In addition, there may be requirements to place handling instructions and information or first aid information on the label. Product manufacturers can also supplement this safety-related information, as needed.

To address other questions, manufacturers should have properly educated customer support personnel to provide answers to customers and effectively communicate issues to the public regarding the safety of enzyme-containing products. Examples of essential information that customer support personnel should be able to communicate are as follows:

1. Composition of the product.
2. First aid information.
3. Use and handling guidelines with detailed examples of correct use and concrete recommendations to steer consumers from misuses.

Further work is needed in risk communication for enzymes to the general public since understanding the use and function of enzymes in products varies. Thus, an important part of risk communication is to ensure that the audience understands what enzymes are and what they do, so that their benefits and risks are understood.

Other stake holders

Other stake holders may be governmental, non-governmental organisations or industry partners. An important route toward gaining acceptance of stake holders is through interaction among experts in the field or the industry, government authorities and the interested stake holders, such as consumer associations, scientific journalists and academia. The goal is to build confidence in the company or industry using the technology.

An attitude of open-ness and willingness to share information and data is essential, while recognising the legitimate needs of companies to protect competitively sensitive information. The amount and detail of information that may be needed in dialogues with some stake holders may be more extensive than what is provided to the general consumer. Position papers and dossiers giving details of the product with particular reference to the enzymes used may be considered. In addition to information relevant to consumers and workers, product manufacturers should anticipate requests from some stake holders related to the production process and containment procedures employed in the production facility, since levels of exposure in the workplace are generally higher than in the product use setting if exposure management steps are not taken. Finally, any document or dossier provided might be more readily accepted if it has been subjected to a peer-review process.

In short, an effective risk communication programme requires the ability to provide useful information in response to worker, consumer and other stake holders' inquiries. By providing this information, it is possible to promote the safe use of the product and in turn, reduce the risks associated with exposure to the enzymes contained in the product.

To sum up, enzymes can bring significant benefits to consumer products, including improved efficiencies and previously unavailable product benefits. However, prior to introducing an enzyme preparation into a product, a risk assessment should be conducted to ensure the safe use of the enzymes by the consumer. The primary challenge associated with enzyme use is preventing the generation of allergen-specific antibodies and the development of symptoms of Type 1 hypersensitivity.

Experience in the cleaning products industry demonstrates that potential risk of adverse effects can be successfully managed by identifying the hazards to be managed, carefully assessing exposure, characterising the risk and then applying appropriate risk management.

Each company that intends to use enzymes in its products can play an active role in understanding and managing the risks associated with enzymes. If the risk is created and not appropriately managed, the consequences may spread beyond a single product or company. This could lead to unwarranted limitations on the use of enzyme technology in other consumer applications. Therefore, it is recommended that companies using enzymes responsibly consider how they are managing enzyme safety and whether appropriate risk assessment and risk management programmes have been employed.

24

Safety regulation of food enzymes

24.1 Introduction

The majority of industrial enzymes available at present are used in food industry. Safety regulations of food enzymes differ among countries, including fundamental aspects, whether a pre-market approval is needed and on the level of details, e.g., what particular information manufacturers have to provide in the course of safety evaluation. Occupational safety concerns focus on allergenic properties as it is well established that enzymes are potent inhalative sensitisers and can cause allergic reactions including asthma. Otherwise toxic substances including bacterial toxins and mycotoxins might also be present in enzyme isolates and might thus constitute a safety risk to consumers. Safety precaution and monitoring measures established by industry have also reduced but not entirely eradicated occupational risks. Challenges to regulators and industry arise from unresolved issues, e.g., whether enzymes might be contact sensitisers and from the lack of harmonisation of both legislation and safety evaluation. In the EU, most food enzymes are not covered by food safety regulations neither on community nor on national level. On top of this the availability of enzymes with new and unusual properties raises questions of safety. In the EU there seems to be a chance that these challenges will be tackled in the course of establishing a harmonised legislation on food enzymes.

In many cultural contexts various applications of enzyme producing microbes in food processing have been practiced for centuries or even thousands of years. The application of purified enzymes of microbial origin is a rather recent invention that dates back to the first half of the 20th century. Improvements in bioprocessing and strain development and especially the advent of genetic engineering have greatly increased the availability of enzymes and tailoring of enzyme properties in order to meet the sometimes very distinct requirements of the technical environments where they are used. As enzymes are often used to replace steps in food processing encompassing harsh chemical or physical conditions (temperature, pressure, chemicals), they are frequently perceived to be in line with both sustainable industrial production and careful processing of food in order to maintain nutritionally important ingredients such as vitamins, etc. The rise of food enzymes as substances applied in food processing has also captured the interests of regulators in most industrialised countries. Newly marketed food enzymes might need a prior market authorisation by the US Food and Drug Administration (FDA) and – if they are considered food additives – also by the European Food Safety Authority (EFSA). In addition, a small number of EU Member States do have national legislations on food enzymes in place. These legislations mainly serve to ensure the safety of enzyme preparations for the final consumers and sometimes include specifications for purity and activity. In certain cases occupational health issues that might arise from manufacturing and handling enzymes are also part of the regulatory overview.

Risks of allergic reactions in workers handling enzymes are not specific to enzymes and have been discussed at great length in the scientific literature, within industry themselves and among stakeholders concerned and even made it to the general press. In fact, the occurrence of severe allergic responses in workers of detergent enzyme manufacturers and allegedly also in consumers of washing powders almost led to the collapse of the enzyme industry between the late 1960s and early 1970s. Safety concerns for consumers largely focus on toxic by-products that might be inadvertently present in the enzyme preparation and might thereby be transferred to foodstuff.

Given the restricted scope of the EU food legislation the former aspect has mainly been an issue in the context of EU chemical legislation. Under chemical legislation occupational hazards have to be considered for both notification of new substances and for classification and labelling purposes as well.

24.2 Safety concerns associated with enzymes

Safety concerns associated with industrial enzymes in general are possible allergenic, irritative and otherwise toxic properties. Allergenic and irritative risks are mainly issues of occupational health in the industrial production and application of enzymes, whereas risks of oral toxicity are especially relevant to consumers of food enzymes.

24.2.1 Allergenic and irritative properties

In occupational contexts enzyme exposure includes mainly to dust or liquid aerosols that are set free while handling enzyme preparation in either manufacturing of the enzyme itself or using enzyme preparations in other industrial contexts. This is generally true for all enzymes regardless of the particular end-use.

The particles are deposited on the skin or on the mucous membranes of the respiratory tract. When an enzyme comes into contact with the respiratory tract or the skin, the body's immune system may be stimulated to produce antibodies resulting in respiratory allergy or contact urticaria, respectively. And because skin has a protein structure, enzymes which catalyse breakdown of proteins such as proteases, are potential skin irritants.

Immunologists differ between allergenic effects and sensitisation. The latter means that individuals who are exposed to a possible antigen for the first time may develop antigen-specific IgG and/or IgE antibodies. The formation of IgG indicates exposure and IgE antibodies indicate allergic sensitisation but not allergic disease. Thus, individuals with a positive skin prick test or specific IgE in their blood would not necessarily be considered ill. If sensitised individuals would, however, subsequently be exposed to the antigen, this might result in an allergic response. So, the question is whether enzymes could be antigens, whether there is evidence that enzymes can elicit allergic reactions and be sensitisers and eventually, if they can act as irritants to the skin. Table 24.1 provides an overview of evidence on these kinds of health effects.

Table 24.1: Health impacts of enzymes in occupational aspects.

Causative agent	*Exposure*	*Health impact*	*Evidence*
Enzyme dust, aerosols	Inhalation	Immediate type allergic reactions (type I), e.g., conjunctivitis, rhinitis, asthma	Good empirical evidence
	Skin contact	Immediate type allergic reactions (type I), e.g., allergic contact urticaria Delayed type allergic reaction (type IV), e.g., allergic contact dermatitis	Good empirical evidence but lack of differentiation from allergic contact dermatitis Sensitisation still contested
Enzyme (containing) powder or liquids		Skin irritation	
Food enzymes	Ingestion		Rare cases following respiratory sensitisation

Many allergens, especially in food and pollen, are proteins. Whether a protein becomes an allergen depends on a set of factors including structural features of the protein. Several common allergens, e.g., from house dust mites, storage mites, ragweed, pollens, *Alternaria*, cat dander and bee venoms, are furthermore proteins with enzymatic activity. And if common antigens are enzymes by nature, enzymes may correspondingly represent antigens. In fact, certain enzymes may also be more prone to cause sensitisation than other proteins because of their intrinsic capacity, e.g., the proteolytic function of many of these allergens has been proposed to be an important factor in the epithelial permeability and the origin of allergy.

Many medical reports on allergenic and sensitising properties of enzymes come from the detergent industry. In fact, it was the handling of enzyme containing washing powders in the second half of the 1960s that led to numerous reports of respiratory illnesses at the workplace. By the early 1970s for every 1000 workers an average of 400 were sensitised and 150 went on to develop respiratory symptoms. Similar reports were made in the enzyme manufacturing industry.

Since then the problem seems to have decreased, because the industry introduced quite extensive measures to diminish and monitor the exposure of workers, including encapsulation of enzymes, using immobilised preparations, avoiding direct contact, introduction of safe working practices and training of workers, replacement of older products by antigenically distinct proteases and exclusion of potentially pre-disposed and especially sensitive workers from directly working with enzyme preparations.

The highest percentage of exposed people now-a-days seems to be affected in the bakery industry and in the enzyme manufacturing industry, but sensitisation may also occur in the animal feed industry, resulting in concerns for meat consumers.

Only recently several reports have been published on allergic reactions in other occupations (pharmaceutical industry, at laboratory work, etc.), and even in consumers after dermal, mucosal, oral and parenteral (injection, infusion) exposure to enzyme drugs, enzymes in cosmetics, soaps, in soft lens fluids, in meat and other products of daily life (Table 24.2).

Table 24.2: Health effects of enzymes in the biotechnical industry.

Industrial sector/enzyme	*Reported effects on*	
	Skin	*Airways*
Starch processing and baking		
α-amylase	CA, CU	RA
Amyloglucosidase	n.r.	RA
Proteases	CA	RA
Dairy industry		
Proteases	CA	RA
Lipases	n.r.	RA
Papain	CU	RA
Trypsin	CU	RA
Chymotrypsin	n.r.	RA
Food industry		
Amylases	CA, CU	RA
Cellulases	CA, CU	RA
Hemicellulases	CA, CU	RA

(Cont'd...)

Industrial sector/enzyme	*Reported effects on*	
	Skin	*Airways*
Protein industry		
Papain	CA	RA
Pancreatic proteinase	n.r.	RA
Animal feed industry		
Cellulases	CA, CU	RA
Hemicellulases	CA, CU	RA
Proteases	CA	RA
Amylases	CA, CU	RA
Phytase	CU	RA
Pharmaceutical industry		
Chymotrypsin	n.r.	RA
Trypsin	CU	RA

Key: CA enzyme causing contact allergy, CU enzyme causing contact urticaria, n.r. not reported, RA enzyme causing respiratory allergy.

In accordance to what is said above, reports from the detergent, the baking and flour industry show that most exposed workers produce IgG antibodies and several develop IgE antibodies (become sensitised). In bread producing factories, exposure rates are high and enzymes have become a major cause of occupational disease. Recent reports have shown that of all the workers in the industry who were sensitised after the exposure, up to 36% work in the enzyme manufacturing, up to 22% in detergent industry, up to 30% in the bakery industry and 31% in the pharmaceutical industry. Lower rates of sensitisation were reported from the animal feed and the textile industry. These numbers are strikingly high despite the fact that industry – especially enzyme and detergent manufacturing industry, has done a lot to diminish worker exposure. However, only a minority of sensitised workers ever get ill. Whereas symptoms are usually mild, they are sometimes bothering enough and hinder individuals to continue working with those enzymes. Beyond possible allergenic properties enzymes are also supposed to be mild skin irritants in the workplace.

The extent of damage will mostly depend on the time and intensity of exposure, the nature of the enzyme (e.g., proteases might attack skin proteins) and its concentration, the integrity of the skin and cofactors such as additional detergent exposure, but controlled studies from the past two decades are missing. Since the concentration of enzymes in end products is very low, irritating properties are not considered a problem for consumers. Validated animal models are available to routine tests for irritative properties of the substances on skin and eyes. A systematic adjustment of these methods for high molecular mass substances such as proteins is still missing.

24.2.2 Toxic properties

Allergenic properties of enzymes are intrinsic to the structure of the enzyme protein (and possibly other proteins present in the enzyme preparation). Toxic concerns are, in contrast, rather focused on by-products and contaminants present in enzyme preparations. In the case of food enzymes, these concerns are especially relevant for obvious reasons. The range of the active enzyme in the final product from the fermentation processes, the enzyme concentrate, varies significantly between 2 and 70% – hence, by-products amount up to 98%. These by-products are varying in terms of quantity and nature, depending on the production organisms and the particular conditions of the bioprocess. Toxic substances could result from contaminants during fermentation or from the production organisms themselves. It is well known that some production organisms or at least phylogenetically closely related strains (same genus or same species) can – under certain conditions – produce toxins (e.g., in the case of certain bacteria and numerous filamentous fungi).

In contrast to the overwhelming evidence about allergic properties of enzymes there seems to be no corresponding evidence of toxic effects of enzyme preparations neither in occupational environments nor in consumers. Generally concluding the absence of toxic risks in enzyme preparations might however be somehow premature. This may rather denote that present safety evaluation procedures have so far been sufficient to detect toxic contaminants before marketing of enzyme preparations.

Nevertheless, the possibility of toxic contaminants in enzyme preparations is still widely acknowledged. Thus, one of the most important factors in safety assurance of enzyme preparations is the toxic and pathogenic potential and a reliable identification (e.g., including DNA fingerprinting) of the micro-organisms used along with the quality assurance system to safeguard against drift in taxonomy, mutations or selection of strains with unwanted abilities. Given the possibility that micro-organisms belonging to the same species but different strains might differ in their ability to produce toxins, both quality assurance and identification have to be on the strain level.

Secondly, changes in bioprocessing conditions such as pH, temperature, purification process or media ingredients might affect the nature and quantity of by-products – including possible toxins. According to Smith and others even micro-organisms not known to be harmful in food applications, might under different conditions turn out to be toxin producers. Thus, there is also a safety reason to thoroughly describe and maintain the process conditions.

Furthermore, enzyme concentrates can be and in fact are, subjected to animal testing for a variety of toxicological endpoints as described above. Besides, microbial toxins that might occur as contaminants can be identified and measured. The latter are chemical analytical tests that have to be conducted for each individual toxin, the former are intended to detect all (known and unknown) toxins that might be present. Beyond toxic substances microbes might contaminate enzyme preparations and even antibiotic substances might be present. Thus, total viable counts are usually measured in enzyme preparations and the presence of coliforms, *Salmonella* sp., *E. coli* and pathogenic micro-organisms is investigated.

24.3 Regulatory contexts and requirements for safety evaluations

In terms of food legislation enzymes have so far been distinguished into food additives and processing aids. The main distinction between food additives and processing aids is that additives have a technological function in the final food, whereas processing aids do not (Table 24.3). This distinction is important because in some jurisdictions a pre-market authorisation, including safety evaluation, is mandatory for additives only. The definitions underlying this distinction vary between jurisdictions. In Canada, US and Japan for instance, all food enzymes are regulated as food additives. In Australia food enzymes are considered processing aids.

Table 24.3: Definitions of food additives and processing aids.

Enzyme category	*Definitions*
Food additives	Means any substance not normally consumed as a food in itself and not normally used as a characteristic ingredient of food whether or not it has nutritive value, the intentional addition of which to food for a technological purpose in the manufacture, processing, preparation, treatment, packaging, transport or storage of such food results, or may be reasonably expected to result, in it or its by-products becoming directly or indirectly a component of such foods
Processing aids	Means any substance not consumed as a food ingredient by itself, intentionally used in the processing of raw materials, foods or their ingredients, to fulfil a certain technological purpose during treatment or processing and which may result in the unintentional but technically unavoidable presence of residues of the substance or its derivatives in the final product, provided that these residues do not present any health risk and do not have any technological effect on the finished product

In the EU food legislation the situation is more complex. Most food enzymes are considered as processing aids. So far, only two enzymes, lysozyme and invertase, are considered as additives. Interestingly, this differentiation was not followed by the European Commission's own Scientific Committee, the Scientific Committee on Food (SCF), which was responsible for the evaluation of food additives: from a toxicological point of view it is not pertinent to distinguish between [enzymes used as processing aids or food additives] since, in both cases, the enzyme preparations may remain in the food.

A few enzymes are covered by the EU specialised legislation, e.g., on winemaking. The majority of about 160 food enzymes produced/marketed in the EU are, however, considered as processing aids, the regulation of which is still governed by national legislation – if there is any legislation in place at all.

National regulations on enzymes used as processing aids differ substantially among the EU Member States. In France, Denmark, Poland and Hungary these enzymes are subjected to an authorisation procedure, in the United Kingdom a voluntary approval system is in place. In many other Member States no national regulation is in place. Furthermore, the range of enzyme products and enzyme applications permitted by national legislation varies among those Member States that have a regulation in place.

The following subsections provide a brief overview on the regulatory contexts of the EU and the US and also touches on the JECFA – given their importance for worldwide standards.

European union

In the EU food additives are regulated by Council Directive 89/107/EEC (28, amended by Directive 94/34/EC) on the approximation of the laws of the Member States concerning food additives authorised for use in foodstuffs intended for human consumption and Directive 95/2/EC (31, amended by Directives 96/85/EC, 98/72/EC, 2001/5/EC, 2003/52/EC and 2003/114/EC) on food additives other than colours and sweeteners. The community legislation on food additives is based on the principle that only those additives that are explicitly authorised may be used. Most food additives may only be used in limited quantities in certain foodstuffs. If no quantitative limits are foreseen, they must be used according to good manufacturing practice, i.e., only as much as necessary to achieve the desired technological effect.

Food additives may only be authorised if there is a technological need for their use, if they do not mislead the consumer and if they present no hazard to the health of the consumer. All food additives have to fulfil purity criteria which are set out in detail in three Commission directives: Directive 96/77/EC (32, amended by Directive 96/86/EC and Directive 2000/63/EC) for additives other than colours and sweeteners.

These guidelines specify conditions of use, requirements for information, documentation and testing and occasionally refer to concrete methods of testing. With respect to hazards, the guidelines mainly focus on toxicological requirements of enzyme preparations, on the safety of the source organism and on unintended reaction products in the food caused by enzymatic reactions in the final foodstuff. Exposure deals mainly with the quantity of enzymes consumed. Allergies or irritative effects are just briefly mentioned by stating that these effects are considered primarily as occupational problems. Whereas the SCF considers enzymes from (edible) plants or animal species as posing no health problems, the toxicological evaluation of enzymes from microbial sources is deemed to be far more important. These toxicological tests should investigate known toxins as well as unknown toxic compounds that might be present in the enzyme preparation.

So far, five food enzymes have been evaluated by the SCF and one by the AFC (thrombin and fibrinogen from cattle and pigs, urease from *Lactobacillus fermentum*, papain from papaya, invertase from *Saccharomyces cerevisiae*, chymosin from *E. coli*, *Kluyveromyces lactis* and *Aspergillus niger*, lacto-peroxidase together with glucose oxidase) and two enzymes, lysozyme (E1105) and invertase (E1103), have been authorised as food additives in the EU. Additives authorised under Directive 95/2/EC are listed in the annexes to this Directive, which are regularly updated.

United states

Enzymes used in food are regulated by the US Food and Drug Administration (FDA) under the Food, Drug and Cosmetic (FDC) Act. Since 1958 the FDA has perceived all enzymes as food additives and therefore subjected to restrictions pertaining to food additives. In general food enzymes may either be classified as: (i) substances that are GRAS, (ii) substances that are not GRAS which are defined as food additives or (iii) substances approved for use in food prior to September 6, 1958 by the FDA or by the Department of Agriculture.

So far, about 50 enzymes have either been approved as food additives or affirmed as GRAS by the FDA.

The regulatory status of food additives or substances affirmed as GRAS is established through a petition process.

In order to evaluate the safety of an enzyme preparation, the petitioner and FDA are comparing the enzyme to be assessed with other enzymes which have already been approved or with those that have been safely consumed as part of the diet throughout human history. According to FDA enzymes that have the same function and that are identified by the same name and EC number often differ slightly in structure and properties when they are obtained from different sources. For example, the structure of an enzyme isolated from one tissue (such as liver) of one animal species may differ slightly from that of the same enzyme isolated from a different tissue from the same species, or from the liver of another animal species. In part, because of this variability, the diet routinely contains many thousands of different protein molecules.

This approach is based on the concept of substantial equivalence (in earlier documents also referred to as substantial similarity), which is also used in the safety evaluation of novel food. According to the definition if a new food or food component is found to be substantially equivalent to an existing food or food component, it can be treated in the same manner with respect to safety. No additional safety concerns would be expected. Consequently, the food or food component can be concluded as safe as conventional food or food component. This concept was further recommended by international expert groups to be applied in the assessment of substances intentionally added to food. For example, a carbohydrase preparation and a protease preparation from *Bacillus subtilis* and *Bacillus amyloliquefaciens* were assessed to be substantially equivalent to carbohydrase and protease enzymes from other micro-organisms that had been evaluated and found to be safe by FDA before (e.g., mixed carbohydrase protease preparation from *Bacillus licheniformis*, carbohydrases from *Rhizopus niveus*, *Rhizopus oryzae* and *Aspergillus niger*).

FAO/WHO

About ten years before the SCF guidelines on enzyme safety evaluation were issued the Joint FAO/WHO Expert Committee on Food Additives (JECFA) had already been conducting safety evaluations of enzymes. The JECFA is an international expert scientific committee that serves as a scientific advisory body to FAO, WHO, their Member States and the Codex Alimentarius Commission, primarily through the Codex Committee on Food Additives and Contaminants and the Codex Committee on Residues of Veterinary Drugs in Foods. It has been meeting since 1956, initially to evaluate the safety of food additives. For food additives, contaminants and naturally occurring toxicants, the Committee: (i) elaborates principles for evaluating their safety, (ii) conducts toxicological evaluations and establishes acceptable daily intakes (ADI) or tolerable intakes, (iii) prepares specifications of purity for food additives and (iv) assesses intake.

In 1991 the General Considerations and Specifications for Enzymes from Genetically Manipulated Micro-organisms were published to supplement the General Specifications. In this supplement the Committee stated that in order to properly evaluate enzymes from GMM it is considered very important to provide adequate information on the source material, the genetic manipulation techniques and the fermentation and recovery process employed. In addition to these enzyme specific guidelines the Principles for the Safety Assessment of Food Additives and Contaminants in Food specified especially toxicological testing requirements.

The requirements to define the strain of the production organism were loosened in 1999 when it was concluded that citation of genus and species of host organisms is usually adequate for those that had been determined to be safe and suitable. The reason given for this amendment was that identification at the strain level might impose unnecessary constraints on the development of production micro-organisms used to produce food-grade enzymes.

24.4 Industry perspective

As the safety evaluation of enzymes used as processing aids is not regulated in most EU Member States it is up to the industry to design an appropriate process of safety evaluations. Industry acknowledges the guidelines for safety evaluations issued by JECFA , SCF and the UK Committee on the Toxicity of Chemicals in Foods, Consumer Products and the Environment (COT) and has come up with their own recommendations as well.

24.4.1 Comparison of safety requirements

Although the requirements for safety evaluations in guidance documents and the scientific literature differ, the overall structure is fairly similar and includes:

1. Basic technical data on the enzyme itself, as the active compound of the preparation.
2. Information on the source material, special attention is given to microbial sources.
3. Substances added to the enzyme isolate (or used during processing).
4. Possible contaminants (micro-organisms, heavy metals, toxins) present in the final enzyme preparation.
5. In the case of GMMs additional information is required on the host organism, the donor organism, the vector, or introduced DNA.

Enzymes are identified by their catalytic activity in all guidelines and provisions. In general, guidelines and provisions do not describe the requirements precisely. Thus, much is left to interpretation. Most detailed physico-chemical information on the enzyme is required by the FDA (enzymatic function, mode of action, substrate specificity, molecular mass, isoelectric point, kinetic properties, specific activity, temperature, pH, inorganic ions).

FDA requires the most detailed information on enzymes from GMM, including detailed technical data on the enzyme itself, on structural modifications, on the GMM (genetic stability, growth properties), vector, introduced DNA, donor organism as well as detailed information on the manufacturing and purification process.

SCF describes basic toxicological requirements, which are toxicological tests to be performed and possible exemptions from testing within the enzyme guidelines. The JECFA laid down general principles for toxicological testing in a separate publication. Particular requirements for toxicological testing primarily depend on the nature of the microbial source, e.g., if it is a microbe that also naturally occurs in food, if it is a GMM, etc. Unlike the contested proposals for toxicity testing in the scientific literature, toxicity endpoints suggested in guidance documents are fairly similar. SCF and JECFA ask for a 90-day subchronic toxicity test along with mutagenicity tests on bacteria and an *in vitro* test for chromosomal aberration.

Other differences are:

1. Only SCF includes requirements on data for the manufacturing process, the usage and stability in food.
2. Detailed descriptions of methods or references for methods to be used in testing and data production are given by both SCF and JEFCA.
3. The AMFEP guidance does not in general include detailed description of requirements. However, AMFEP does acknowledge the guidelines from JECFA, SCF and COT.
4. Safety approvals are often confined to particular enzyme preparations (SCF). Changes in the production process or application of genetic engineering lead to a re-evaluation on a case-by-case basis. Waivers from toxicological testing are also dealt with on a case-by-case basis.
5. FDA takes into account long-term experience with certain enzymes. A GRAS status may be assigned to such enzymes.

Toxicity relevant information would include homology comparisons to proteins known to cause adverse effects, information on stability of the protein under processing and storage conditions and the expected treatment of the food and testing for resistance to proteolytic enzymes. Toxicity testing would include a 28-day oral toxicity study – additional studies would depend on the outcome of the 28-day study. Exemptions from toxicity testing might be justified if both the GMM and the protein have a history of safe consumption by humans. Compared to the other guidance documents, the EFSA Draft would put much more emphasis on allergenicity assessment, including homology comparison, digestibility testing and might even include *in vitro* tests with serum from allergic patients.

This guidance is, however, not laid out for industrial enzymes and certain enzyme specific information requirements, such as for additives in the enzyme preparation, hygiene (total viable counts, production strain), contaminants including thresholds (heavy metals, microbes, antibiotic activity in enzyme preparation, presence of mycotoxins) are not included. Hence, it is likely that for food enzymes additional guidance needs to be provided by EFSA.

24.5 New methods in enzyme production

The application of genetic engineering techniques in enzyme manufacturing is pushing up the exploitation of new enzymes and the development of new enzyme properties.

Enzyme yield has been improved – among other factors – by the use of strong expression or multi-copy systems. Enzymes that were not accessible before, e.g., because it would not have been feasible to set-up a production process with their source organisms, can now be cloned into and produced from a well-known host organism. Thereby, enzymes from almost any source in nature become accessible exhibiting unusual properties such as extreme thermostability. This includes thermophilic and psychrophilic microbes and even non-culturable microbes.

Combinatorial approaches of rational protein design and directed evolution methods turn out to efficiently alter the properties of enzymes, enzyme stability, catalytic mechanism, substrate specificity and range, surface activity, folding mechanisms, cofactor dependency, pH and temperature optima and kinetic parameters have been successfully modified. Protein shuffling and related techniques dramatically increase the variability of enzymes and might lead to enzymes not present in nature so far. Apart from manufacturing enzymes from micro-organisms plants are also investigated for the production of enzymes. Furthermore, enzymes can be chemically modified, e.g., by incorporation of cofactors, or chemical glycosylation.

Since the marketing of the first genetically modified industrial enzyme about 15 years ago and the subsequent development of these methods for modifying enzyme structure during the last 10 years, the number of available enzymes has dramatically increased and enzyme properties have been enhanced significantly.

24.6 Challenges for food enzyme regulation and safety evaluation

From the above discussions several challenges for both further development of food enzyme regulation and for safety evaluation can be identified. These challenges can be attributed to differences in both existing legislation and in requirements for safety evaluation, to prevalent scientific uncertainties and to the introduction of new production technologies, especially genetic engineering.

24.6.1 Regulatory aspects

Whether enzymes are categorised as processing aids or as additives and whether the definitions for both categories differ among countries is not important only to bureaucrats. In the EU, for instance, only food additives are covered by harmonised legislation and are required to undergo a pre-market approval procedure. This might not only be relevant for trade liberalisation but also in terms of consumer safety and perhaps of occupational safety as well. For, if no pre-market approval is mandatory – neither is safety evaluation. Consequently, safety evaluation of food enzymes used as processing aids would be a voluntary task depending on the particular manufacturer only. Although internal industry guidelines for safety of enzyme preparations and safe handling of enzymes exist, this might not be considered sufficient as both consumer awareness of food production methods and scepticism towards the food industrial system have been increasing considerably. In the EU, this picture might soon change though, if a new regulation on food enzymes along with a mandatory safety evaluation is actually established.

Another question is whether occupational health risks in manufacturing and handling of food enzymes are actually covered by harmonised EU legislation. In case of technical enzymes occupational health risks are dealt with under harmonised chemical legislation. If a new enzyme is marketed, the pre-market application will also include information on occupational health issues. Similarly, certain occupational health aspects have to be considered if applying for an authorisation of a new feed enzyme under harmonised EU legislation. In the absence of harmonised EU legislation for food enzymes that are used as processing aids only, they would thus be subjected to hazard classification and labelling according to chemical legislation. Whereas the manufacturing processes and associated risks of most industrial enzymes are similar it would have to be clarified whether this regime establishes different degrees of scrutiny of occupational risks. Given that about 90 enzymes presently manufactured by AMFEP member companies are solely used in food industry, this might be a relevant question to clarify.

Occupational and consumer safety

Industry has achieved a lot in reducing health risks for workers. However, there are still reports on occupational health problems. A survey revealed that enzyme sensitisation and work-related respiratory symptoms were positively correlated with airborne enzyme exposure. With bakers the number of workers that have to leave their job because of work-related problems is still increasing. This kind of reports keep the discussion ongoing and urge for a critical review and improvement of the safety precaution measures to be established.

Following contradicting evidence on skin sensitising properties of enzymes, this calls upon further research.

With respect to enzymes from GMM it has to be mentioned that there are currently no reports on specific problems caused by these enzymes, they appear to have the same sensitising potential and are capable of sensitising exposed employees at the same rate as traditional enzymes. However, special attention might be appropriate. Changes in the amino acid sequence of enzymes, their structure, or properties such as thermostability might change allergenic properties. This is probably less important for occupational health as industry voluntarily labels and accordingly handles the enzymes as respiratory sensitisers. As enzymes might also be present in food it could nevertheless be important for consumers. So far, a possible allergenicity of food enzymes was a rather neglected issue in enzyme safety evaluations. A recent draft guidance document suggests a change in perception – at least for enzymes from GMM. If this guidance document is to be used for enzyme evaluations the allergenic properties of new enzymes will have to be evaluated in light of their similarities to known enzymes, as well as food and environmental allergens. Such assessment procedures were originally introduced for foods derived from genetically modified crops. These procedures are increasingly contested for evaluating food from GMOs but might nevertheless still be appropriate for food enzymes. At least well characterised and purified enzyme isolates would perhaps not allow for a high percentage of other proteins to be present.

As the use of enzymes expands into new areas there is also a risk of exposure of the general population, which differs from the current situation. And if enzymes are increasingly used for cosmetic and other consumer product uses, it becomes imperative to understand how people can be exposed to enzymes, the level of exposure and the risk for sensitisation.

Further research on understanding how enzymes act as allergens, linked with an understanding of how individuals become sensitised to enzymes is therefore important for the continued control of occupational and non-occupational disease caused by enzymes.

With respect to otherwise toxic properties of enzyme preparations, it seems that present safety evaluation approaches seem to work well to identify both critical amounts of toxins inadvertently present in the enzyme isolate and possible unknown toxic effects including the enzyme protein itself. Furthermore, it has to be acknowledged that the use of GMMs has the potential to improve product safety by reducing the number of production organisms to a small set of well characterised strains that sometimes have safely been used in enzyme manufacturing and even in food processing for decades. This seems to be important for both occupational risks and for consumer safety.

The application of genetic engineering techniques resulted in increased variability and modified properties of enzymes such as thermo- and pH-stability. Furthermore, enzymes from exotic sources might be used in food for which there is no prior experience of exposure to humans. For this reasons it might be premature to generally relax the requirements for toxicity testing as demanded by Pariza and Johnson. Furthermore, as shown in the sections above, there are differences in the requirements for safety evaluations of food enzymes among countries and scientific committees alike, e.g., toxicity endpoints. These differences are indicators of scientific uncertainties and point to a need for either clarification and/or scientific research.

This would be helpful to clarify the relevance of endpoints and test methods contested in the scientific literature and in guidance documents for industry. Thereby, consistency in safety evaluation would be improved and unnecessary testing avoided.

The challenges mentioned above will most likely concern regulators and industry alike. The introduction of an EU harmonised legislation on food enzyme would offer an opportunity to reconsider regulatory approaches and scope and requirements for safety evaluations as well. Legislation and the implantation process will take place under critical observation and perhaps involvement of well-informed consumer and environmental groups. Scientists should also be aware of this process and might be interested and willing to contribute in order to arrive at a sound and reasonable regulatory framework. Ideally, further research will be conducted to clarify the open questions indicated.

References

Arbige, M.V., *Industrial Enzymes*, Applied Science Publishers, London.

Bailey, J.E., *Biochemical Engineering Fundamentals*, McGraw-Hills, New York.

Bilitewski, V., *Biosensors and their Application in Enzymes*, Marcel Dekker, New York.

Burtis, S.M., *Enzymes and their Role in Cereal Technology*, Prentice-Hall Inc., Englewood Cliffs, New Jersey, USA.

Connell, W.C., and Miller, G.J., *Chemical Aspects of Enzyme Biotechnology*, John Wiley & Sons, New York.

Cornish-Bowden, A., *Enzyme Catalysed*, Cambridge University Press, Cambridge.

Drauz, K., *Enzyme Catalysis in Organic Synthesis*, Peragam Press, London.

Dugan, P.D., *Industrial Enzymes*, Plenum Publishing Corporation, London.

Edelson, S.S., *Commercialisation of Engineered Industrial Enzymes*, Academic Press, London.

Eisenthal, R., *Safety Evaluation of Genetically Engineered Enzymes*, Academic Press, London.

Fernie, T., *Industrial Enzymes*, Marcel Dekker Inc., New York.

Fersht, A., *Enzyme Structure and Mechanism*, Academic Press, London.

Guilbault, G.G., *Analytical Uses of Immobilised Enzymes*, University of Wisconsin Press, Madison.

Henzle, E.B., *Enzyme Technologies,*, Wiley, Hoboken.

Holton, J.B., *Enzyme Biotechnology*, Humana Press Inc., New Jersey.

Jackwerth, N.P., *Principles of Enzyme and its Activity*, McGraw Hill, Columbus, Ohio, USA.

James, L.A., *Enzyme Engineering,*, John Wiley & Sons, New York.

Koch, T.R., *Principles of Enzymatic Analysis*, Pergamon Press, Oxford, New York.

Kraut, J., *Analytical Methods for Measuring Enzymes*, Cold Spring Harbour Press, UK.

Lewis, R., *Enzymes in Wheat, Floor and Bread*, Science Publishers, New York.

Linderstrom, K., *Enzymic Hydrolysis of Food Proteins*, Tata McGraw Hill, New York.

Mann, J., *Biosynthesis of Enzymes*, Oxford University Press, Oxford.

McCaull, J., and Crossland, J., *Biotechnology of Microbial Enzymes,*, Harcourt Brace Jovanovich, New York.

Nelson, D.L., *Lehninger Principles of Biochemistry*, Reston Publishing Co., Reston, Virginia.

Richard, O.M., *Carbohydrate-degrading Enzymes in Food Industries*, Prentice-Hall of India Pvt. Ltd., New Delhi.

Rudd, D.F., *Fundamentals of Enzyme Biotechnology*, Prentice Hall, Englewood Cliffs, NJ.

Smith, L., *Enzyme Activity and Effectiveness*, Pearson Education, Singapore.

Snell, E., *Enzymes: Biochemistry, Biotechnology and Chemical Chemistry*, Affiliated East-West Press Pvt. Ltd.

Tyler, E.G. and Wayne, A.P., *Advances in Enzymes Biotechnology*, McGraw-Hill, New York.

Umetsu, H., *Industrial Enzymes and their Characteristics*, Plenum Publishing Corporation, London.

Watanabe, M., *Legislation and Regulation in Industrial Enzymology*, John Wiley & Sons, New York.

Watson, J.D., *Molecular Biology*, Routledge, New York.

Wilbur, A., *Handbook of Enzyme Biotechnology*, Ellis Harwood, New York.

Index

H

I

K

L

M

N

O

P

For Product Safety Concerns and Information please contact our EU representative GPSR@taylorandfrancis.com Taylor & Francis Verlag GmbH, Kaufingerstraße 24, 80331 München, Germany